Condition Monitoring, Troubleshooting and Reliability in Rotating Machinery

Scrivener Publishing
100 Cummings Center, Suite 541J
Beverly, MA 01915-6106

Rotating Machinery Fundamentals and Advances

Scope: Rotating machinery represents a broad category of equipment, which includes pumps, compressors, fans, gas turbines, electric motors, internal combustion engines, etc., that are critical to the efficient operation of process facilities around the world. The objective of the "Advances in Rotating Machinery Series" book series is to provide industry practitioners a time-saving means of learning about the most up-to-date rotating machinery ideas and best practices. To meet this intent, this series covers industry-relevant topics, such as design assessments, modeling, reliability improvements, maintenance methods and best practices, reliability audits, data collection, data analysis, condition monitoring, and more.

About the Series Editor:
Robert Perez is a mechanical engineer with more than 40 years of rotating equipment experience in the petrochemical industry. He has worked in petroleum refineries, chemical facilities, and gas processing plants. He earned a BSME degree from Texas A&M University at College Station, an MSME degree from the University of Texas at Austin and holds a Texas PE license. Mr. Perez has written numerous technical articles for magazines and conference proceedings and has authored 5 books and coauthored 4 books covering machinery reliability. He is also the technical editor of Kane's Rotating Machinery Dictionary.

Publishers at Scrivener
Martin Scrivener (martin@scrivenerpublishing.com)
Phillip Carmical (pcarmical@scrivenerpublishing.com)

Condition Monitoring, Troubleshooting and Reliability in Rotating Machinery

Volume 3

Edited by
Robert X. Perez

Scrivener Publishing

WILEY

This edition first published 2023 by John Wiley & Sons, Inc., 111 River Street, Hoboken, NJ 07030, USA and Scrivener Publishing LLC, 100 Cummings Center, Suite 541J, Beverly, MA 01915, USA

For more information about Scrivener publications please visit www.scrivenerpublishing.com.

Wiley Global Headquarters
111 River Street, Hoboken, NJ 07030, USA

For details of our global editorial offices, customer services, and more information about Wiley products visit us at www.wiley.com.

Library of Congress Cataloging-in-Publication Data

ISBN 9781119631545

Cover images: Left to right
Natural gas compressor station, GTS Productions I Shutterstock.com
Gas turbine burner, Red_Shadow I Shutterstock.com
Tech using thermal imaging camera, Joyseulay I Shutterstock.com
Lubricant and Gears, Andrey VP I Shutterstock.com
Cover design by Kris Hackerott

Set in size of 11pt and Minion Pro by Manila Typesetting Company, Makati, Philippines

Printed in the USA

10 9 8 7 6 5 4 3 2 1

Dedication

This book series is dedicated to rotating machinery professionals around the globe who have devoted their careers to repairing, evaluating, and optimizing their equipment. It is through their diligence that critical machines are able to operate safely, efficiently, and reliably between scheduled outages.

Contents

Preface

"Good reliability engineering is not the search for perfection—rather, it is the search for pragmatic solutions to business problems."
H. Paul Barringer, Reliability Consultant

Rotating machinery represents a broad category of equipment, which includes pumps, compressors, fans, gas turbines, electric motors, internal combustion engines, etc., that is critical to the efficient operation of process facilities around the world. These machines must be designed to move gases and liquids safely, reliably, and in an environmentally friendly manner. To fully understand rotating machinery, owners must be familiar with their associated technologies, such as: machine design, lubrication, fluid dynamics, thermodynamics, rotordynamics, vibration analysis, condition monitoring, maintenance practices, reliability theory, etc.

The goal of the three volumes of the "**Advances in Rotating Machinery Book Series**" is to provide industry practicioners a time savings means of learning about the most up-to-date rotating machinery ideas and best practices. This three-book series will cover industry-relevant topics, such as design assessments, modeling, reliability improvements, maintenance methods and best practices, reliability audits, data collection, data analysis, condition monitoring, and more.

In my 40 year career in the industry, I have trained myself to keep H. Paul Barringer's quote (shown above) in mind when dealing with real-world machinery issues. I approach every reliabililty problem as a business problem. I begin by asking: What is the cost of the problem? Then, I ask: What are the possible practical solutions to the problem? Hopefully, we can find a solution that is both effective and economically justified, but this is not always the case. This book series' mission is to offer proven, cost effective solutions and best practices that can help users manage their machinery problems.

Volume 1 begins the series by focusing on machinery design and analysis. Volume 2 in the series covers machinery reliablity concepts and practical machinery reliability improvement ideas. Volume 3 continues the series by covering:

1. Machinery monitoring concepts and best practices
2. Lubrication best practices
3. Machinery troubleshooting
4. Reliability improvement ideas
5. Professional development advice

Readers will find a good mix of theory and sage experience throughout this book series. My hope is that practicioners of machinery reliabilty technologies will use the wisdom contrained in this book series to achieve best in class reliablity performance at their facilities.

Robert X. Perez, Editor
Summer 2022

Acknowledgements

I would like to thank all the contributors for their expert advice and their clear and insightful prose. Without them, this book series would not have been possible. I would also like to thank the publisher for believing in me and allowing me to develop this comprehensive book series. Finally, I would like to thank my wife for reviewing my drafts and for her encouragement.

Part 1

CONDITION MONITORING

1

An Introduction to Machinery Monitoring

By Robert X. Perez

The aim of employing predictive maintenance technologies in process facilities is to assess the condition of equipment by performing periodic inspections such as vibration analysis, temperature monitoring, oil analysis, ultrasonic analysis, etc. or by using permanently installed equipment, such as vibration or temperature sensors. The primary tenant of the predictive maintenance philosophy is that it is more cost-effective to perform maintenance when degradation or distress is detected than to risk running equipment until it loses performance capability and adversely affects the process (Figure 1.1). Operating personnel hope to identify and address

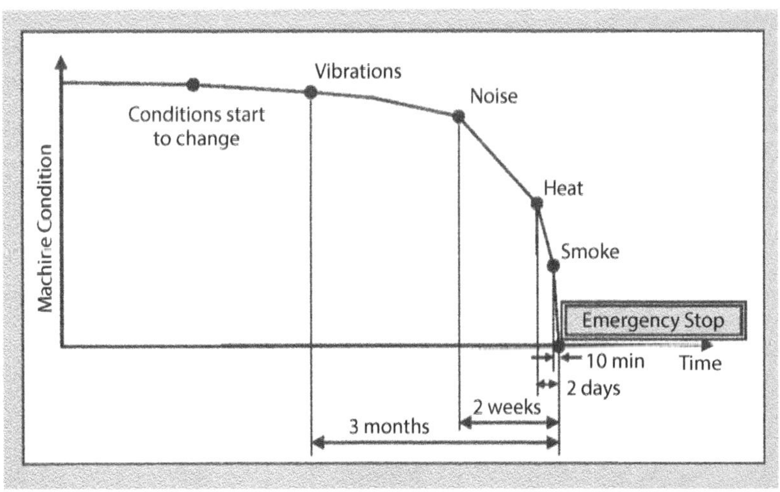

Figure 1.1 As a machine begins to fail, it begins showing signs of distress. First, vibration levels increase and then noise is detected. In the final stages of failure, heat and smoke are experienced before a catastrophic failure occurs. It makes economic sense to invest in condition monitoring technology that can detect early signs of failure before secondary damage can occur.

Robert X. Perez (ed.) Condition Monitoring, Troubleshooting and Reliability in Rotating Machinery, (3–14) © 2023 Scrivener Publishing LLC

machinery issues in the primary state before costly secondary damage is experienced.

There are two types of data collected by condition monitoring systems: a) Dynamic data which is composed of electrical signals that rapidly change versus time, as seen in Figure 1.2. Dynamic data requires some type of signal processing to convert it into a user-friendly format. b) Static data which are signals that do not change rapidly versus time, therefore signal processing is not required. The most common condition indicators used for monitoring machines and piping in critical processes are:

- **Vibration**—This is dynamic data is collected by measuring the motion of a vibrating surface, such as a bearing housing or shaft. Analysis of this type of data requires complex signal processing and pattern recognition. More on vibration analysis later.
- **Pressure**—This can be either in the form of static or dynamic data collected by inserting a pressure traducer into a fluid stream. Trending static field pressure data can be used to spot changes in rotating machinery performance or to ensure operating conditions are normal.
- **Temperature**—This is usually in the form of static data. Thermocouples (TC) or resistance temperature devices

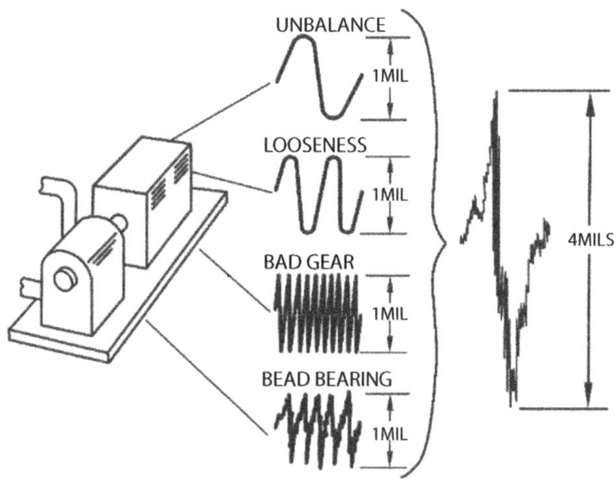

Figure 1.2 Mechanical vibration levels are commonly used to assess the condition of vital process machinery. Complex vibration waveforms measured in the field are a combination of multiple machinery phenomena such as unbalance, looseness, etc.

(RTD's) are often inserted into a fluid stream or on or below bearing metal surfaces to measure temperature. Portable infrared temperature guns and contact thermometers can also be used to monitor machine surface temperatures. In some applications, thermography is employed to visualize temperature distributions across a machine in order to identify component issues such as failing bearings, etc. Thermography can also be used to spot electrical problems in the field on motors and control panels.

- **Oil Analysis**—Oil analysis requires that oil samples be collected in the field and then sent off-site for lab testing. Although most oil properties are usually determined by lab testing, some oil properties can be monitored in real time.
- **Piping, Duct Work, and Structural Vibration**—The vibration of piping, vents, duct work, and supporting structures connected to machinery can signal problems, such as critical speeds issues, resonances, and unwanted flow conditions. On new installations, excessive piping vibration may be indications of poor installation and/or design practice.

Vibration Analysis

In the simplest terms, mechanical vibration in rotating machinery is simply the back and forth movement or oscillation of machines and components, such as drive motors, driven devices (pumps, compressors, and so on), and the bearings, shafts, gears, belts, and other elements that make up mechanical systems. Vibration in industrial equipment can be both a sign and a source of trouble. With a basic understanding of vibration and its causes, the maintenance professional can quickly and reliably determine the cause and severity of most machine vibration and provide recommendations for repair.

Elements of Vibration

Machinery vibration as a repetitive movement around a point of equilibrium characterized by its variation in amplitude and frequency. Vibration can be the dynamic motion of a bearing housing or piping system or the dynamic motion of a rotor relative to a bearing or stator. Both the amplitude and the frequency are used to assess and analyze vibration issues. **Amplitude** is the maximum extension of the oscillation and it is measured from the lowest point to the highest point of the waveform. We can say the amplitude is the total movement of a surface or object during a cycle which is used to quantify the intensity of the vibration. **Frequency** measures the rate at which movements in vibration occur per second (Hz) or cycles per minute (CPM). For example, every piano note is tuned to a unique frequency. If you examined the vibration waveform of each note, they would each have a unique frequency corresponding to a defined note. The piano frequencies and amplitudes of each note are combined to create a complex signal. Similarly, vibration can be a composition of multiple frequencies that are the result of different machinery phenomena (Figure 1.2). Every machine will have its own vibration signature related to many factors such as its construction, installation, and condition. It is the job of the monitoring system to faithfully detect and display the vibration that is occurring. The vibration analyst must have complete trust in the vibration monitoring system before they can begin the vibration analysis process.

When Vibration is a Problem

Most industrial devices are engineered to operate smoothly and avoid vibration, not produce it. In these machines, vibration can indicate problems or deterioration in the equipment. If the underlying causes are not corrected, the unwanted vibration itself can cause additional damage. In critical process machinery, smoother operation is generally better and a machine running with vibration levels close to zero is ideal.

Effects of Vibration

The effects of excessive vibration can be severe. If unchecked, machine vibration will accelerate wear rates in mechanical seals, internal seals, and bearings and potentially lead to catastrophic equipment failures. *High machine vibration levels usually lead to a shortened machine life and a higher probability of catastrophic failure.*

Vibrating machinery will also:

- Create more background noise
- Create safety issues due to flammable product leaks to the atmosphere
- Lead to degradation in plant working conditions due to external product and oil seal leaks
- Lead to excessive power consumption due to the wear of internal seal clearances
- Affect product quality by damaging seals and allowing oil, water, and other contaminates to enter the process

In the worst cases, vibration can damage equipment so severely that it will fail rapidly and potentially halt plant production. Yet, there is a positive aspect to machine vibration. Measured and analyzed correctly, vibration can be used in a preventive maintenance program as an indicator of machine condition and help guide the plant maintenance professional to take remedial action before disaster strikes.

Monitoring Systems

Predictive maintenance programs rely on either portable or permanently installed monitoring systems that can accurately sense and report one or more key equipment condition and performance indicators. For example, vibration monitoring systems, which are commonly used to assess the mechanical condition of process machinery, have several distinct components (Figure 1.3) that work together to deliver a useful output. Other examples of monitoring systems are those that measure bearing temperature by using embedded resistance temperature detectors (RTDs) or thermocouples (TC) with outputs that are connected to temperature monitors. The role of a temperature monitor is to convert the input from a temperature sensor into an output voltage proportional to the temperature that can then be displayed, monitored, or stored for later use. Pressure monitoring systems employ pressure transmitters to measure vital pressures, such as suction and discharge pressures, oil and seal system pressures, and process pressures. The intent of every monitoring system is to sense physical measurements occurring in the field and display them in real time so that it can be analyzed and acted upon as required.

Vibration monitors all have some type of motion sensitive sensor that detects and transmits a motion signal, usually a current or voltage, to the signal processor. To select the proper sensor, the user must ask: What am I trying to measure? To answer this question, you need to know the expected amplitude and frequency range of the vibration phenomena. For example, Figure 1.4 shows spectral analysis bands from a machine with rolling

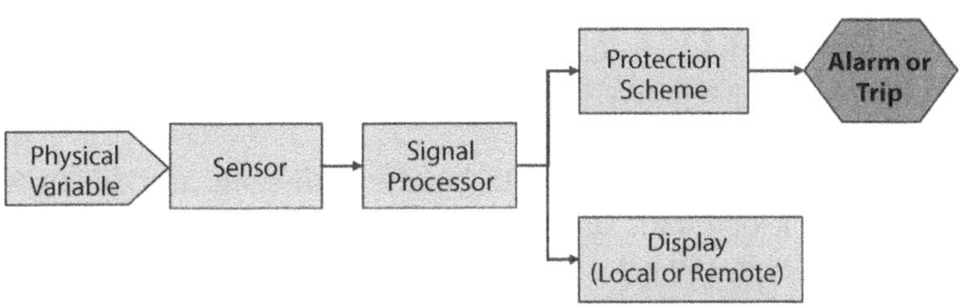

Figure 1.3 Most monitoring systems are composed of a sensor, signal processor, and a local or remote display of some kind. Critical machinery measurements can have one or more protection schemes, which contain local or remote alarms and trip settings.

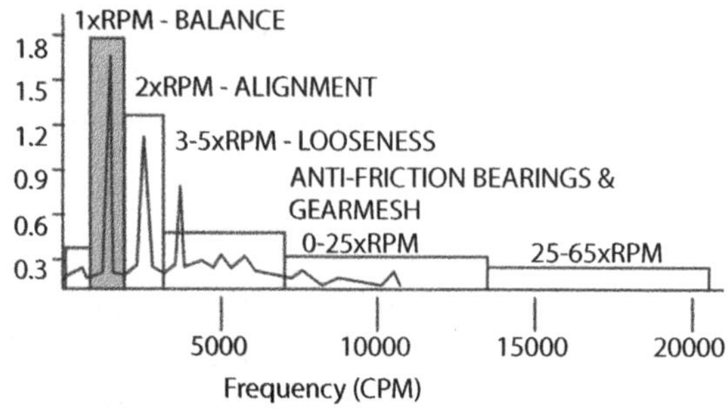

Figure 1.4 This hypothetical vibration spectrum illustrates some possible frequency content. Notice that the various machine issues show up in different frequency bands.

element bearings. Notice there are various potential vibration issues identified based on experience, such as imbalance, alignment, etc. By inspection, we can see that to adequately monitor the vibration phenomena shown in Figure 1.4, we would need a sensor capable of detecting vibration in the range from 0 to 20,0000 cycles per minute (CPM). The lesson here is: The expected machinery defects expected will dictate the type of sensor used. If we look at the various sensor options listed in Table 1.1, we see that we should use velocity measurements to monitor this machine. (Velocity would be measured using an accelerometer with single integration.)

The next element in a monitoring system is a signal processor that receives the signal and converts it to a usable output format. Signal processing can include filtering out unwanted portions of the input signal, converting the signal to a digital set of values, or calculating the average, maximum, or minimum value of a series of inputs. One key parameter of the signal processor is that it must be capable of sampling data at a higher rate than the highest frequency transmitted by the sensor. To avoid aliasing, the type of signal distortion and signal processor must be capable of sampling at more than twice the highest frequency contained in the input of the signal. The designs of signal processors are numerous and varied in purpose. In the end, you want the output of the signal processor to be free of processing errors and in a usable format that that can be displayed and/or used in a protection scheme.

Signal processors are designed to handle both static and dynamic signals. An example of a static signal is temperature. If you plot a temperature over time, you typically get a gradually changing series of points that can visually be studied and analyzed. Static signals, such as temperature and pressure signals, do not carry any rapidly changing, i.e., dynamic, components. On the other hand, dynamic signals can vary rapidly with time,

Table 1.1 Summary of frequency capabilities of various sensors.

		Frequency Range[1]		
		0-10 Hz	**10 to 1000 Hz**	**1000+ Hz**
Type of Measurement	**Displacement (accelerometer with double integration)**	Detection of common mechanical defects in low speed machines		
	Displacement (accelerometer with double integration)	Piping vibration analysis[2]		
	Displacement (proximity probes measuring relative motion between bearing and shaft)	Detection of common mechanical defects in high speed machines with fluid film bearings		
	Velocity (accelerometer with single integration)		Detection of common mechanical defects in 1800 to 3600 rpm machines	
	Velocity (accelerometer with single integration)	Piping vibration analysis[2]		
	Accelerometer (case mounted)			Detection of high frequency phenomena, i.e., Impacts, rolling element bearing defects, gear defects

1) Notice that an accelerometer with single integration can be used to detect most common mechanical defects in the 1800 to 3600 rpm machines. 2) Proximity probes are used to detect common mechanical defects in high speed machines with fluid film bearings.

as seen in Figure 1.5. Dynamic signals, also called dynamic waveforms, require more complex signal processing to determine their properties. Typical waveform properties are frequency, peak amplitude, root mean square amplitude, and phase. In some cases, both static and dynamic information are extracted from the raw sensor data.

A spectrum analyzer can deconstruct a complex input waveform (A) into its fundamental sine waves (B1, B2, B3). These constitute that sine waves provide insight into the mechanical condition of a machine being analyzed.

Most in-depth analysis of machinery vibration is done in the frequency domain or using spectrum analysis. Spectral analysis is the process of transforming a signal from the time domain to the frequency domain (see Figure 1.6). It is often done using a spectrum analyzer. The signal is analyzed to determine any substantial frequencies coming from the machine's components. Where there is a peak in frequency signal, that is the likely source of vibration. Common applications for spectral analysis include the rotational speed of a shaft or how often tooth meshing occurs on a pair of gear wheels.

Care must be taken to ensure all vibration measurement locations provide a true indication of bearing vibration. On larger machines, horizontal and vertical bearing housing vibration data collection is recommended (Figure 1.7). A comparison of horizontal and vertical vibration levels can be used to identify machine specific problems related to support stiffness, bearing loading, bearing fits, etc.

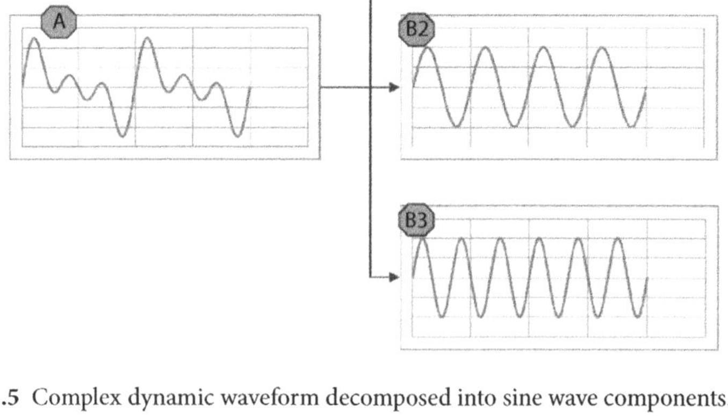

Figure 1.5 Complex dynamic waveform decomposed into sine wave components.

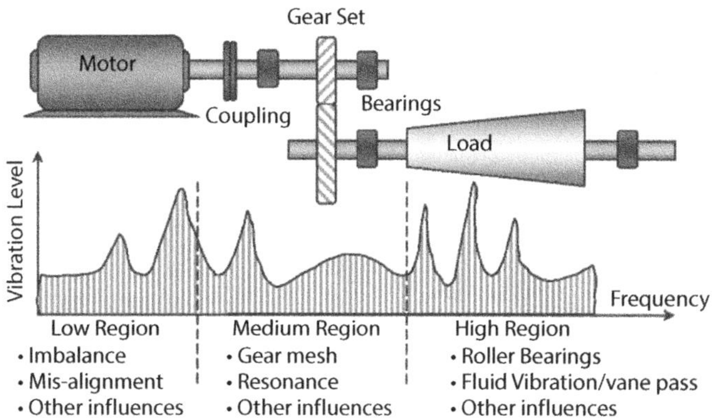

Figure 1.6 Here is a hypothetical vibration spectrum for an electric motor, gearbox, and compressor machine train. By transforming vibration signals from the time domain into the frequency domain, characteristic machinery phenomena can be identified. Notice that the spectrum here contains several discrete frequency components. Every machine has a set of unique spectrums or signatures that are based on their construction and operating conditions.

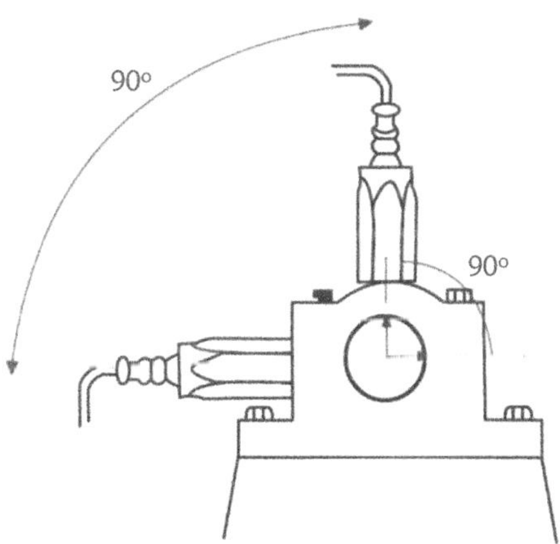

Figure 1.7 It is recommended that both horizontal and vertical vibration data be collected on critical machines.

DCS Systems

It is common that critical field machine parameters such as overall vibration levels, bearing temperatures, field pressures, and flows are relayed to a control room for real-time monitoring and display (see Figure 1.8). If properly conceived and installed, key machinery data is displayed in ways

Figure 1.8 Control room operators depend on a distributive control system to provide them with vital real-time process information in a visual format.

that is easy to observe and interpret. Displays can use dials, scales, simple numerical displays, or other visual interfaces to communicate the status of variables being measured. Today, most plants have DCS, or distributed control systems, that allow field vibration and temperature information to be easily transmitted from the field to a centralized computer for monitoring. Most DCS systems also have storage capabilities that provide a means of trending and comparing the present status with the past, which is a must in critical machinery applications.

Data Presentation

Figure 1.9 illustrates two trend plot examples. One plot is that of a gradually increasing value, whereas the other shows a step change in a measured value. Trend plots are useful because they provide visual representations of the measured parameter over time and this representation can help in the troubleshooting process. Suppose a step change occurred at the same time as a change in the process, there may be a correlation between the two events that should be investigated. A gradually increasing trend plot may indicate either a deteriorating internal or external component.

Critical monitoring systems may also have built-in protection schemes, which can be configured to serve the unique needs of specific machine trains. These systems can provide either remote or local alarms whenever an undesirable condition has been detected. However, it is recommended that all safety critical trip points, such as machine speed, be handled with independent, dedicated instrument loops to ensure their reliability. To

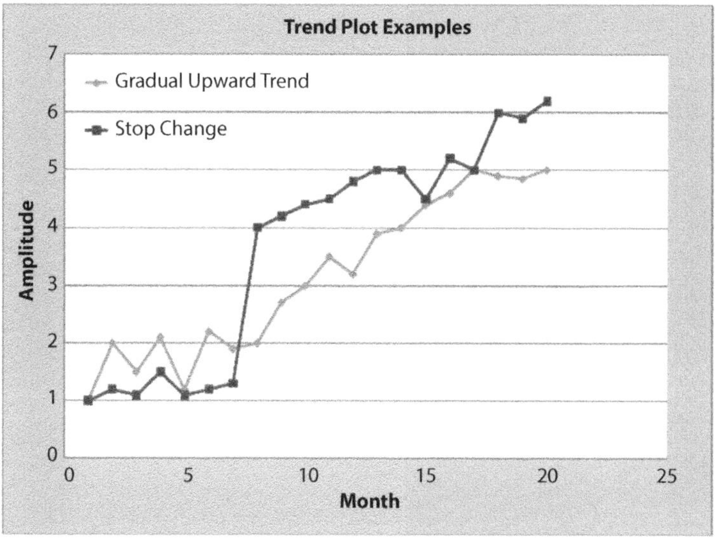

Figure 1.9 Examples of trend plots.

ensure reliable machinery monitoring systems, they must be regularly calibrated and maintained to ensure all monitoring points are working properly.

Finally, for a monitoring program to be complete, assessment criteria are required to determine when the machine owner should be concerned or if action must be taken immediately. Without assessment criteria, there would be no need for monitoring systems because their outputs would be meaningless. If assessment criteria are set too low, then time and money are wasted. However, if assessment criteria are set too high to avoid nuisance alarms or trips, then human, environment, and equipment health may be placed in jeopardy.

2

Centrifugal Pump Monitoring, Troubleshooting and Diagnosis Using Vibration Technologies

By William D. Marscher

Introduction

This chapter outlines the current and developing practice of condition monitoring and diagnostic troubleshooting for centrifugal pumps and their systems through application of vibration-related technologies. Key concepts and methods will be described in understandable terms. Modern vibration-based data acquisition and data analysis options are presented, along with physics-based pump condition evaluation strategies. There will be discussion concerning the reasoning behind international guidelines and standard available from HI, ISO, and API and their vibration acceptance requirements.

Circustances for which test results should be augmented by rotordynamics analysis or structural finite element analysis will be discussed in the context of what information these analyses can provide to an End User or OEM that could be informative to reliable and trouble-free operation. Mechanical reliability problems typically involve either tribology (friction and wear) or metal fatigue (crack formation and possible fracture). The likelihood and severity of these tribological and fatigue problems will be considered in terms of physical phenomena and forces associated with them, so that by applying techniques to detect these phenomena and their forces the problems they may cause can be avoided or at least delayed.

The most common sources of force that may result in excess vibration include the rotor being out of balance, the presence of excessive misalignment between the pump and its driver and excessive hydraulic

Robert X. Perez (ed.) Condition Monitoring, Troubleshooting and Reliability in Rotating Machinery, (15–76) © 2023 Scrivener Publishing LLC

off-design-flow forces such as those from inlet recirculation, stall, or surge. Other common vibration-causing hydraulic forces that may be excessive include the stator vane or impeller blade pass pressure pulsations that are present at some level at all flow rates.

When forces such as these become large, vibration will increase more or less in proportion to any force increase until running clearances are used up. Once radial clearance is gone, vibration will increase at some frequencies such as integer harmonics of 1x and maybe 1/2x, but ironically may sometimes instead actually *decrease* at 1x rpm, because the vibrating rotor is jammed against the much stiffer and massive casing. Like many vibration-related phenomena, this situation can be very confusing unless a systematic investigation is followed. The best troubleshooting methods will include as much key information as possible, rather than focusing on only one type of data. This chapter will discuss how x is typically the most important vibration component but should be evaluated in the context of other frequencies, as well as process information, bearing temperature, and (if available) oil monitoring analysis information.

Like many other aspects of rotating machinery, evaluating vibration is often about perceiving "wheels within wheels", in that the end result may not be from a single cause but rather a combination of several factors. One example is the amplifying phenomenon known as resonance, which can result in high vibration even if exciting forces are reasonably low. This is due to motion being magnified by the presence of a rotor or structural natural frequency, which occurs when a machine operates at a "critical speed", as will be explained later. Another example is subsynchronous vibration due to rotordynamic instability, in response to which the rotor vibration becomes unbounded, with radial motion limited only by rubbing at the clearances. If circumstances are right, such unstably high vibration can occur even if the rotor had been nearly perfectly balanced and aligned. To understand what can lead to phenomena such as resonance and instability, as well as unexpectedly high excitation forces, it is important to be familiar with key vibration concepts, as presented below. In support of this, the instrumentation and data analysis options to best monitor and evaluate vibration will be presented.

Vibration Definitions

The reader may be familiar with vibration concepts and term definitions from other references, but for convenience some of the most important issues are defined or illustrated below:

The foundational concept of vibration is oscillating motion, such that there is at least one complete back-and-forth cycle of a mass supported relative to another mass (or to a reacting "ground", such as a foundation) by a flexible element, typically behaving like some form of spring. If energy is not removed from the oscillation, typically by some form of "damping" (like an automobile shock absorber), then the vibration oscillation "waveform" (motion in space over a single oscillation period of time) becomes a forever repeating "cycle", vibrating without end. A typical waveform over the "period" of one cycle is shown in Figure 2.1, along with several measure numbers for vibration waveform amplitude.

Key terminology concerning the vibration waveform or cycle is as follows:

- Vibration: The oscillation of an object about its position of rest
- Cycle: Movement of a mass through all its positions back to its point of rest, on an oscillatory basis
- Period: The time it takes for one cycle to complete
- Frequency: Number of cycles in a given amount of time (e.g., 1 cycle/s, known as a "Hertz", Hz), which is also the inverse of the vibration's period
- CPM: Number of cycles in one minute
- CPS: Number of cycles in one second (Hz)
- RMS: Root Mean Square, which is the square root of vibration level squared, summed over calculations step-by-step

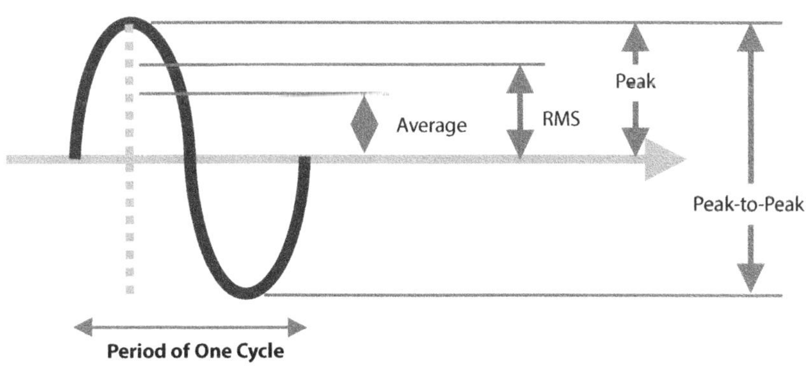

Figure 2.1 Cyclic nature of vibration considered as waveform. Courtesy Mechanical Solutions, Inc.

throughout a selected time interval, covering one or more complete cycles
- Fourier Transform: the mathematical relationship between an overall waveform and the component waveforms at differing frequencies that add together to form the overall waveform
- FFT: Fast Fourier Transform, a computer method to quickly determine a vibration vs. frequency "spectrum" from a vibration vs. time waveform however complicated it may be

How Vibration vs. Time Relates to a Vibration vs. Frequency "Spectrum"

The relationship between vibration is considered as a function of time versus vibration which is considered as a function of frequency, is illustrated in Figure 2.2. The figure shows how each approach is presenting the same actual total vibration. If the vibration is plotted only in the time domain, the waveform can become quite complicated, as seen in the upper right figure. However, as shown in the left view of the left figure, when considered one frequency at a time, the vibration breaks into simple sine waves of different frequencies and amplitudes. Seen in the right view of the left figure, the amplitudes are straight vertical lines, each line being of a given height (the amplitude) at a given frequency, with frequency increasing left to right. This is called a frequency spectrum. The math that relates the left view with the right view for Figure 2.2 is what is known as the Fourier Transform, with its calculations simplified by the so-called Fast Fourier Transform, FFT, as performed by modern vibration signal analyzers.

The following are some considerations about whether it is best to display vibration in the "time domain" or the "frequency domain":

- Using the frequency domain, specific defects show up as peaks at discrete frequencies, which gives important clues concerning what is causing the vibration
- Small (but maybe important for diagnostics!) repetitive signals typically are not hidden in the frequency domain, even among stronger signals at other frequencies
- The time domain is ideal for observing and quantifying a one-time event (a "transient") or an impulsive excitation, which can occur during cavitation

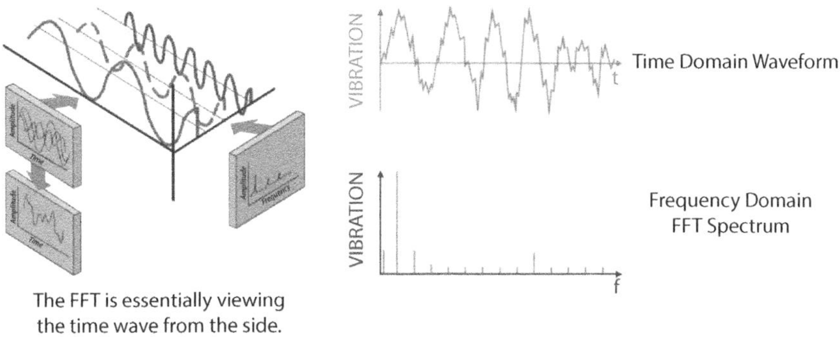

Figure 2.2 How vibration vs. Time can relate to vibration vs. frequency. Courtesy Mechanical Solutions, Inc.

What are Reasons for Excess Vibration?

A wise man once said, "If a pump is not vibrating, it's not running!" Any operating machine produces some degree of waste energy and this waste energy shows up primarily as vibration, noise, and heat. So, the issue is not whether or not the presence of vibration is acceptable, but rather how much is acceptable. More will be discussed about how to determine this when we discuss vibration standards below.

In preparing to evaluate vibration, it is useful to consider the physical reasons why some of a machine's energy ends up as vibration. An outline of the process of how a pump's operation will lead to some level of vibration is as follows:

- Forces internal to the pump due to the process of pressurizing and moving fluid react against the pump rotor and together with imbalance forces (no rotor is perfectly balanced) and misalignment forces (no pump and driver are perfectly aligned), cause motion ("response") of the rotor.
- The force leads the response in time
- There is a phase lag (timing delay) between force and response
- The rotor response is an oscillating motion that we call vibration (called "rotordynamics", for a rotor) that causes the rotor to react against the pump casing through its bearings and bearing housings to transfer some of the vibration from the rotor (such that it now becomes "structural vibration").

Some structural vibration may also be caused directly by fluid pressure pulsations or "acoustics"

- The vibration of either the rotor or the stationary structure may be amplified by a natural frequency "resonance", in which the natural unforced oscillation of the vibrating component (like a guitar string after it is plucked) is at or near to the forcing frequency
- Based on what potential damage issue is being evaluated, the vibration may be considered in terms of the direct motion ("Displacement"), the motion's rate of change during an oscillation cycle ("Velocity"), or the rate of change of the velocity ("Acceleration"). Displacement emphasizes lower frequency components of vibration, acceleration emphasizes high frequency components, and velocity is in the middle. Most vibration standards are in terms of, or at least emphasize, velocity.
- If either the rotor or structural vibrations, and thereby their differential, are excessive, then rubbing or fatigue damage can occur. The most important factor in rubbing is rotor vs. bearing housing or seal interface and relative displacement.

Relationship of Vibration to Centrifugal Pump Acceptability and Reliability

Pumping machinery vibration provides diagnostic information, particularly relevant for the bearing and seal failure problems that vibration or excessive displacement may cause. These and other problems typically associated with vibration are responsible for a significant amount of the maintenance budget and lost-opportunity cost at many refineries, chemical plants, electric utilities, water/wastewater plants, and other facilities where industrial pumps are used. Applying effective vibration-based condition monitoring and troubleshooting methods therefore makes not just technical, but also financial sense. This chapter will leverage industry standards such as ISO 10816-7 (ISO, 2009), API-610 (API, 2021), and API-684 (API, 2010) and the combined HI 9.6.4 (HI/ANSI 2017), 9.6.5 (HI/ANSI 2017), and 9.6.8 (HI/ANSI 2019), so that the reliability-conscious End User or his consultant can use them as a guide, and at least a starting-point in setting up a condition monitoring program and its criteria for alarm and trip. It will be emphasized that when specific vibration values are compared to the values in the industry standards, this must be in the context of where a centrifugal pump is operating on its head versus capacity curve, as well as its speed. Lowest vibration will occur during operation close to the Best Efficiency Point "BEP" for any given speed. In addition to this, as already mentioned, it is also very important to determine how close the rotor critical speeds and rotor-support structural natural frequencies are to running speed or other strong forcing frequencies, such as not only 1x running speed, but also (often but with lesser effect) 2x running speed and vane or blade passing frequency. The vibration levels allowed should also account for machine maximum power level, relative stiffness of the machine mounting, and how much radial clearance is available within bearings and annular seals or the thrust balance device, as will be discussed in some detail. In certain circumstances (such as vertical pump "reed" frequency vibration discussed later), height of the measurement location above the foundation is also a factor, such that higher vibration levels may be permitted at locations farther above the foundation, as discussed in ISO 10816-7.

Vibration Standards, Informal and Formal: Intent and Basis

The fundamental basis for international vibration standards is that experience and intuition each indicate that increased vibration is likely associated with diminished life of a pump. This principle has two complementary aspects to it. The first aspect is that more vibration implies higher hydraulic and/or mechanical (e.g. imbalance) forces within the pump. A large portion of these forces must be reacted through bearings, which can only tolerate limited load, and the forces also lead to relative motion of the rotor versus the casing, thereby utilizing radial clearances (e.g. at wear ring seals, mechanical seals, and balance drums), resulting in likely increase in rubbing wear and therefore loss of some portion of the pump's remaining useful life.

The second aspect is that increased vibration indicates that the pump has already been damaged or degraded in some manner and may imply a catastrophic event will follow at some point before maintenance is planned. So, the first reason for concern over vibration level is that its increase will decrease reliability, while the second reason for concern is that it already signifies a problem exists. Both reasons are why vibration is evaluated as part of condition monitoring, to at least periodically determine and be prepared to properly interpret centrifugal pump vibration.

The direct loss of pump life due to high vibration involves either wear of clearances or metal fatigue. Wear results from higher vibration displacement at close running clearances, such as in fluid film bearings and seals, because vibration displacement increases the opportunity for contact (fortunately lubricated to at least some degree). This is a "tribology" problem and is the primary reliability issue associated with excessive vibration (Marscher, 1987). The other issue, metal fatigue, can take place in rolling element bearings, on mechanical seal faces, or in structural components such as bearing housings or their bolts, as well as in foot bolts, support pedestals, and suction or discharge nozzle fillets (Marscher, 2002). In shafting, fatigue can also occur due to unusually high oscillating axial thrust loads or due to high torsional vibration. However, shaft fatigue is seldom caused by lateral vibration with the exception of near the driver coupling due to coupling problems (such as severe imbalance or misalignment of the coupling) since radial clearances inside of centrifugal pumps typically restrict shaft lateral motion internally to modest levels that do not permit excessive bending stresses to develop (Marscher, 1997).

Vibration Measurement Form

As discussed in the various standards, such as those of ISO, there are several common forms by which to measure how much a location moves in an oscillating fashion (Figure 2.3). Each has advantages and disadvantages depending on the type of sensor that is affordable or convenient and the particular frequency range likely to be of interest. These vibration forms are as follows:

> **Displacement:** This is the distance the object is vibrating ("mils peak-to-peak", or "microns peak-to-peak). Displacement is most relevant to machine damage at lower rpms and frequencies.
> **Velocity:** This is "how fast" the object is vibrating ("inches per second peak" or rms, or "mm/s peak" or rms).
> **Acceleration:** This is the force on the attachment of anything attached to the vibrating part minus rate of change of velocity over time (expressed in "Gs peak"). This is most relevant to machine damage at higher frequencies.

These three different forms of the same vibration are related to each other through an "equation of motion", in which the forces that develop in reaction to the vibration (for example, at the bearings) all sum up to an overall net force level. The equation of motion for forced vibration is:

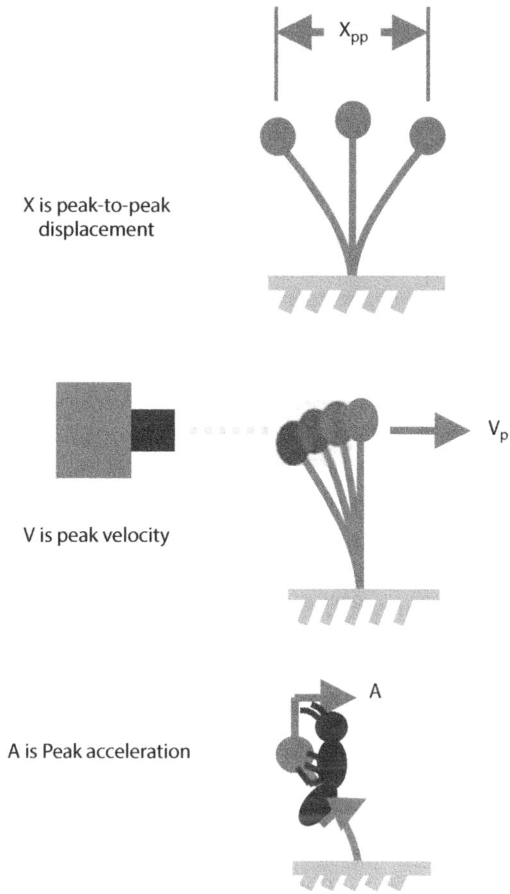

Figure 2.3 Illustration of common forms in which vibrational motion is quantified and considered.
Courtesy Mechanical Solutions, Inc.

Force = Mass X Acceleration + Damping X Velocity + Stiffness X Displacement

Damping can be considered in terms of the damping constant, c, which when multiplied by motion velocity is the force that an automobile shock absorber pushes back with when the auto goes over a bump. In pumps, most of the damping comes from the bearings and a smaller amount from the annular seals like wear rings, and a still smaller amount from deflection of stationary structural members. It is this damping that restrains vibrating motion from becoming infinite when a pump operates with a strong excitation force (like imbalance at 1x rpm) frequency being at a rotor natural frequency. Damping force, i.e., c times velocity, is also different from the acceleration and displacement reaction forces because it acts at right angles (is "perpendicular" or "orthogonal") to the acceleration and displacement forces, rather than acting in the same direction as they do.

Because acceleration is the rate of the rate of change of displacement, the ratio of acceleration versus displacement increases proportional to the square of the frequency of vibration. Similarly, the ratio of acceleration to velocity goes up directly proportional to the frequency and the ratio of velocity to displacement goes up directly proportional to the frequency. For this reason, acceleration is not a very sensitive measurement of vibration energy at low frequencies, while displacements are tiny for relatively high velocity or acceleration at high frequency. Therefore, it is best to use displacement probes where lower frequencies (e.g., less than about 10 Hz) are likely to be important, especially at locations where the clearance is tight. ISO 10816-7 develops some of its vibration criteria for vertical pumps based on this principle.

On the other hand, for very high frequencies, nearly all the reaction force from vibration will be in the vibrating mass times the acceleration and an accelerometer will provide the best measurement of potentially damaging vibrational energy (Marscher, 1987). No present vibration standards for pumps are based on this since pump accelerations are seldom high at frequencies above the range where vibration velocities have been found relevant to reliability.

The situation in which vibration velocity best represents vibration energy is at or near a natural frequency resonance, where a pump is supposed to be designed and installed to avoid. Nevertheless, velocity is what is typically used by modern standards to evaluate vibration severity level since it is considered a "best compromise" across the largest range of frequencies where machines like pumps are likely to experience strong excitation forces, i.e., typically from about 20 Hz to 500 Hz for a broad range of pump types and speeds.

Nearly all major standards today base vibration acceptance criteria upon RMS vibration velocity, with sometimes tighter standards at specific important frequencies such as 1x rpm, e.g., API-610. In cases where running speed is very low (e.g., below 600 rpm) or in locations where clearance may be less than the vibration displacement at reasonable levels of vibration velocity (e.g., near fluid film radial bearings), displacement may be (and should be) used either in place of or in combination with vibration velocity. In any of these cases, when velocity is measured, it is typically determined using an accelerometer, the output of which is mathematically "integrated" to produce a velocity signal, essentially dividing the acceleration by its frequency. This approach is used rather than a probe engineered to detect velocity directly since no commercially practical method has been discovered to do the direct measurement of velocity.

Vibration Detection Sensors

Given the above, regardless of whatever the vibration source is or its details are, the resulting vibrational motion can be characterized in the three different, but inter-related forms. These forms are the displacement range (e.g., peak-to-peak mils or microns), velocity (in/sec or mm/sec rate of displacement change versus time, RMS), and acceleration (rate of change of velocity versus time, peak), as illustrated above. Most vibration standards are based on RMS (root mean square) velocity, but peak-to-peak displacement is used in certain situations, as described in the various standards.

Some locations where instrumentation is commonly applied to pumps, although typically not to the comprehensive extent shown, is provided by the example in Figure 2.4.

Note that bearing and/or lubricating oil temperatures are typically measured to complement vibration and that subsequent evaluation should account for speed in a variable speed pump. Lubrication oil monitoring for

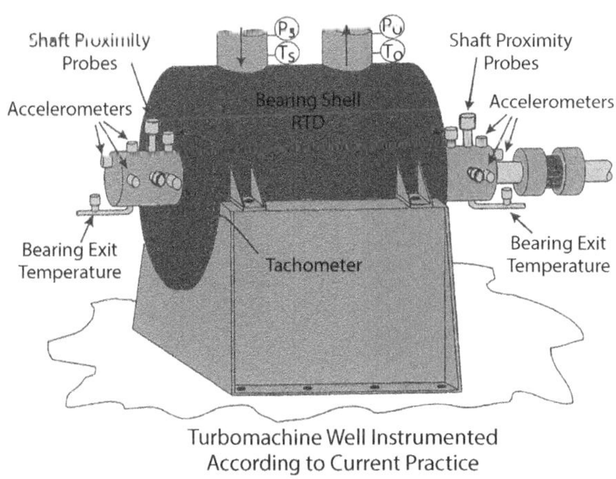

Figure 2.4 Types and locations of instrumentation in high value centrifugal pumps. Courtesy Mechanical Solutions, Inc.

particles, oxidation, and water contamination is also often applied to high value equipment. Depending upon the nature of a developing problem, sometimes bearing temperature or oil quality provides an earlier indication, or more sensitive indication, of the problem. This is sometimes true of severe pump/driver misalignment, for example. The best method to detect this may be a thermographic camera, observing temperature differential from one side of the bearing housing to the other (Marscher, 1997).

Vibration levels are typically evaluated in one of two groups. The most common is motion of a non-rotating structure (usually a bearing housing) in terms of its vibration velocity, as indirectly detected by integration of the RMS electrical signal output of an accelerometer (i.e., a sensor that directly detects acceleration, not velocity) firmly attached to the structure. The second form of vibration detection is direct measurement of lateral motion of a rotating shaft, typically by observing the peak-to-peak electrical output of a proximity probe, such as an eddy current or capacitance sensor, which is calibrated so that its signal correlates with the displacement of the shaft versus the probe mounting point and, therefore, the gap between the probe tip and the shaft surface.

Accelerometers

Accelerometers come in various sizes, from the size of a small insect to the size of a small soda can. The most typically used size is a cylinder roughly 1 inch (25 mm) long and a diameter about half its height. Examples of uniaxial (cylindrical shaped and measures vibration in only one direction) and triaxial (box shaped and measures vibration in three perpendicular directions) accelerometers are shown in Figure 2.5.

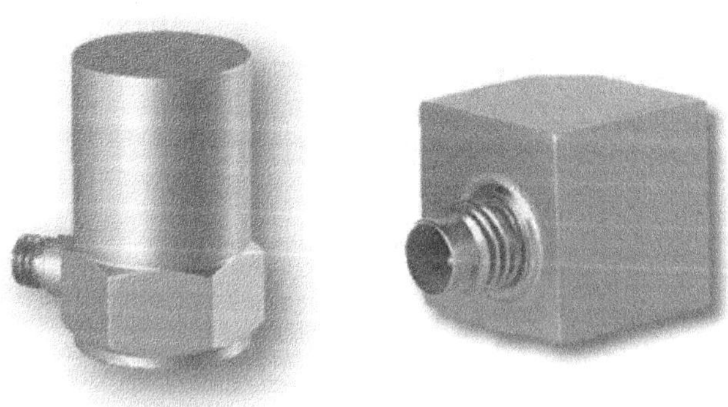

Figure 2.5 Typical Uniaxial (left) and Triaxial (right) Accelerometers (approximately twice actual size).
Courtesy Mechanical Solutions, Inc.

Generally, the smaller the accelerometer, the higher the frequency range it can detect. This also depends upon how it is connected, however. Magnets, wax, glue, and double-sided tape have been successfully used, in order of frequency transmission capability (magnets being the worst). In troubleshooting, hand-held accelerometers, with or without a small diameter "stinger" stick, in the hands of an expert can also be used to over 1 kHz, although low frequency (less than 5 Hz) amplitudes may become artificially high due to an unsteady hand. ISO Standard 5348 (ISO, 2021) provides good guidelines for accelerometer attachment and use.

Proximity Probes

Proximity probes are used to detect displacement directly. They may be of the older "shaft riding" type (similar principle to a dial indicator) or older still, "shaft vibrometer" which is a reed on a stick with an inertial mass on its end. Capacitance probes and light-based (e.g., laser or "photonic") probes are also sold for specialty applications. However, the most commonly used proximity probes in the machinery industry are eddy current probes, developed in the 1950's and 1960's by Don Bently, and independently by the R&D company MTI in Albany NY. An example of the most common implementation of these probes is provided in Figure 2.6.

A very small diameter wire is coiled in the eddy current probe's head and the cigarette shaped probe is attached to the bearing housing by a stiff holder bracket. The head of the probe is held about 50 mils (1.25 mm) from the shaft surface and the electricity coursing through its coil couples with currents and the coil induces locally in the shaft. The electrical current is sourced in a "proximitor box" that forms a circuit including the probe head and its eddy current "coupling" with the shaft, the transmission wire resistance and capacitance, and the power supply (typically 24 volts DC). As the gap changes, the coupling changes and the proximitor box signal output changes voltage proportional to the gap over about an 80 mil p-p (2 mm p-p) range. Proximity probe systems are more expensive to buy and more complex to install than accelerometers, so typically they have only been used on high value equipment. Some key factors concerning proximity probes include the following:

- Usually thread mounted
- Typical Calibration: 200 mV/mil
- Typical Gap DC Voltage: 10 volts (50 mils)
- Has poor sensitivity against non-magnetic steels and materials such as austenitic 300 series steel

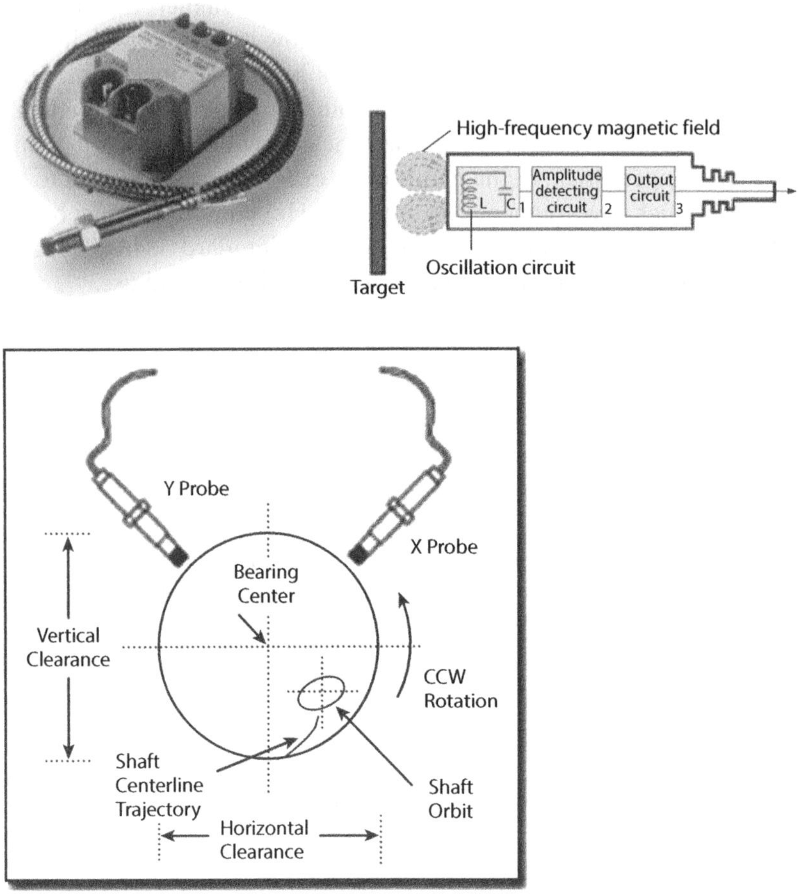

Figure 2.6 Typical eddy current proximity probe construction and implementation. Courtesy Mechanical Solutions, Inc.

Motion Magnified Video (aka Vibration Video Amplification)

A new non-contact methodology for detecting both rotor vibration and non-moving structural vibration has entered the scene in the last several years, involving optical measurement of the vibration displacement and then using this motion detection either directly or converted to vibration velocity. This technology has been independently developed by several groups and has been named Motion Magnification by the pioneering team at MIT and by the author's company, Video Vibration Amplification by the Vibration Institute, Motion Amplification® by one of the first firms to commercialize it, and in similar forms is known as Digital Image Correlation

(DIC), or Laser Holography. Each of these developments has a somewhat different approach, with each specific approach having benefits and drawbacks. With the exception of some laser systems, they all attempt to translate field-of-view light intensity variations location-by-location in a scene into vibrational displacement perpendicular to the direction of the view. This approach has the advantage of being able to simultaneously evaluate hundreds or even thousands of locations without the need to install large numbers of expensive probes and signal analysis channels, as well as associated cables and transmitters/receivers.

The arrival of these optical vibration detection and measurement methods, in practical and cost-effective forms, is a very useful development. The MIT (Massachusetts Institute of Technology) method that they named Motion Magnification is the procedure that the author is most familiar with. This procedure is able to use high speed high resolution video, obtained with a (reasonable cost) mass-produced camera, to shoot time periods of scenes in the machinery room that include entire machine trains, or groups of smaller machines, and their piping, attachments, and baseplates/foundations. Any oscillating motion that occurs in the scene is magnified, typically on the order of a factor of 1000x. In this manner, a vibration of 1 mil p-p (25 microns p-p) is shown in the replay as motion of 1 inch p-p (25 mm p-p). This makes normally "invisible" but diagnostically significant vibration easily visible and puts the vibration of a given machine or location on a machine in the context of the motion of other machines, locations, or components (e.g., piping or a soft foot's gap) around it. In many plants, this is becoming a useful troubleshooting tool, but is now moving into the continuous monitoring field as well. It is able to provide acceptably accurate non-contact vibration detection on equipment regardless of its temperature, height, or accessibility. The main drawbacks are that adequate lighting is required and camera characteristics must be properly accounted for by the software programming if accuracy is desired.

Also, for higher frequency phenomena (i.e., greater than about 500 Hz), most optical/video methods are challenged to provide useful results because displacements associated with specification vibration velocity limits get very small at these higher frequencies. In such situations (e.g., certain piping acoustics, or perhaps vane passing frequency), an acceleration or velocity-based detection method rather than most displacement-based video detection methods should be considered.

However, for typical frequencies of interest with regard to centrifugal pumps and their drivers, Motion Magnification and associated video techniques are a modern method of quickly determining the overall vibrating pattern of the machine, including its support, piping, and other system components. The traditional method of obtaining this information was either to gather data at many locations and manually plot it in cartoon fashion or to do this in a computerized method know as Operating Deflection Shape (ODS). ODS will be discussed in detail in a later section and the benefits of Motion Magnification versus ODS will be discussed at that time.

In addition to vibration displacement detection, some video systems marketed can also detect and magnify/display thermal growth and thermal or pressure distortion. This can be very useful in diagnosing piping load changes and how they correlate with vibration during start-up or under various operating conditions, as well as for observing pump casing key/base assembly hang-up during warm-up of boiler feed pumps, for example.

In general, as previously discussed, vibration standards are based on vibration velocity with peak-to-peak displacement possibly being a criterion for low speed equipment. However, acceleration is often useful in troubleshooting, particularly to detect bearing faults, as discussed below, and in detecting and quantifying damaging cavitation. In the instance of rolling element bearings, the acceleration pulses are consistently spaced with the spacing being related directly to faults in specific components within the bearing, as discussed later. In the case of cavitation, the acceleration also shows up as spikes, but in this case the spike amplitude and spacing is relatively random. The cavitation energy occurs generally in the range of 10 kHz to 120 kHz because of the extreme rapidity that the damaging cavitation bubbles collapse with (order of 10 to 100 microseconds). Such cavitation spikes should be observed with equipment capable of at least 50 kHz frequency response and preferably 100 kHz. Accelerometer probes should be small and rigidly attached to the pump casing, preferably on a casing wall external location with internal side line-of-sight to the impeller suction eye. Research has concluded that cavitation spikes measured in this manner are not causing damage if they exhibit less than 50 G's peak and are very likely to cause surface erosion damage if they exhibit over 100 G's peak, with more rapid erosion occurring roughly proportional to the rate of high acceleration pulses per second and roughly proportional to the G level once past 100 G's (Marscher, 2017).

International Vibration Acceptance Standards

Over the last several decades, the science of machinery condition monitoring and predictive maintenance has developed some very sophisticated approaches for reliable diagnostics. However, implementation of this technology has been only slowly entering the rotating machinery industry at the plant level. End Users are very busy with producing product, which is their core business. It is inefficient for them to develop deep expertise in this technology, or in the details of machinery design and the complex physics of vibration beyond what is required to optimize their plant's production. Therefore, until any new sophistications can be cast in a sufficiently user-friendly manner (the market perception is that this is a work-in-progress), nearly all vibration-based condition monitoring is presently performed by simple comparison of measured vibration levels to conservatively established norms. ISO does this in the most detailed manner, breaking vibration norms into various levels, including A-new, B-acceptable (usually Alarm is set at 125% of this value), C-may be a problem, check soon (usually Trip is set to 125% of this value), and D-shut it down quickly to avoid probable damage. Unfortunately, the "simply check the vibration level" approach will sometimes red-flag pumps that do not in fact possess reliability problems and in other (less frequent in the author's experience) cases, will green-flag pumps that are actually operating with serious present or impending issues. A more detailed description of these evaluation zones and the acceptance limits associated with them are provided in Tables 2.1 and 2.2.

To make vibration assessment more reliable for those who need to get deeper into the science (e.g. unnecessary down-time is expensive), in addition to the typically conservatively low vibration overall-level acceptance criteria, international standards are being developed for more detailed predictive maintenance as well. In addition to such global standards, many OEMs also provide useful troubleshooting procedures and guidelines. Typically, OEMs will rightfully encourage pump operators to use information from sources in addition to vibration sensors, such as temperature

Table 2.1 ISO vibration evaluation zones.

Zone A: The vibration of newly commissioned machines would normally fall within this zone.
Zone B: Machines with vibration within this zone are normally considered acceptable for unrestricted long-term operation.
Zone C: Machines with vibration within this zone are normally considered unsatisfactory for long-term continuous operation. Generally, the machine may be operated for a limited period in this condition until a suitable opportunity arises for remedial action.
Zone D: Vibration values within this zone are normally considered to be of sufficient severity to cause damage to the machine.

Table 2.2 ISO Vibration Velocity Limits, Units of mm/s RMS, for Centrifugal Pumps Over 1 kW and >2 Blades.

Category	**Category I**		**Category II**	
Power Level	≤ 200 kW	> 200 kW	≤ 200 kW	> 200 kW
Zone A Newly commissioned machines in preferred op range	2.5	3.5	3.2	4.2
Zone B Unrestricted long-term operation in allowable op range	4.0	5.0	5.1	6.1
Zone C Limited operation	6.6	7.6	8.5	9.5
Zone D Hazard of damage	> 6.6	> 7.6	> 8.5	> 9.5
Alarm setting (≈ 1.25 times the upper limit of zone B)	5.0	6.3	6.4	7.6
Trip setting (≈ 1.25 times the upper limit of zone C)	8.3	9.5	10.6	11.9

(especially in bearings or on bearing housings) and lubricant quality sensors (may evaluate samples off-line), and to account for process operating point values (e.g. discharge pressure, flow rate, and rotational speed).

The primary international standard for acceptability of a given level for vibration both in terms of bearing housing velocity and shaft relative displacement for centrifugal pumps is ISO 10816-7 (in the process of being incorporated into ISO 20816). Its recommended vibration acceptance levels are presented in Table 2.2. Note that Category I consists of pumps whose proper operation is critically required, while Category II is for less critical pumps.

There are other important public standards developed for pumps in certain specific industries and services, such as API-610 (the primary pump standard for the oil and petrochemical industry) and ANSI/HI 9.6.4 (the primary standard for water and wastewater services and often used for commercial building services). In addition to these acceptance standards, ISO is developing a 13373 series of standards aimed at condition monitoring and diagnosis/prognosis ("predictive maintenance") and there is ANSI/HI 9.6.5 for condition monitoring, as well as 9.6.8 for reliability enhancement of pumps and their systems through upfront analysis and detailed post-installation mechanical evaluation (including determination of as-installed natural frequencies). Certain countries also have their own standards (the German VDI-2056 was one of the earliest and best, and became a template for ISO), but most of these have become incorporated into, and/or replaced by, the ISO 10816-7 standard.

Pump Components Playing Key Roles in Vibration Diagnostics

The reason for monitoring vibration is either because certain components may become damaged by high vibration or that certain vibration frequencies and/or amplitudes imply some components are already seriously degraded and therefore in imminent danger of being degraded at a higher than normal rate. The concept of "higher than normal rate" is important versus "not degrading at all" because as one sage has said, "if it's not wearing out, then it's not running". So, as any pumps operate, it is expected its operating life will be limited. Depending upon the service, the life limit in terms of expected Mean Time to Failure, or recommended Mean Time To Repair, may be one year in a very difficult (e.g., aggressive chemical or slurry) application or 15 years in a typically mild application (e.g., clean water at constant speed and flow). In any of this range of cases, the probability of eventual failure may be based on specific components within the pump that gradually wear, erode, or experience fatigue. The components most typically at risk are the bearings, seals (wear rings and balance drums as well as mechanical seals or packing), the driver's flexible coupling, the impeller and other primary flow-path surfaces, and the casing load-carrying members, such as pump bearing housings, feet, and pedestal, as well as nearby foundation and mounting bolts. There are specific vibration characteristics associated with each component's mechanical integrity and/or vibration-driven degradation. We will discuss the vibration characteristics and issues with each of these key components now, beginning with the bearings.

Rotor Support by Bearings: Fluid Film Journal Bearings vs. Rolling Element Bearings

Rotor shafts must be supported by some type of bearing to keep elevated while resisting gravity and other loads (e.g., imbalance). A bearing must accomplish this while still allowing the shaft surface to rotate without the rotating interface wearing out. The two types of bearings practical for industrial pumps are fluid film journal bearings and rolling element bearings. Journal bearings develop support stiffness across the lubricant film entrained between the shaft cylindrical journal and the cylindrical bore of the bearing shell, with a stiffness of typically somewhere between 100,000 lbf/in, and 1,000,000 lbf/in, depending upon the size of the bearing diameter and length, with larger sizes possessing more stiffness. As part of developing this stiffness, the journal bearing also develops considerable damping (acts to discourage motion and absorb energy, like a shock absorber in an automobile chassis, and discourages vibration, especially near resonance) and cross-coupling (acts to potentially increase vibration in a possibly unstable, uncontrollable way, discussed later). More complicated forms of journal bearings, like fixed lobe or tilting pad bearings, are designed to maximize the ratio of damping to cross-coupling stiffness, while maximizing the "direct" stiffness in the direction that the load is applied. If vibration of a pump or motor is primarily below running speed ("subsynchronous", usually a little less than half running speed when it occurs), it suggests that the cross-coupling is winning out over the damping, which can occur, for example, when a bearing is underloaded (e.g., as caused by unloading due to misalignment). On the other hand, if the bearing is over-loaded, then it can wear out more quickly and maybe even be immediately damaged by a rub once its clearance has been taken up over a given arc of the shaft vibration "orbit". The vibration response to such a rub is typically strong multiples of 1x running speed, perhaps up to as high as the 10th harmonic, but more usually up to about the 4th harmonic. Sometimes, exact ½ running speed and up to its first 10 or so harmonics will also be significant, especially in vertical turbine pumps, as discussed later. Another factor to consider with the degradation of a journal bearing is that its stiffness and damping can significantly increase as clearances increase, leading to unexpected resonance where there was none before (Bowman, 1990).

A ball bearing or cylindrical (sometimes spherical) roller bearing contains a set of balls or roller cylinders, trapped between an inner race attached to the rotating shaft and an outer race attached to the stationary bearing housing. The ball or roller rolls as the shaft rotates, while the bearing "inner race" rotates with the shaft. Many assume that such bearings are nearly infinitely stiff because (except for a very thin lubricant film on the surfaces of the rollers and races) the shaft load is transmitted to the ground in a nearly metal-to-metal contact manner. However, besides the ultra-thin lubricant film, the balls or rollers also have finite stiffness and the bearing housing itself is not infinitely stiff. The net result is that the overall stiffness of rolling element bearing support is typically about 2 to 3 times that of journal bearing support. This is significant but not high enough that the bearings should be considered rigid mounts for the shaft. On the other hand, both damping and cross-coupling are typically very low in rolling element bearings.

The primary aspect of bearing type that is relevant to vibration diagnostics is what frequencies indicate that one or more bearings are exhibiting

$FTF = \frac{N}{2}(1-\gamma)$ Fundamental Train Frequency

$BSF = \frac{D_P}{D_b} * \frac{N}{2}(1-\gamma^2)$ Ball Spin Frequency

$BPFO = N_b * \frac{N}{2}(1-\gamma)$ Ball Pass Frequency Outer Race

$BPFI = N_b * \frac{N}{2}(1+\gamma)$ Ball Pass Frequency Inner Race

N = RPM

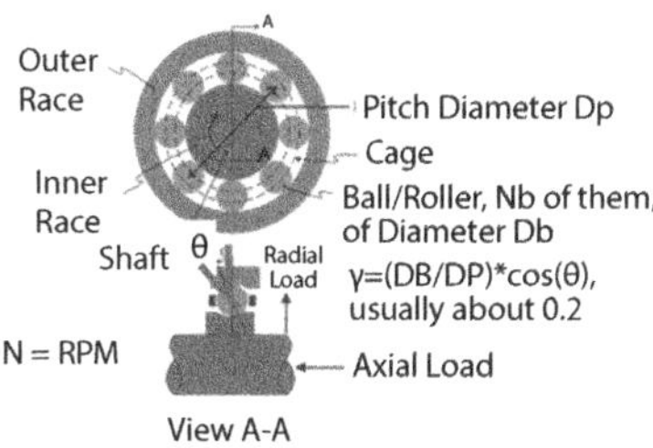

Figure 2.7 Rolling element bearing defect frequencies.
Courtesy Mechanical Solutions, Inc.

a problem relevant to reliability. In the case of journal bearings, lubricant breakdown or bearing overload which results in rubbing typically leads to harmonics (i.e., integer multiples) of running speed and sometimes in exact ½ running speed, multiples as well. In the case of rolling element bearings, so-called defect frequencies become important (although usually still not dominant) spikes in the vibration versus frequency spectrum. These defect frequencies may be listed as shown in Figure 2.7.

The defect frequencies are always present, even in new well-manufactured bearings. However, they become significantly stronger when the bearings have become degraded and tend to impulsively excite very high frequency "ringing" in the bearing races. This latter situation is what enables the effectiveness of the acceleration-squared diagnostic offered by some instrumentation vendors. In the author's experience, such G-squared parameters are useful for tracking bearing health, but primarily on a relative basis, not on an absolute basis unless the bearing is already badly damaged. The quantitative G-squared value depends upon bearing static load and frequency of the vibration excitation forces, lubricant quality and viscosity, and bearing housing mass. Therefore, percentage change in G-squared over time is what is of significant value as a diagnostic parameter.

Rotor Support by Seals: Annular Seal "Lomakin Effect"

Annular seals (e.g., wear rings, interstage bushings known as rear wear rings, and balance drums) in centrifugal pumps accidentally act as bearings as well. As such, annular seals can greatly affect rotordynamics by changing the rotor support stiffness and therefore, the rotor natural frequencies (Black, 1979 and Childs and Moyer, 1985). This can either avoid or induce possible resonance between strong forcing frequencies, particularly at one and two times the running speed and one of the lower natural frequencies (Marscher, 1989). This effect is so strong for multistage pumps that API-610 requires that annular seals be taken into account when calculating critical speeds for pumps of three or more stages and that their clearances be assessed for both the as-new and 2x design clearance "worn" conditions. This provision by API is because the stiffness portion of this "Lomakin Effect" (first noticed by the Russian pump researcher A. Lomakin) is roughly inversely proportional to radial clearance. It is also directly proportional to the pressure drop and (roughly) the product of the seal diameter and length. An illustration of how the Lomakin Effect occurs is given in Figure 2.8.

Keep in mind that Figure 2.8 is a simplification of what occurs in a liquid annular seal. It illustrates the primary effect of flow velocity decreasing in the area being closed down and speeding up in the area opening up. Because of this, Bernoulli's equation indicates that static pressure increases on the closing side and drops on the opening side, giving a net restoring "bearing" force. However, if significant relative swirl (i.e., swirl other than the most typical, but not guaranteed, half shaft rotational speed) is present at the seal inlet, the physics gets more complex. In such a case, besides flow patterns that depend strongly on shaft eccentricity, a given stream tube of flow that starts at one end of the seal must actually speed up as it encounters reduced gap because of incompressible flow continuity, rather than the flow slowing down at the narrower gap due to friction factor. This can lead to negative Lomakin stiffness when otherwise positive stiffness would be predicted. In addition, swirling flow at the inlet of these seals tends to increase harmful cross-coupling and lessen beneficial effective modal damping of the rotor. This increases vibration response near resonance and, in the extreme, can lead to roughly half speed whirl turning

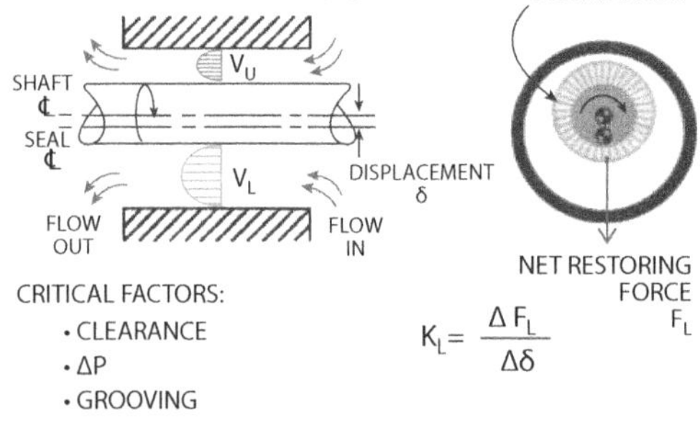

Figure 2.8 Illustration of lomakin effect stiffness KL in Annular sealing passage. Courtesy Mechanical Solutions, Inc.

into shaft "whip", with strong and unstable shaft orbit pulsations that are present at the subsynchronous natural frequency (usually the first bending mode of the rotor). Appropriate patterns of holes in the inner surfaces of annular seals have been found to increase their damping and decrease their cross-coupling with no harmful reduction in stiffness.

As rule of thumb, for short plain annular seals (e.g., ungrooved wear rings) in water, the Lomakin Effect stiffness is approximately equal to 0.2 times the pressure drop across the seal times the seal diameter times the seal length, divided by the seal diametral clearance. For grooved seals or long L/D (greater than 0.5) seals, the coefficient 0.2 diminishes by typically a factor of 2 to 10. In highly viscous fluids, the coefficient can decrease very significantly, i.e., by a factor of 10 or more.

In troubleshooting the vibration of multistage pump rotor systems, determining the likely value of the Lomakin Effect and how its control can affect pump vibration level (particularly by avoiding resonance) is an important factor that should not be ignored. Essentially, each wear ring and the balancing device are acting as additional bearings, individually weak but collectively changing a multistage pump's first natural frequency (typically U-shaped) by as much as a factor of 2 to 3. Therefore, in troubleshooting, this should be accounted for.

The other types of seals common to pumps are mechanical seals and wound packing in a "stuffing box", with such seals being placed between the pump flow path and the bearings and environment. The effect on vibration of a mechanical seal is negligible. The reverse is not necessarily true, since vibration of more than about 2 mils (50 microns) p-p have been reported to cause face wear, or chipping and fatigue of seal components such as o-rings and springs. Similar to mechanical seals, packing wear will increase with higher vibration, and it have little effect on shaft support stiffness and therefore, rotor natural frequencies. However, unlike mechanical seals, packing strongly increases the damping of any rotor mode that would have significant motion at the packing location if the packing was not present. This can cause a potentially resonant natural frequency to respond with negligible vibration, even when the excitation frequency should be resonant with it, and then this becomes a problem if the packing is replaced with a mechanical seal. Another important difference with packing is that it must be adjusted to allow a small amount of leakage to promote "boundary lubrication". If the packing is adjusted too tightly, it can cause the shaft to chatter as its surface rubs against the packing coil material's interior surface, causing "stick-slip" with unexpected high broadband vibration, especially at frequencies below running speed. In fact, over-tightened packing can cause the pump to visibly "shudder" or shake.

Couplings

Sometimes, industrial pumps are connected to their driver with a rigid coupling, such as metal clamshell halves or rigid flanges at the end of the driver and driver shaft, or sometimes with axial face teeth on each flange ("curvic" or "Hirth" coupling), in either case, then bolted together without any intermediate element such as an elastomeric bushing or (in the case of jawed flanges) a spider element. Such rigid couplings do not introduce any new vibration phenomena or frequencies, as long as they are nearly perfectly aligned. However, if misalignment exists statically or is induced dynamically by pump vs. driver vibration, especially out-of-phase with each other, then certain frequency components get emphasized, possibly to the point of violating the acceptance standards. The most common frequency induced is 2x running speed, but additional 1x or other integer harmonics are possible depending upon the orbit that results as the misaligned shafting interacts with the inboard bearings, as discussed with examples in the alignment section below.

To minimize the effect of misalignment on vibration and on coupling and bearing wear, various designs of flexible couplings are typically implemented. These can be as simple as a hollow rubber connector between shafts for very small pumps or can be an expansion of the rubber tube to form a large "donut" that looks like an automotive tire or rubber spiders trapped between jawed flanges at the end of each shaft. Entirely metal flexible couplings, most typical for larger pumps, include male-female gears at opposing shaft ends forming a "gear coupling", a flexible disk or diaphragm where the I.D. is attached to one shaft and the O.D. is attached to the other shaft end, or a "disk pack" coupling in which a stack of thin disks is bolted around the circumference, where first the driver shaft coupling hub is attached and then 30 degrees later (for example), the pump coupling hob is attached and so on alternating around the circumference. For certain pump services, a "U-Joint" or "Hooke's Joint" coupling is common, especially in mining or wastewater dry pit pump applications.

In the case of all flexible couplings, it is also possible (even desired for maintenance purposes or to better accommodate lateral misalignments) to place a "spacer piece" or "drive shaft" between opposing sets of flexible couplings and their hubs. This reduces the strength of the 2x vibration component or other harmonics when misalignment occurs, but can lead to 1x vibration due to spacer imbalance or eccentricity if sufficient care is not taken in spacer balancing and mounting or if the end joints (e.g. U-Joints) wear out.

An excellent machinery alignment standard is the ANSI/ASA S2.75, "Standard Shaft Alignment Methodology", as will be discussed further when misalignment effects are presented in more detail.

Bearing Housings and Attachment Bolts

Bearing housings and their attachment bolts and flanges are intended to very rigidly react forces between the shafting and the pump casing. As such, the primary impact of significance by them on vibration is to either somewhat reduce the shaft to lower natural frequencies (because even "rigid" does not imply infinite stiffness) or to exhibit looseness, such as might occur if the housing is cracked or the housing bolts are loose or broken. In the case of cracked or broken housing components, the vibration spectrum typically will show increased 1xRPM harmonics, together with some broadband vibration well below running speed and sometimes together with exact 1/2x harmonics as well.

One other important issue that can be associated with bearing housings is a cantilever (i.e., like a diving board) mode with natural frequency close to the impeller vane passing frequency. Bearing housings attached to multistage pump casings through seal-access "drip-pockets" are particularly prone to this and in recent years are often replaced by aftermarket "360 degree" attached bearing housings, driving the bearing housing natural frequency far up and away from the vane pass. If bearing housing resonance occurs, it typically does not harm fluid film journal bearings or seals because the vibration displacement is low due to the relatively high frequency involved. However, fatigue of instrumentation wires and oil lubrication or drain piping has been experienced and velocity-based vibration criteria are violated, causing chronic alarms that, from a reliability standpoint, may be false and therefore are a considerable annoyance.

Finally, bearing housings are important to condition monitoring and troubleshooting because vibration is measured on only the bearing housings in most machinery standards. The exception is when the shaft is observed directly with proximity probes, but in even that case the displacement measurement is the displacement of the vibrating shaft *relative to* the vibrating bearing housing.

Pump Casing, Feet, and Foot Attachment Bolts

Faults in the casing proper, or in its support feet or the attachment of these feet to the baseplate or foundation, have two primary effects. The first is to decrease the rigid body rocking natural frequencies, possibly by a significant amount (e.g., up to 25%). The second is that cracks opening and closing, or a "soft foot" tapping, will lead to clipping of what would otherwise be sinusoidal motion at the primary forcing frequency, nearly always 1x RPM. This clipping results in strong integer harmonics at 2x, 3x, etc. of running speed. The number of harmonics that get strongly excited depends upon the length of the clipping of each sine wave, i.e., the percentage of time that the crack or space under the lifted foot is open versus closed. The shorter the closure time, the greater the number of harmonics.

Pump Pedestals, Baseplate, and Foundation

Flaws in the primary support under the pump casing can shift the rigid body rocking modes of the pump and/or motor, thereby possibly shifting one of these natural frequencies into resonance. In larger pumps, such as boiler feed pumps, the rocking modes are often designed to be 25 to 50% above running speed, since lower natural frequencies are associated with what is considered too flexible a support, while higher natural frequencies require significantly more extent (and therefore expense) in the foundation and baseplate. Therefore, as pedestals, baseplates, and foundations age over decades (e.g., through corrosion, fatigue, etc.), many power plants have experienced a gradual onset of resonance where there was none when the pump was first installed. This type of issue is easily detected by a properly performed impact test and then a full modal test or a Motion Magnification high-speed video evaluation can be performed to determine the faulty component (e.g., a cracked pedestal) so that it can be repaired.

Piping, Suction, and Discharge

While piping is not a component of the pump per se, other than the pump itself, it typically is the most influential component of the "pump system". The piping has mechanical effects, hydraulic effects, and acoustic effects. The primary mechanical effect is that if it is rigidly connected to one or both pump nozzles, it adds both mass and stiffness to the assembly so that pump structural natural frequencies get shifted to unexpected values. Because of Murphy's Law, there have been many instances where one of these shifted natural frequencies has become close enough to an important excitation force frequency (like 1x RPM) to result in a serious resonance. Such cases can be resolved by stiffly bracing the piping at the pump nozzles or by attaching the piping with flexible joints. Tie rods are typically required between pump and piping nozzles to carry the hydraulic load (pressure times pipe cross-sectional area) without over-stressing the flexible element (typically some kind of rubber or other elastomer, possibly in "donut" or bellows form). However, if these tie rods are torqued tightly against each flange, as is often erroneously done, this undoes much of the benefit of the tie rods because it relatively rigidly couples the pump and piping again, at least in the direction of the pipe centerline. This overtightening can also pull the pump out of alignment with its driver or even distort the pump casing (especially in Vertical Turbine Pumps). On the other hand, if the tie rods are loose, then as the piping pressurizes it places a large and unexpected (by the OEM) piping load on the pump nozzle, equal to the nozzle pressure times nozzle cross-sectional area, which can distort the pump casing using the nozzle as a lever. If this occurs, pushing the pump casing as it reacts against its foundation can lead to significant pump/motor misalignment, with the resulting frequency harmonics of misalignment, as discussed in the alignment section below. Therefore, tie rods are best adjusted "finger-tight" but not torqued when the pump is not running and there is no pressure in the piping. If chronic misalignment symptoms (like 2x RPM) persist after multiple cold alignments, improperly tightened tie rods should be investigated.

Loose attachment of tie rods also enables stronger vibration excitation by oscillating hydraulic or acoustic forces. Whatever the travel of the pump nozzle that occurs due to the static pressure in the unsupported pipe/pump nozzle, the vibration due to pressure oscillation will increase as the ratio of the pressure pulsation to the average hydraulic pressure.

However, regardless of the presence or not of tie rods and whether or not they are properly tightened, hydraulic forces as well as piping acoustics can lead to high pump vibration. Proper piping attachment typically attenuates vibration sourced in such fluid oscillation but does not eliminate it. Acoustic pulsations may occur if the pump running speed or, in particular, the impeller vane passing frequency of the pump is close to one of the first several piping system acoustic natural frequencies. The most powerful of these acoustic natural frequencies is the so-called quarter wave, which is equal to the speed of sound in the pipe (normally only about 90% of the speed of sound in the fluid as listed in a handbook because of the effects of pipe wall flexibility) divided by 4 times the distance from the pump impeller O.D. to the first major interruption in a straight suction or discharge pipe, such as an elbow or tee, or entry into or from a header. Other important acoustic natural frequencies potentially prone to resonance are 2x, 3x, and 4x the quarter wave frequency depending upon piping details that affect boundary conditions at each pipe end. Acoustic pressure pulsations that can cause vibration can also be caused by stand-pipe or tee-pipe quarter waves or by vortex shedding from tee-openings or valve components, if the resulting frequencies are at impeller vane pass. These acoustic generators are beyond the scope of this chapter but are covered in Blevins, 1977. The simple acoustic modes that most commonly cause vibration problems, and their frequencies, are shown in Figure 2.9.

Hydraulic forces, other than acoustics, can cause high vibration at the vane pass, particularly in pump impellers with only one to three vanes, especially if they are single volute tongue pumps (like dry pit wastewater pumps) or if there are certain combinations of impeller blades and diffuser vanes (Bolleter, 1988). Sometimes, cavitation will increase vibration at vane pass as well, but surprisingly sometimes vibration at vane pass actually decreases in the presence of cavitation, apparently because the bubbly fluid smooths out the vane-to-vane pressure gradients a bit. In any event, the main effect of cavitation on vibration is to excite the pump casing at very high frequencies. Cavitation vibration is most evident in the impact-like impulses that it causes in high-frequency content (at least 10 kHz) time domain plots of vibration, as evidenced by 100 Gs or higher vibration acceleration spikes on the suction casing with impeller line-of-sight if the cavitation is damaging. This is presented in Figure 2.24 and presented later in the detailed discussion of cavitation-produced vibration.

Sometimes, cavitation can occur as a pulsating surge in which high flow through the impeller inlet causes decrease of static pressure to below the vapor pressure, which results in extensive cavitation bubble and vapor formation in the passage, blocking it. This makes the flow drop dramatically, which eliminates the static pressure drop and leads to the vapor being washed out or re-absorbed, freeing the passage once again for high flow. This behavior continues in a cyclic manner, with bursts of high-G

a = Speed of Sound in Fluid

$$fn_{A_{1/4}} = \frac{a}{4L}$$

$$fn_{A_{1/2}} = \frac{a}{2L}$$

For Natural Frequency Number "i":

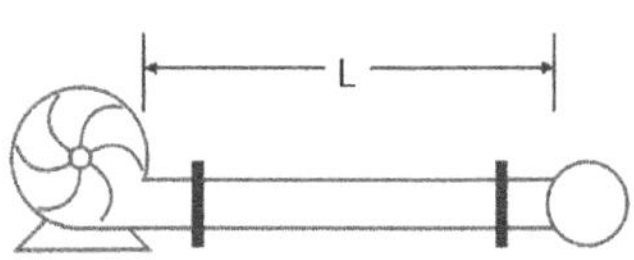

$$fn_{A_i} = \frac{(2i-1)\,a}{4L}$$ for quarter wave (closed one end)

or $\frac{ia}{2L}$ for half wave (both ends open or both ends closed)

Figure 2.9 Acoustic modes in constant diameter piping the conventional speed of sound of the liquid must be adjusted down (usually about 10%) to account for the flexibility of the metal pipe walls, as shown in the references.
Courtesy Mechanical Solutions, Inc.

cavitation occurring periodically, with a frequency typically in the range of 7 to 11 Hz in those pumps prone to this phenomenon, especially double suction pumps or other pumps that must deal with low suction pressure.

A more common cause of high vibration by hydraulic forces is the operation of the pump not close enough to its Best Efficiency Point (BEP), as discussed later with the assistance of Figure 2.20. In the extreme (say, at flows outside the range of 70% to 120% of BEP), fluid stalling in the impeller can occur, leading to suction or discharge internal flow recirculation and possible rotating stall. The resulting vibration tends to feel to nearby personnel like pulsing or rumbling and most of the vibration is "broadband" (no single frequency) below running speed. If rotating stall gets fully established, then more coherent pulsating vibration occurs, typically exhibiting vibration frequencies of 80% running speed, +/- 15%.

Pump Drivers

The vibration of pump drivers, such as AC motors (induction and synchronous, with and without VFDs) and steam turbines, with or without intermediate gear boxes are outside of the scope of this chapter. That being said, the primary frequencies to be concerned with in the context of pump diagnosis are x running speed for imbalance or bent shaft and possibly 2x running speed for driver-pump coupling misalignment, as discussed later with the assistance of Figures 2.17 and 2.18. In the case of motors, 2x slip frequency or 2x line frequency, as well as VFD pulse construction frequencies (typically 6x and 12x motor electric feed frequency) have also been problematic. The slot passing frequency of induction motors and the blade passing frequency of steam turbines are seldom a vibration problem, although they can cause excessive noise, especially if they are coincident with a "bell" mode of the casing or frame.

Evaluating Causes of Excess Vibration: Excitation vs. Amplification

All of us know by intuition that excessive vibration can be caused by shaking forces ("excitation forces") that are higher than should be expected. As mentioned above, various pump and system components could cause or facilitate vibration at a given frequency. For example, as shown in Figure 2.10, maybe the pump or driver rotor imbalance is too high, leading to elevated 1x RPM vibration. Other shaking forces commonly include misalignment, impacts from looseness, and rubbing. As discussed above on a

Frequency	Source
0.05 – 0.45 x	Diffuser Stall
0.41 – 0.49 x	Instability
0.500 x	Rubbing
0.55 – 0.95 x	Impeller Stall
1 x	Imbalance
1 x + 2 x	Misalignment
#Vanes x	Vane / Volute Gap
#Blades x	Blade / Diffuser Gap

Figure 2.10 Common vibration frequencies and their sources. Courtesy Mechanical Solutions, Inc.

component basis, sometimes shaking forces are hydraulically based, such as from blade passing forces, suction recirculation at low flow, or piping pressure pulsations. Motor-sourced excitation forces can be electrically based, such as from uneven rotor/stator air gap, a broken rotor bar, a stator phase-to-phase or phase-to-ground short circuit, or from VFD harmonic pulses reacting electromagnetically across the rotor/stator gap.

Vibration-causing force may occur at just one frequency ("narrow-band") or over a range of frequencies ("broadband"). In either case, we can pick apart the force's effects one frequency at a time with the aid of a Fast Fourier Transform (FFT) analyzer. This is very useful because vibration occurring at a given identified frequency implies a short list of potential problems physically associated with that frequency--a condition monitoring dream! As a limited example, a list of some of the most common vibration frequencies in pumps and the physical phenomena associated with them is given in Figure 2.10.

Vibration at any of these frequencies can be greatly amplified by the phenomenon known as natural frequency resonance, illustrated in Figure 2.11 and discussed by Blevins, 1979, Bowman, 1990, and Marscher, 2013. At frequencies well below the lowest or "first" natural frequency of the rotor (the frequency of simplest shape vibration that occurs when a rotor or structure is struck like a tuning fork), nearly all the force is reacted by displacement acting against the support stiffness. Therefore, this displacement becomes a reaction force when multiplied by the "spring" stiffness, for example some combination of the rotor stiffness and bearing plus bearing-support stiffness. Conversely, at frequencies well above the natural frequency, nearly all of the force is reacted by the acceleration, which becomes a reaction force by multiplying it by the rotor mass ("F = ma"), weighted by the zones where the rotor moves a lot compared to non-moving vibration "node points".

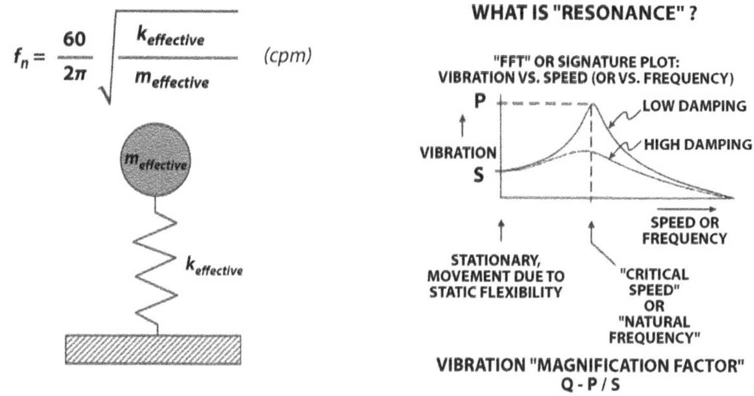

Figure 2.11 Illustration of natural frequency "fn" resonance and effects of damping. Courtesy Mechanical Solutions, Inc.

Therefore, for a given shaking force at a given frequency, the force reacts (per one of Newton's laws, "action equals reaction") against the combination of a) support stiffness, which leads to a reaction force proportional to displacement, b) the vibrating mass that the force is causing to move, which leads to force proportional to the acceleration (in vibration, acceleration always acts in the opposite direction of displacement), and c) any energy-absorbing damping associated with the vibrational motion, which leads to reaction force proportional to the velocity. Observing displacement from a given reference frame where the displacement starts out as negative, once the negative oscillating displacement crosses zero displacement and becomes positive, the acceleration needs to push in the opposite (now negative) direction to slow the vibrating component down or it would take off into space instead of oscillating. Exactly at a natural frequency, the displacement for the natural frequency's vibrating pattern in space (the "mode shape") times the stiffness discouraging that motion cancels the mass times acceleration for the node shape. In fact, this equivalency of spring reaction force versus mass inertial reaction force is what causes a "natural frequency" of oscillation to occur.

Process of Resonant Amplification due to Coincidence of Excitation and Natural Frequencies

The mathematical description of the physics of vibration indicates that in the absence of damping, the vibration amplitude equals the amount of displacement if the excitation force was applied once, statically, and divided by the difference between the spring reaction effect minus the mass reaction effect. Therefore, at a natural frequency, with the displacement effect exactly canceling out the acceleration effect, the vibration due to an oscillating force of the same frequency would become magnified to infinity if it were not for the presence of damping. Any damping eventually consumes all the initial vibrating energy, bringing the rotor to a gradual stop if the force stops. More importantly, as shown in Figure 2.11, it limits the amount of magnification (or "amplification", Q) if the force is continuously supplied in an oscillating manner. In other words, at resonance, whereby definition exciting frequency equals natural frequency, only damping times velocity is available to react the force at a frequency equal to the natural frequency. This is why damping in a rotor system is something to be maximized if practical.

Impact Test Method of Determining Natural Frequencies

In the performance of vibration diagnostic testing, it is possible to perform tests to specifically determine the natural frequencies of the pump and its driver, as well as the damping at those natural frequencies. Typical vibrating patterns in space, or "mode shapes", are shown along with simplified formulas to predict the natural frequency values in Blevins, 1979.

As discussed earlier, operating a pump at one of its natural frequencies would theoretically cause infinitely high vibration if it were not for the presence of energy-absorbing damping. The damping is quantified in terms of percent of critical damping zeta (i.e., the amount of damping to quash a vibration after just one oscillation) that is present or alternatively as a "log dec", i.e., logarithmic decrement, which is approximately 2 times pi (=3.14) times zeta. One other measure is the amplification factor Q that predicts the amount of motion that will be achieved at resonance versus the deflection that would have occurred if the force was just slowly applied statically. Q = 1/(2*zeta).

The most straightforward type of test to determine natural frequency and damping, using modern instrumentation of modest cost, is the impact modal test, sometimes called the "bump" test, the force vs. time and force vs. frequency of a typical such test being shown in Figure 2.12.

The damping can be determined from the natural frequency peak obtained in a vibration response vs. frequency plot, known as a Frequency Response Function (FRF) plot. The width of the peak corresponds to the damping. Specifically, two times the width in delta-frequency at the "half power point" of amplitude, divided by the natural frequency that the peak is center at, is equal to the damping ratio, zeta. Since the vibration power is

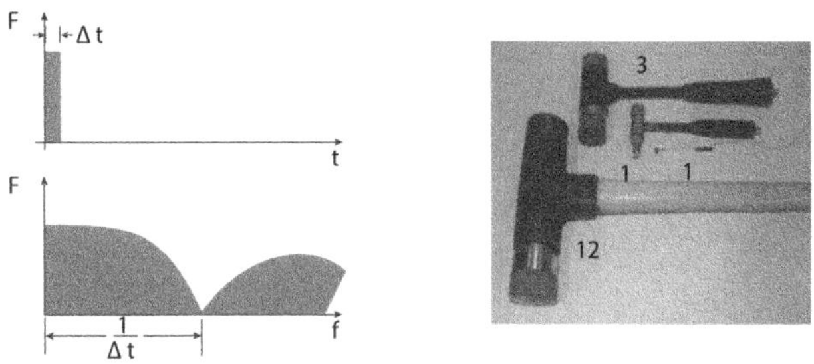

Figure 2.12 Impact test exciters and principle: A brief impact force excites a broad frequency range.
Courtesy Mechanical Solutions, Inc.

proportional to amplitude squared, if the vibration FRF plot is amplitude vs. frequency, then the half power point is the natural frequency amplitude peak divided by the square root of 2.

Once natural frequencies are determined by either impact testing or other means, such as waterfall plots of frequency spectra at ever increasing speed (see Figure 2.22 later as an example of a waterfall or "cascade" plot), the troubleshooter should assess the potential for them to cause a resonance by being too close to an important excitation frequency. An excellent graphical method for presenting this is the Campbell Diagram, as discussed in API 684, 2010 and HI/ANSI 9.6.8, 2019. The Campbell Diagram plots both excitation and natural frequencies on the vertical axis and running speed variation on the horizontal axis. Angled lines emanating from the zero-zero origin of the plot represent important excitation frequencies such as 1x, 2x, and vane pass, while the natural frequencies are gently sloping, nearly horizontal lines, typically sloping upward because of extra stiffening present, especially in rotors, as speed and therefore (usually) bearing stiffness increases. On this plot, if the any of the slanted excitation force lines cross a natural frequency line within the speed range of the pump, then a resonance is predicted. Typically, at least a 10 to 25% margin is desired relative to such a crossing, depending on how much damping is in the natural frequency (refer to API 610, 2021) and how accurately the natural frequency has been determined to refer to HI/ANSI 9.6.8, 2019). An example of a Campbell plot is shown in Figure 2.13.

Rotor natural frequencies can be difficult to determine during operation and because of the Lomakin Effect, cannot be practically determined unless the pump is operating at the load of interest. Marscher, 1999b, provides a practical method based on time domain averaging of the rotor vibration response to a large number of random impacts. Besides the Lomakin Effect,

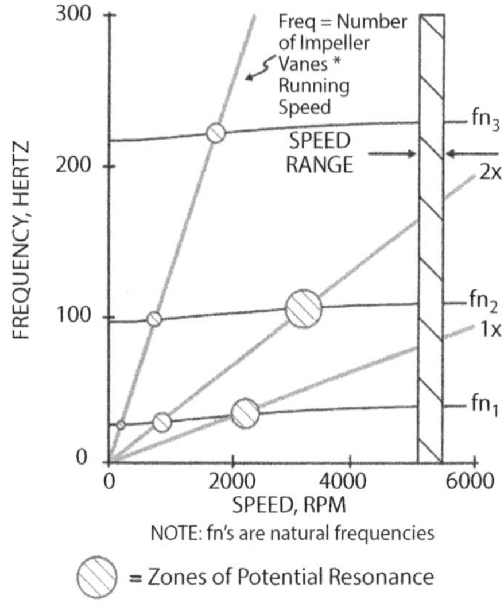

Figure 2.13 Using a campbell diagram to predict resonance problems. Courtesy Mechanical Solutions, Inc.

rotor natural frequencies are sensitive to fluid added mass and shaft/sleeve or shaft/impeller press-fit (Marscher, 2013), causing shaft natural frequencies to be much different in water than in air and are likely to change with operating speed because of centrifugal effects of fit-ups, respectively.

Specific Forces in Centrifugal Pumps

Figure 2.10 gave a limited list of some common excitations in pumps. Figure 2.14 shows the locations or origin of a larger sample (still not ALL the possibilities) of typical potentially major excitation forces in a centrifugal pump. The example is for a between-bearings multistage pump, but the forces shown conceptually apply to cantilevered and/or single stage pump as well. Some of the forces shown are mechanically sourced and some are hydraulic in nature. A number of the forces listed in Figure 2.10 are illustrated.

Mechanical Excitation Forces

Balance

Based on End User surveys by EPRI (Electrical Power Research Institute) and other qualified groups, imbalance is the most common cause of

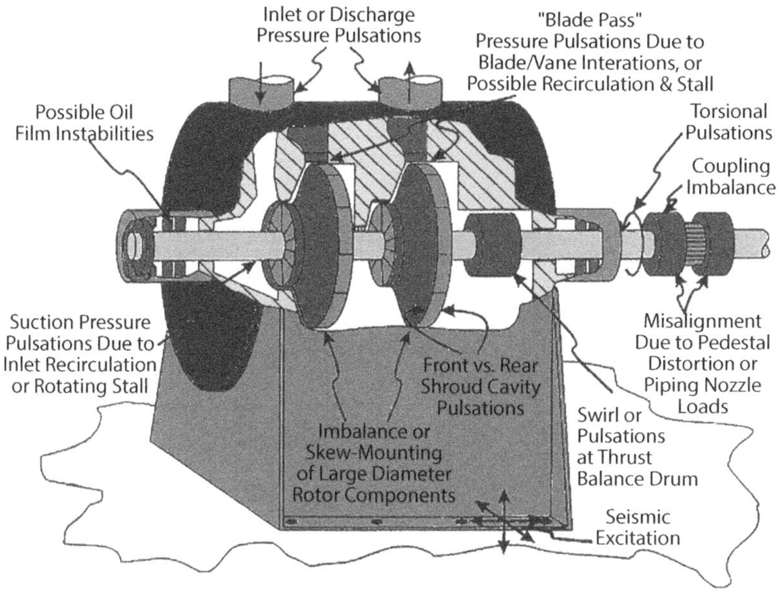

Figure 2.14 Typical excitation force sources in centrifugal pump.
Courtesy Mechanical Solutions, Inc.

excessive vibration in pumps and their drivers, followed closely by misalignment. As illustrated in Figures 2.15 and 2.16, imbalance leads to forces and vibration response, acting at exactly 1x running speed.

The 1xN is because the heavy side of the rotor is rotating at exactly rotating speed and forces vibration movement at exactly that frequency.

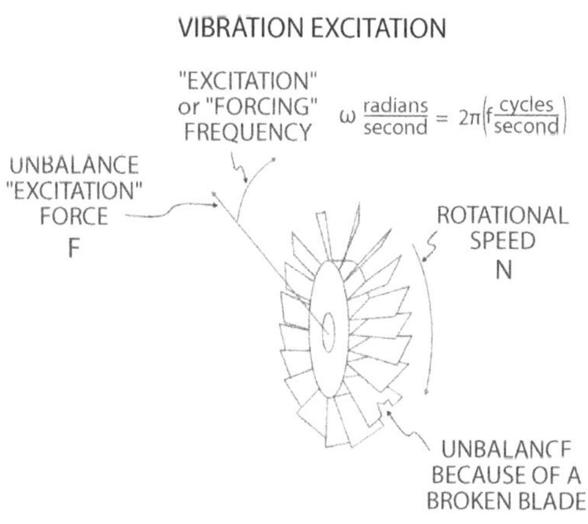

Figure 2.15 Description of reason for imbalance and 1x RPM frequency associated.
Courtesy Mechanical Solutions, Inc.

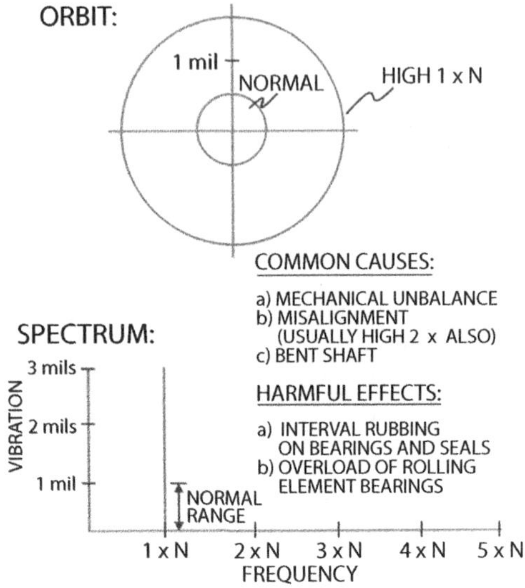

Figure 2.16 Imbalanced example of orbit and FFT.
Courtesy Mechanical Solutions, Inc.

Typically, this also results in a circular shaft orbit (the orbit is what the shaft centerline traces out in space at a given point along the shaft axis, typically near the rotor bearings). However, the orbit may be oval if the rotor is highly loaded within a journal bearing or may have spikes if imbalance is high enough that rubbing is induced.

One way to confirm that imbalance is the cause of 1x RPM vibration is that, as speed is varied, imbalance force and, therefore, response is proportional to rotating speed squared.

Misalignment

Next to imbalance, misalignment is the most common cause of vibration problems in rotating machinery. Misalignment is usually distinguished by two forms: offset and angular, as shown in Figure 2.17. Offset is the amount that the two centerlines are separated radially from each other. Specifically, this is the distance between the centerlines when extended to be next to each other at the plane of the coupling. Angular misalignment is the differential crossing angle that the two shaft centerlines make when projected into each other when viewed from first the top and then in a separate evaluation from the side. In general, misalignment is a combination of both offset and angular misalignment.

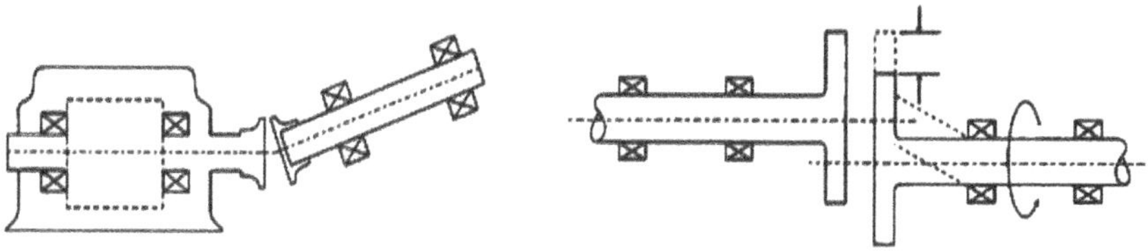

Figure 2.17 Illustration of angular and offset misalignment.
Courtesy Mechanical Solutions, Inc.

The latest thinking and recommendations concerning machinery alignment are addressed very well in ANSI/ASA S2.75 (ANSI, 2017). An excellent classic reference is Dodd, 1974.

When misalignment is a problem, it typically causes primarily 2x running speed, because of the highly elliptical orbit that it forces the shaft to run in, pinched on the misaligned side. Sometimes the misalignment load can cause higher harmonics (i.e., rotor speed integer multiples), especially 3x RPM and may even decrease vibration level because it may load the rotor unnaturally hard against its bearing race or shell. Alternately, misalignment may actually cause increased 1x vibration in pumps with journal bearings by lifting the rotor out of its radially-loaded "bearing hydrodynamic wedge", to result in the bearing running relatively unloaded (this can also cause shaft rotordynamic instability, as discussed later). Figure 2.18 shows

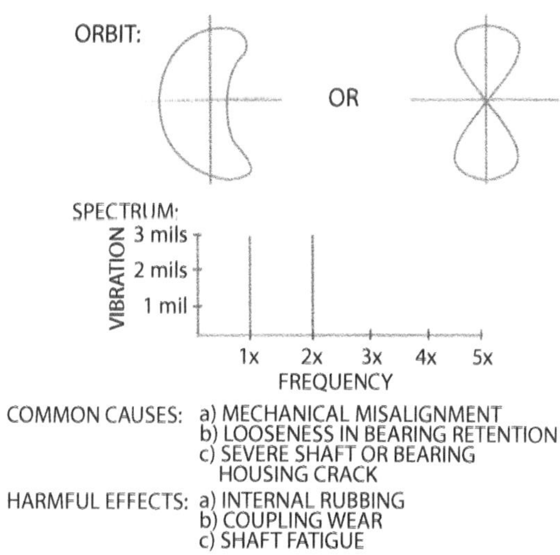

Figure 2.18 Misalignment example of shaft orbit and FFT spectrum.
Courtesy Mechanical Solutions, Inc.

a typical orbit and FFT spectrum for misalignment, in which 2x running speed is the dominant effect. This is often accompanied by relatively large axial motion, also at 2x, because the resulting dynamically-misaligned coupling experiences a non-linear "crimp" twice per revolution, resulting in an axial direction force at 2x RPM.

Because the rotor vibration effects from imbalance and misalignment are typically present at some combination of 1x and 2x running speed and because studies show that imbalance and misalignment are by far the most common source of excessive rotor vibration, API 610 requires that 1x and 2x running speed be accounted for in any rotordynamics analysis and that any critical speeds close to 1x or 2x be sufficiently damped out. A damping ratio as high as 0.15 is required if a natural frequency is close to 1x or 2x running speed (API, 2021).

Mechanical Forces Due to Dry Running Pump, Dry Running Seal, Overtightened Seal

When a dry rub, or poorly lubricated rub, occurs, the frequency spectra associated with this tends to be in two forms. The first is high integer or half-integer (e.g., exactly 1/2x rpm) harmonics that tend to gradually get lower at higher frequencies, forming "rolling hills" of harmonics that periodically reach lows that depend upon the arc length of the rub. The shorter the arc length is, the wider the hills of harmonics are.

The second type of spectrum effect is that there is a general broad-band amplitude increase, especially below running speed (can look like a ski-slope, running left to right, with the peak at zero Hz), and sometimes with broad-band sidebands forming +/- the lower integer harmonics of running speed.

Hydraulic Forces and Blade Passing Frequency

Hydraulic force determination, including the frequencies such forces act at, is typically more difficult than assessing mechanical forces. However, the blade or vane passing force, as illustrated in Figure 2.19, is an exception and is the most commonly encountered type of hydraulic force (Bolleter, 1988). The worst case zero-peak amplitude vane pass levels for an impeller are typically (in the author's experience) between five and fifty percent of the product of the pressure rise for that stage times the impeller OD times the exit flow passage width. Near BEP, the five percent value is a best guess

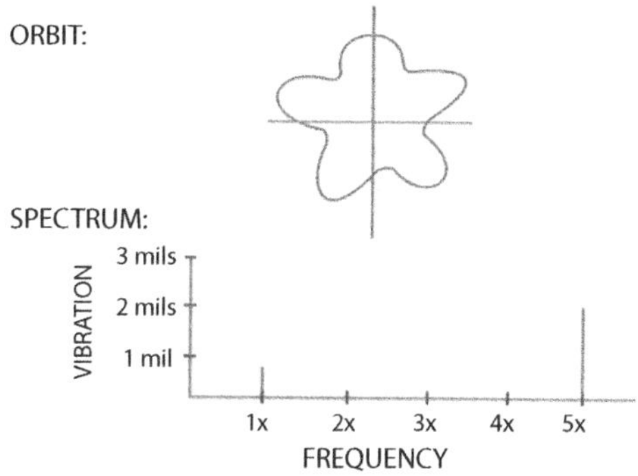

COMMON CAUSES:
a) "GAP B" TOO TIGHT
b) DISCHARGE RECIRCULATION
c) FLAT OR DAMAGED VOLUTE TONGUES
d) INTERNAL RESONANCE OF DIFFUSER WALLS OR VANES

HARMFUL EFFECTS:
a) FATIGUE IN INSTRUMENTATION WIRE CONNECTIONS OR DRAIN PIPE CONNECTIONS
b) IF INTERNAL RESONANCE IS THE CAUSE, FATIGUE CRACKING OF THE RESONATING PART

Figure 2.19 Vane pass vibration.
Courtesy Mechanical Solutions, Inc.

in the absence of OEM or field test data, while close to the minimum continuous flow, fifty percent is a worst case estimate.

Hydraulic forces, at vane pass and otherwise, are minimized in the vicinity of the Best Efficiency Point (BEP). Even close to the BEP, hydraulic force prediction is very design and installation dependent however, and especially depends upon the impeller-blade-to-diffuser-vane (or volute tongue) radial gap, "Gap B" (Marscher, 2007). The strength of vane pass forces also often depends upon interaction with the entire fluid system that the pump operates in. Such system interactions can include acoustic natural frequencies, water hammer, system surge due to capacitance within the suction or discharge lines and being forced by the system operating characteristic impedance curve to operate far from the pump BEP, as illustrated in Figure 2.20 and discussed by Marscher, 2008.

Operation well below the BEP at any given speed, as well as operation at flows well above that point, causes a mismatch in the incidence angle of incoming flow versus the impeller vanes and the diffuser vanes or volute tongues of the various stages. This loads up the vanes considerably in an unintended manner and may even lead to "airfoil stalling" with associated formation of strong vortices (miniature tornadoes) that can severely shake

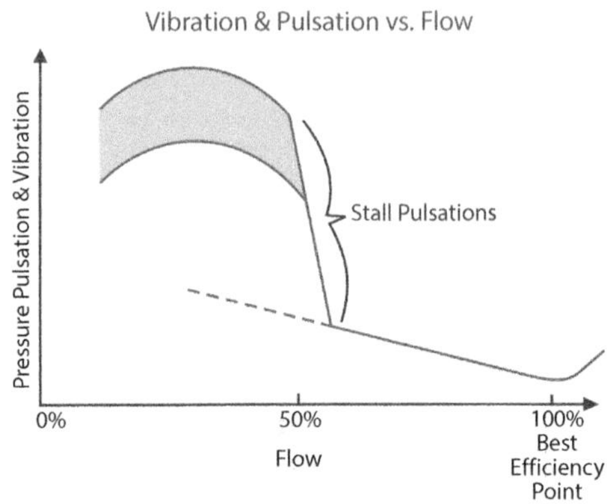

Figure 2.20 Effects of vibration on Off-BEP operation.
Courtesy Mechanical Solutions, Inc.

the entire rotor system at subsynchronous (i.e., below running speed) frequencies. This process can result in vibration, which is high and at frequencies below running speed, but not unbounded like a rotor instability, discussed below. The strongest steady side-loads and shaking oscillations from this mechanism occur at flows at or below the onset of suction (inlet) or discharge internal flow recirculation (Fraser, 1985). The typical effect on rotor vibration (and pressure pulsation) of the operation of a pump at off-design flows is shown in Figure 2.20.

Hydraulic Vibration Forces Below Running Speed, Including Subsynchronous Whirl

Shaft whirl is a forced response at a frequency, usually below running speed ("subsynchronous"), driven by a rotating fluid pressure field. The fluid rotational speed typically forces the whirl speed of the rotor. The most common cause of whirl is fluid rotation around the impeller front or rear shroud side passages, average fluid rotational speed in the close clearances of journal bearings, or swirl within the wear rings or balance drum clearances. In journal bearings, such fluid rotation is typically about 48 percent of running speed because the fluid is stationary at the stator wall and rotating at the rotor velocity at the rotor surface, such that a slightly less than half running speed flow distribution is established in the running clearance. In grooved wear ring "labyrinth" clearances and balance drum clearances, other fractions of rotational speed have been observed, even occasionally supersynchronous (i.e., higher than running speed) rotation, although this is rare. An example of subsynchronous vibration due to whirl at less than rotating speed is given in Figure 2.21.

Subsynchronous vibration due to rotating stall that can set up during suction recirculation tends to be in the vicinity of about 80% running speed, which is the rotating frequency of the stall field. In a waterfall presentation of frequency spectrum vs. rotor speed, the vibration caused by the stall shows up as a constant percent of the 1x rpm vibration "spine" in the plot, as shown by the example of the 4 stage boiler feed pump operating at about 50% of BEP, in Figure 2.22.

Rotordynamic stability (API 610, 2010) refers to phenomena whereby the rotor and its system of reactive support forces are able to become self-excited, usually subsynchronously, leading to potentially catastrophic vibration levels even if the active, otherwise stable excitation forces are quite low. Instability can occur if a rotor's natural frequency is in the range where fluid whirling forces (as explained, usually about ½ running speed) are able to synchronize with the rotor whirl. In pumps, this normally occurs only for relatively flexible multistage rotors and, in general, rotordynamic instability is much more common for compressors or turbines than for pumps. In addition to the need for a "subsynchronous" (i.e., below running speed) rotor natural frequency for this instability to occur, at that same frequency as the whirl, the effective damping associated with the vibration at this natural frequency must somehow drop below zero. Typically, the mechanism for this is that the rotor will go unstable if the cross-coupling stiffness force

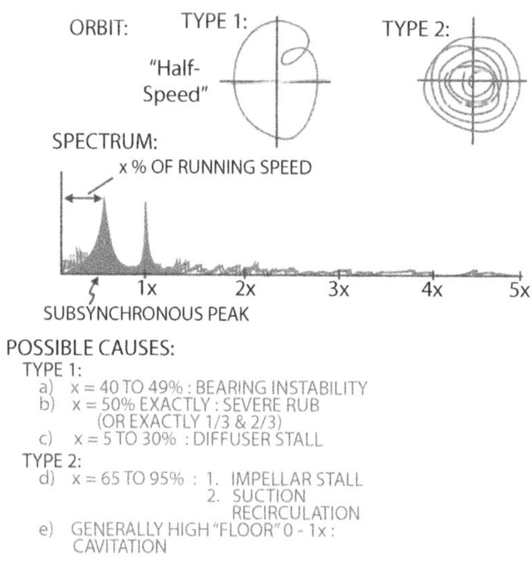

Figure 2.21 Subsynchronous vibration.
Courtesy Mechanical Solutions, Inc.

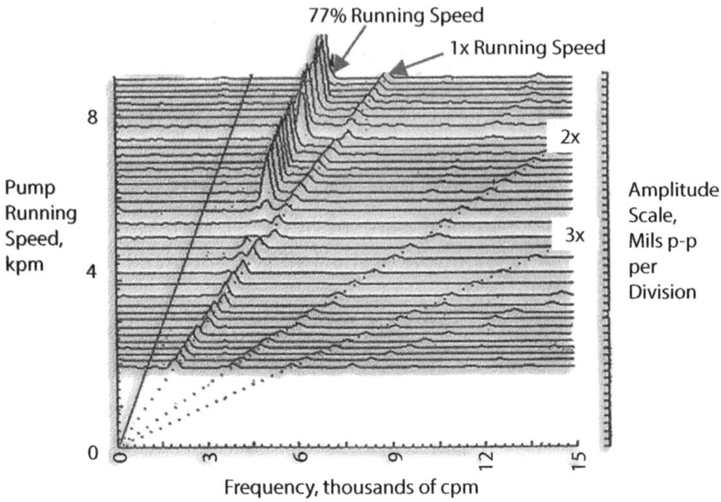

Figure 2.22 Example of Field Data (waterfall plot) from inboard proximity probe of four stage boiler feed barrel pump when operating at about 50% of BEP point. Courtesy Mechanical Solutions, Inc.

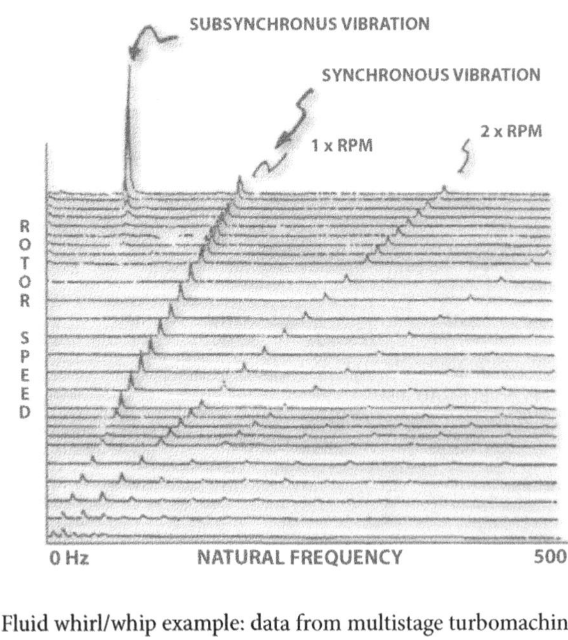

Figure 2.23 Fluid whirl/whip example: data from multistage turbomachine. Courtesy Mechanical Solutions, Inc.

(of the combined bearings and close-clearance seals) is greater than the damping force. This situation is called by many a "shaft whip" and is highly destructive. The frequency spectrum and orbit for shaft whip is shown as the "Type 1" picture in Figure 2.21. The nature of a shaft whip is that once it starts, the rotor is excited by its own motion and all self-excitation occurs at the unstable natural frequency of the shaft, so the vibration response frequency "locks on" to that natural frequency, unlike the constant percentage of running speed as is the case for the stable vibration of a rotating stall. An example of this situation is shown for a multistage high speed compressor in Figure 2.23.

Detection of Effects of Cavitation

Figure 2.24a exhibits dynamic pressure transducer output at a pump suction during the occurrence of strong cavitation, while Figure 2.24b exhibits acceleration response over a period of time on the surface of the suction casing. The data of Figure 2.24b shows the second stage of the cavitation process. In the first stage, cavitation bubbles may form through low-temperature "boiling" in local low pressure regions (if they are below vapor pressure of the fluid), typically in the pump suction near the impeller inlet. This may cause some blockage of flow and typically will reduce the discharge head able to be produced by the pump. If this decrease is 3% or more relative to the expected discharge head at a given speed and flow rate, a Hydraulic Institute "head drop" test would red-flag that significant cavitation is occurring. However, before this 3% head drop point, the cavitation noise will be noticeable by the human ear (sounds like grinding rocks) and may already be strong enough to indicate that stage 2 of the cavitation is causing serious damage. In stage 2, the bubbles are swept by the flowing liquid along a streamline of quickly increasing pressure, such that the bubbles can no longer exist as vapor according to the thermodynamic properties of the fluid being pumped. At that point, the bubbles implosively collapse, causing a narrow sonic jet to be directed through and out of the bubble center and potentially impact the impeller or casing sidewall surface with tremendous impact stress on the order of the fatigue strength of the metal. This eventually cause cavitation erosion.

Stage 2 results in a stress wave that quickly propagates as "structure-borne noise", essentially a vibration stress wave, as discussed by Marscher, 2017. The strength of the associated acoustic pulse is strong enough to result in a very large acceleration pulse on the pump casing exterior walls, a typical result of which is shown in Figure 2.24b. This acceleration pulse is strongest when there is a line-of-sight to where the cavitation bubbles are collapsing.

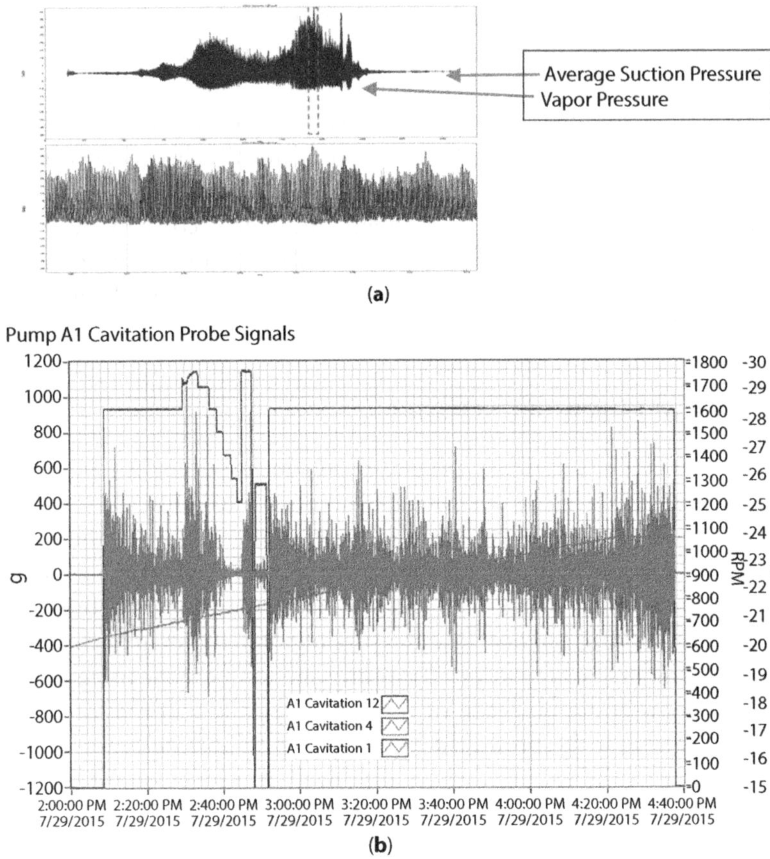

Figure 2.24 (a) Negative pressure amplitude clipped-off below vapor pressure indicates cavitation. Without cavitation the time waveform is symmetric above and below average. The lower plot is showing a time-expanded view between the red vertical lines of the upper plot. Plots are showing +/- 200 psia.
Courtesy Mechanical Solutions, Inc.
(b) Acceleration amplitude versus time for a large axial flow water pump operating variable speed up to 1800 rpm. At 1400 rpm and below (as seen in the top solid line), the cavitation was nearly eliminated when NPSHa (suction head available) was greater than the NPSHr (suction head required to avoid significant cavitation, per OEM HI/ANSI test stand data). Note that the cavitation pulses are typically over 200 Gs peak and occasionally reach 1000 Gs. The CA6NM stainless steel impellers eroded away in a period of months. Accelerometers 4 and 8 were on the suction casing with line-of-sight to the impeller and Accelerometer 12 was downstream on the discharge pipe.
Courtesy Mechanical Solutions, Inc.

The strength levels consistent with the likely occurrence of damage, versus the bubble collapse occurring harmlessly midstream in the fluid, for example, have been observed to be about 100 Gs peak. Levels of 300 to 1000 Gs peak are not unusual if the cavitation is particularly strong and in such cases, the erosion rate of the impeller and/or casing would be expected to be rapid (e.g., 1000 to 5000 hours until unacceptable amounts of erosion damage take place). Measurement levels of 50 Gs may still be problematic for less erosion resistant materials or if cavitation is taking place in a non-line-of-sight location.

Torsional Excitations

Oscillating torques can be sourced in mechanical phenomena, such as they can be produced at 2x RPM by kinking in many styles of shaft couplings. However, torsional excitations more often tend to be due to static pressure or fluid velocity and variation around the periphery of the stator (such as a volute or diffuser). Typically, such circumferentially varying pressures or velocities have strong components at 1x running speed (typical of impellers being off center of the volute center of pressure) or 2x running speed (typical of twin volutes or "pinched" distortion of pressure or velocity versus a simple sine wave around the circumference). Sometimes, the most important torsional excitation is the number of impeller blades times running speed (so-called "blade pass" or "vane pass" vibrations) from discharge pressure pulses as the impeller blades move past a volute tongue or diffuser vane. Figure 2.25 shows typical worst case torsional excitation levels for various common excitation frequencies and, in cartoon fashion, shows an example of how these excitations often relate to the first several torsional natural frequencies for centrifugal pumps and motors. If a strong excitation occurs at the same frequency as a torsional natural frequency, strong resonance and associated damage can occur. Keep in mind that the rotor system torsional natural frequency is not the same as the pump torsional or motor torsional natural frequencies by themselves, but depends upon how their rotary inertias and torsional stiffnesses, along with those of their coupling, combine (Marscher, 2008). They are also impacted potentially by press-fits. However, fluid added mass has little effect on torsional natural frequencies, so they can be determined by circumferential impact testing of the dry system rotor (Marscher, 2013). There may be some operational effects, however, in certain couplings whose stiffness is affected by torque level.

The excitations in Figure 2.25 are shown in terms of ratio of oscillating torque (peak-to-peak, or p-p) to steady torque. This ratio is the so-called

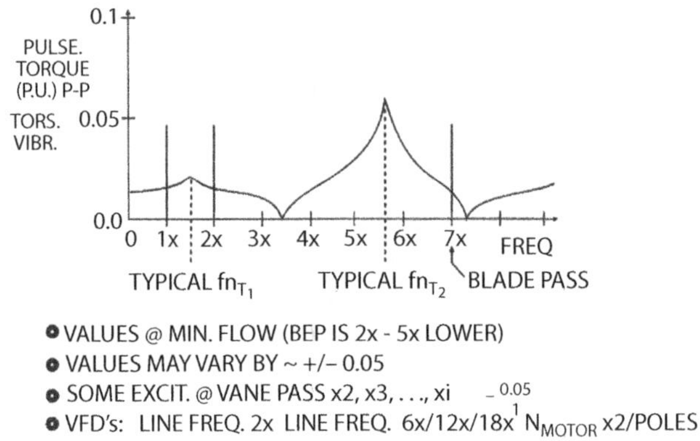

Figure 2.25 Typical torsional critical speeds and typical worst case per-unit excitation level.
Courtesy Mechanical Solutions, Inc.

per unit factor, or p.u. As an example, Figure 2.25 predicts that a worst case 1x or 2x rpm excitation would be 0.05 p.u., i.e., about 5% p-p of the steady torque. However, this figure is given only as an estimate and such expectations should be confirmed by testing. The estimated levels will typically increase a factor of two or more at flows outside of the pump's Preferred Operating Range (POR, see HI/ANSI 9.6.8).

Per Unit (P.U.) means oscillating torque (peak-to-peak) given as a fraction of steady state operating torque. The values shown are based on typical high-side excitation levels at the pump's BEP. In the case of 2x slip excitation, for induction motors, worst case p.u. should be set to 0.01 p-p, while for synchronous motors 0.05 should be applied when slip is occurring during the start-up unless OEM or test estimates can be obtained.

For an electric motor operated through a VFD, API and HI caution the need to anticipate response at 1x line frequency and 2x line frequency, as well as n x RPM, where n is a set of integers defined by the drive and/or motor manufacturer. Besides n = 1 and 2 signifying 1x and 2x running speed, VFD waveform harmonics can be significant. For example, because they simulate a true sine wave with six stepped voltage levels, older VFD's had strong torsional harmonics at 6x, 12x, 18x, and sometimes 24x the simulated line frequency. However, modern adjustable speed drives, or pulse-width-modulated VFDs, are typically designed to have relatively weak harmonics. In the case of lack of other information, it can typically be assumed that worst case p.u. factors are 0.02 p-p for the VFD 6x simulated line frequency and are less for each succeeding 6x harmonic based on 0.02 times the ratio of 6x/nx, where n is an integer multiple of six.

Torsional excitation harmonics for reciprocating engine drives are considerably more complex than those for electric motors or turbines as drivers. For 4-cycle diesel or gas engines, strong torsional harmonics exist at m x running speed, where m is an integer multiplier of running speed. Running speed harmonics of torque for m greater than twice the number of pistons are most typically small enough to ignore, with regard to potential to cause damaging torsional resonance. Even with this limitation, this may result in quite a few harmonic peaks (beyond those of the driven machine) that need to be avoided with regard to coincidence with torsional natural frequencies. In fact, the situation is even more difficult. In addition to the high 1x RPM harmonic content, often the strongest torsional harmonics of a reciprocating engine are "half-harmonics" of the number of pistons times running speed, given that 4 cycle engines fire every other stroke. This includes ½ running speed, particularly strong for a single mistuned cylinder. Therefore, in reciprocating engine drivers, m is not necessarily a whole number "integer" but can be a half integer. Often, reciprocating engine torsional harmonics have been found strongest at ½ x RPM as well as ½ x cylinder number.

The detection of high torsional vibration is most directly performed with on-shaft RF telemetry, available at modest cost from several instrumentation manufacturers. The telemetry radio signal voltage strength is designed to be proportional to the torque level, as detected by a strain gage assembly glued to the shaft and wired to the transmitter. In some designs, the strain gage bridge voltage is converted to digital form instead and then broadcast. In any event, the biggest challenge in such assemblies is keeping them safely attached to the shaft. Tape and hose clamps are common temporary mounting methods, but are often problematic. The centrifugal forces are high, especially since the telemetry and its batteries or induction power unit must be attached on the shaft O.D. The assembly becomes hot from frictional air heating, degrading the tape or any glue, and hose clamps have stress concentrations in their threads that can lead to sudden failure. Designs that completely encircle the shaft with a collar arrangement are recommended, especially for permanently installed units. A different problem in permanent installations, however, is that strain gages tend to have limited fatigue life, unless the torsional oscillations are very low and "uninteresting". Non-telemetry methods of determining torsional oscillations also exist, such as demodulation of axial stripes placed on the shaft observed by an encoder or use of a "torque cell" coupling extension. These are more difficult to implement in a troubleshooting situation, however.

An indirect method of detecting torsional oscillations, at least qualitatively, is to measure the vibrational motion of each of the pump and motor feet, especially on the inboard (driven) end. As the torque reacts on the machine casings, the reacted torque transfers from the casing feet to the baseplate and foundation. By observing one side of the pump lift up and the other compress and observing opposite motion on the motor, the effect of strong torsional oscillation can be detected. This is especially useful as part of an evaluation using an Operating Deflection Shape or a Motion Magnification video approach as at least part of the troubleshooting toolbox.

Vibrations Particular to Various Centrifugal Pump Types

Vertical Turbine Pump Evaluation

The vertical turbine pump, or VTP, is very different from other pumps and turbomachinery because of its less stringent balancing, shaft straightness, and motor shaft alignment tolerances. This lower precision is typically tolerable because of the VTP's long flexible casing and the casing's flexible attachment to ground and because of the peculiar spaghetti-like lineshafting which connects the motor to the below-ground liquid-end "bowl assembly" of the pump. In spite of these peculiarities, for the VTP like other pumps, it is the bearing loads, the seals, and the bearing and wear ring clearances where problems are most likely to occur, as discussed by Kovats, 1962.

The flexibility of the VTP structure and shafting result in many closely spaced natural frequency modes within the range of frequencies for which strong exciting forces are expected, which for VTPs are typically 1x and 2x running speed. An average of one mode per 100 cpm is not unusual for deepwell VTP's. VTP pumps also exhibit nonlinear shaft dynamics because of the large shaft excursions which occur in the lightly loaded long length/diameter ratio bearings (Marscher, 1986).

The most important factor for VTP rotordynamics is the statistical character of the support provided by any given lineshaft bearing. The normally lightly loaded lineshaft bearings exhibit a rapid, nonlinear increase in bearing stiffness as the lineshaft gets close to the bearing wall. Given the flexibility of the lineshaft and the relatively flexible support provided by the pump casing "column piping" and given the relatively large assembly tolerances and misalignments in the multiple lineshaft bearings of these machines, the contribution of each bearing to the net rotordynamic stiffness is a nearly random and constantly changing situation, as explained conceptually in Figure 2.26. The result is that in practice there is no single value for each of the various theoretically predicted shaft natural frequencies, but rather the natural frequencies of the lineshafting and the shaft in the bowl assembly must be considered on a time-averaged and location-averaged basis, so that their net effect is "smeared" across a broad frequency range (Marscher, 1986). This is a good thing in that it makes developing a strong resonance and associated damaging vibration of the VTP shafting very difficult, even over a wide frequency range (Marscher, 1990). This is not to say, however, that shaft vibrations may not become high if lineshaft or bowl bearing clearances become too large due to running dry during start-up or due to unexpectedly high abrasive content in the pumped water. This can be a rapidly deteriorating situation, such that bearing wear leads to higher vibration, which leads to higher bearing forces from the whirling shaft, which leads to an even more rapid wear rate, especially if abrasives are in the mix.

An important advancement in the monitoring of VTP pumps was the development decades ago of the underwater proximity probe by a major instrumentation supplier. Studies reported in the literature which have made use of such probes to observe actual shaft motion during various conditions of interest include Marscher (1986, 1990) and Spettel (1985).

Figure 2.26 for VTP lineshafting bearing alignment as a function of downthrust implies that the degree of loading of individual bearings

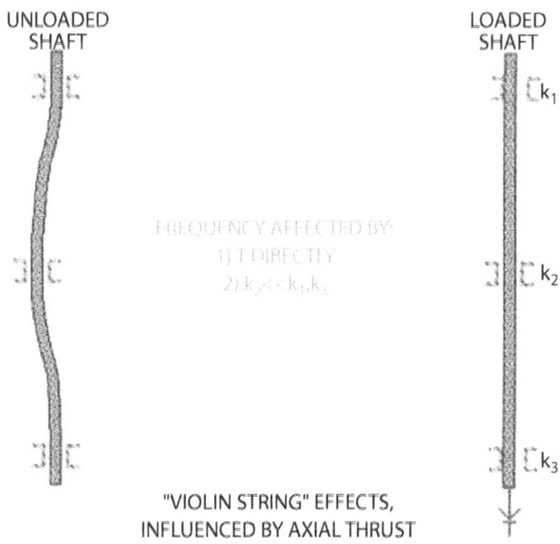

Figure 2.26 Vertical pump lineshaft rotor behavior example. Courtesy Mechanical Solutions, Inc.

depends not only on lateral loads but also strongly upon the amount of downthrust on the lineshaft. Likewise, the shaft natural frequencies also depend upon this downthrust. As the operating point flow rate changes, the discharge pressure changes in accordance with the pump head-capacity characteristic curve. The decreased discharge pressure at flow rates above BEP (moving toward "runout condition") decreases downthrust, permitting the shaft to more easily follow a curved shape to decrease the bearing loads. Lower bearing loads imply much lower bearing stiffness (so they act more like "bumpers" than bearings), thereby reducing the shaft natural frequencies, especially the first several modes. Therefore, when testing VTP pumps for rotor natural frequencies and/or general vibration behavior please be aware that operating point matters very much (Marscher, 1986).

In the case of VTP structural (i.e. non-rotating component) natural frequencies and overall vibration, operating point is typically much less important. The greatest number of VTP structural problems have been associated with the first bending natural frequency (the "reed frequency") of the motor and pump as an assembly, in which the motor tips back and forth as a nearly rigid beam, anchored to ground through the pump's relatively flexible motor stand/discharge head (HI/ANSI 9.6.8, 2017). When this natural frequency mode (usually between 500 and 1200 rpm for most VTPs) is within 5% or so of rotating speed, very large vibrations can result from resonance. Occasionally, the second "above ground" bending mode instead is the natural frequency in resonance and it tends to be in the range of 700 to 1800 rpm. This natural frequency's mode shape is like a coat hook, with the motor top moving in the opposite direction of the motor bottom, where the bottom may even move significantly more than the top. This mode is particularly sensitive to motor coupling misalignment and may be excited strongly by 2x rpm misalignment forces.

Common VTP vibration issues include:

- Above ground "Reed Frequency" resonance
- VFD as a "shaker" finding a natural frequency as speed changes through a large turn-down range
- Line shaft behaving like a "violin string", sensitive to downthrust as discussed
- Shaft enclosure tube (when used to isolate bearing lubrication from the pumpage) natural frequencies
- Discharge nozzle can possess improper tie-rod tightening, encouraging discharge head distortion
- Sump vortices or asymmetrical inlet flows cause subsynchronous fluid excitation

- Column piping acoustics, especially at vane pass
- Sub-synchronous vibration (at exactly 1/3 or ½ running speed) due to lineshaft rubbing
- Can have "shudder" vibration due to overtightened packing, with broadband vibration below running speed

Vertical Dry Pit Pump Vibration Issues

The most common vibration problems encountered in vertical dry pit pumps involve balance, alignment, vane pass resonances (typically not suspected or evaluated by the OEM, in the author's experience), and a reed frequency 1x RPM resonance, often due to an excessively flexible foundation or casing support or unsupported piping mass. In general, the troubleshooting shortlist should include:

- Excitation of structural natural frequencies
  - 1X- Biggest resonance issue, usually the 1st bending reed frequency
  - 2X- Less likely a resonance problem, but can be caused by misalignment or driveshaft Hooke's joints
  - Vane Pass- Can cause strong resonance
- Packing vs. Mechanical Seals - overtightening causes shuddering, with broadband frequency below running speed.
- Non-integral motor can have intermediate shafting dynamics and resonance, at either 1x rpm or 2x rpm.

Submersible Pump Vibration Issues

The most common vibration problem encountered in submersibles is some sort of structural support issue, causing an unexpected resonance. This resonance can be driven by imbalance at 1x running speed, but also can surprisingly be caused by stable whirl of the fluid within the pump and motor cavities, typically at a bit less than ½ rotor speed. This is sometimes misdiagnosed as a dangerous rotordynamic instability, but in the author's experience always turns out to be the non-symmetry of the whirling fluid driving a simple and stable reed frequency resonance of the mounted pump/motor/structural support. This can be verified by increasing the motor speed to "run through" the resonance, such that vibrations come down at speeds above the whirling resonance frequency. Rotordynamic instability does not improve at higher speeds and in fact is typically more severe. An alternate test would be to more stiffly support the structure of the pump/motor, which (if stiff enough) would shift the reed frequency up enough to cure the resonance and associated elevated vibrations.

End Suction Overhung Single Stage Pump Vibration Issues

The most common vibration problems encountered in end suction single stage horizontal pumps are piping load effects on alignment and casing distortion, as included in the first three items in the list below:

- Nozzle loads lead to casing distortion and can cause rubbing (look for ½ x rpm subharmonics)
- Nozzle loads affect alignment of (and damage of) the mechanical seal (look for ½ rpm and/or 2x rpm)
- Baseplate flexibility affects alignment (look for 2x rpm)
- Bearings are low precision and lead to higher vibration of rotor if measured (look for high 1x rpm)
- Packing is susceptible to over-tightening, which causes shudder (look for broadband vibration below 1x rpm)
- Product lubricated bearings in mag drive and canned motor pumps are susceptible to fluid flashing and dry rub (look for ½ x rpm subharmonics and elevated harmonics above 1x rpm)

Between Bearing Double Suction Single Stage Pump Vibration Issues

The most common vibration problems encountered in double suction single stage horizontal pumps is likelihood of cavitation because these pumps are used in low suction head applications, a low-frequency (about 5 Hz) movement toward and away from the coupling called "axial shuttling", sensitivity to casing distortion due to piping load effects, and bearing housing tilting due to casing distortion from pressure oscillations. Specifically:

- Used when suction pressure is very low. Therefore, suction recirculation and cavitation have been issues, with strong subsynchronous vibration peaks.
- Double suction impeller typically equalizes thrust side-to-side, but at off-design is prone to axial shuttling at 5 to 10 Hz.

- 180 degree "drip pocket" allows bearing housing and stuffing box to flex downward due to pressure pulsations, e.g., at 1x or vane pass or due to piping loads

Horizontal Multistage Pump Vibration Issues

The most common vibration problems encountered in horizontal multistage pumps of various types include sensitivity to the Lomakin Effect, which often shows up as a 1x rpm resonance after the pump wear rings and balance device clearances increase with wear and deterioration of pedestals and/or foundation pads as they crack and/or crumbled over the years, again typically resulting in an unexpected 1x rpm resonance gradually developing. Additional problems specific to the various pump sub-types are listed in outline form below.

Potential Issues Specific to Barrel Pump with Balance Drum or Disk and (Typically) Stiff Shafts

- 180 degree "drip pocket" allows bearing housing natural frequency near vane pass
- Diffuser "strong-back" plates can resonate at vane pass
- Because of an intended broad turn-down range in flowrate, it can be susceptible to onset of suction or discharge recirculation due to stalling of the impeller or diffuser (or volute) vanes, respectively. In the case of suction recirculation, subsynchronous vibration at 70 to 90% rpm is typical and in the case of discharge recirculation, a strong 10 to 30% rpm peak is typical. In the case of the latter, beware that vibration velocities may meet ISO criteria, even though shaft vibration levels may be taking up most or all of the available clearance since displacement goes up as frequency decreases for a given velocity (for this pump, ISO is based on velocity).

Note:

For Barrel Pumps with thin shafts, the rotor critical speeds are particularly sensitive to the Lomakin Effect and therefore wear ring and drum clearances.

Potential Issues Specific to Axially Split-Case Pumps:

- Acoustics in crossover, excited by vane passing frequency
- Top casing half "humps" allowing inter-stage jet leakage with possible vibration "whistle" at frequency above vane pass

- Axial shuttling at off-design conditions, typically at 5 to 10 Hz
- High sensitivity of 1st bend mode to center bushing clearance with problems showing up at 1x rpm

Potential Issues Specific to Ring Section "Donut" Pump with Balance Disk:

- Less robust than barrel pump, but fewer natural frequency problems, based on experience
- Designs where thrust disk totally replaces oil lubricated thrust bearing have encountered rub and binding problems during start-up or process upsets with expected ½ x harmonics and high 1x harmonics during the rub and "ski-slope" in spectrum below running speed, peaking near DC

Steps in Pump Evaluation through Vibration Monitoring

As discussed at the beginning of this chapter, key factors to be determined during troubleshooting involve where the vibration may be damaging components or representing damaged components and what the forces and natural frequencies that may be involved in causing any excessive vibration levels are. The inter-relationship of any exciting forces, the natural frequencies, and ultimate measured vibration has been discussed throughout the chapter, but it may be useful to bring the combination into focus using the example (with unusually strong force harmonics, for the purpose of illustration, not typical reality) in Figure 2.27 below.

Also important in most circumstances is the timing, or "phase relationship", between when (and the direction which) forces are applied versus the

Figure 2.27 Example of how vibration of equipment depends upon interplay between natural frequencies and excitation forces across Frequency Spectrum.
Courtesy Mechanical Solutions, Inc.

timing and direction of the vibration response to the force, or the so-called phase angle. This is pictorially illustrated in Figure 2.28 below:

Before the advent of computer-assisted vibration evaluation, troubleshooters would use a combination of vibration increase at a given frequency along with a 90 degree phase angle change to identify natural frequencies of rotors and structures. This can still be performed if computer hardware and software are not available to the person doing machine diagnosis. For example, for the rotor, a keyway can be focused on so that it stays "stationary" with a strobe light. As the speed of the machine drifts through the natural frequency range, the phase angle (timing) change between the strobe trigger versus peak lateral motion can be observed. The direction of the peak motion should stay the same during this process. This may be visible motion if it is large enough or (much better) a prox probe or accelerometer mounted on a housing or at the end of a shaft-rider stick, discussed for steam turbines in ISO 10816-2.

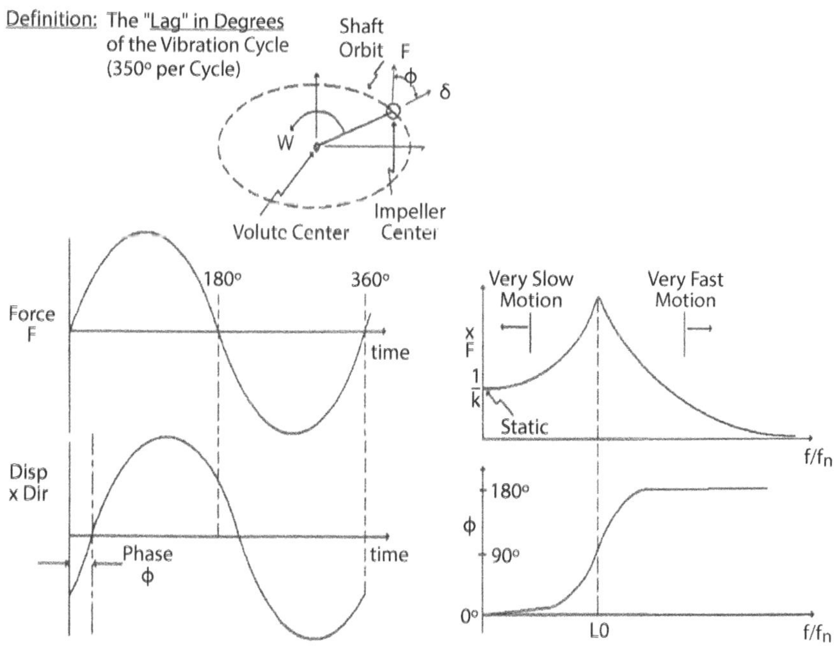

Figure 2.28 Example of force vs. displacement phase angle.
Courtesy Mechanical Solutions, Inc.

Use of the Bode and Nyquist Plots to Confirm Natural Frequencies

A particular method for accounting for amplitude and phase change simultaneously with frequency is provided by the Bode plot, which originated with electrical engineers many decades ago. This is shown at the bottom of Figure 2.29, along with an alternate approach known as a Nyquist plot, a type of polar plot shown at the top of Figure 2.29. In the Nyquist plot, the amplitude is the length between the plot center and any given point of the plot and the phase angle is the angle between the positive x-axis (horizontal axis) and the line that connects the center to the plot point. Each point has its own frequency associated with it and typically, a cursor from zero to the plot point would start at zero amplitude and phase and gradually increase in amplitude as phase decreases past -359 degrees and eventually through to -180 degrees (minus means the vibration displacement lags the force rather than is ahead of it, as would be expected since cause leads to effect, not the other way around).

What the Nyquist plot helps with is distinguishing actual natural frequencies from peaks that "masquerade" as natural frequencies but are actually hydraulic force peaks over a range of amplitude plus phase change. It also helps to better distinguish closely spaced natural frequencies. This is a major benefit if there is noise in the vibration signal or if other equipment is operating nearby. True natural frequencies of the component being monitored will show up as portions of a circle in the Nyquist plot, while the

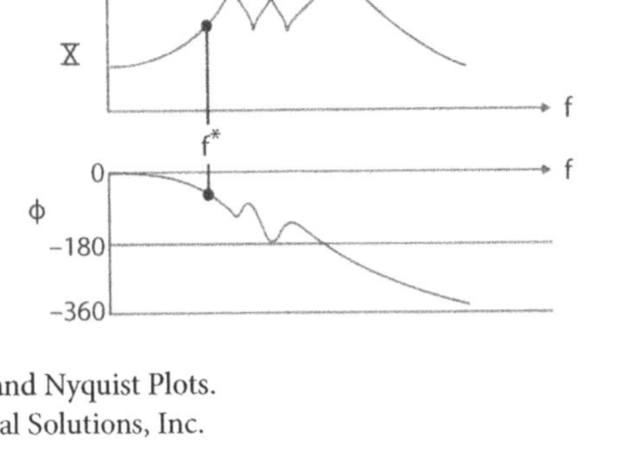

Figure 2.29 Bode and Nyquist Plots.
Courtesy Mechanical Solutions, Inc.

"gulls wings" of the Bode plot can take on non-circular shapes if caused by hydraulic force flow surges or turbulence, as might occur during suction recirculation for example.

Operating Deflection Shapes (ODS)

The phase angle becomes very evident when the vibration of the entire structure is plotted as the complete assembly of tested points, which can be presented in a cartoon animation. If a specialty computer program is used, hundreds of points can be combined in an overall moving display of what is vibrating, how much, and in what direction. This is known as an Operating Deflection Shape, or ODS, as discussed versus Motion Magnified Video (MMV) methods earlier. Before the advent of MMV and similar video-based methods, this was considered by most troubleshooters as the "gold standard" of vibration evaluation, when combined with vibration vs. time or vibration vs. frequency spectra at key locations. ODS

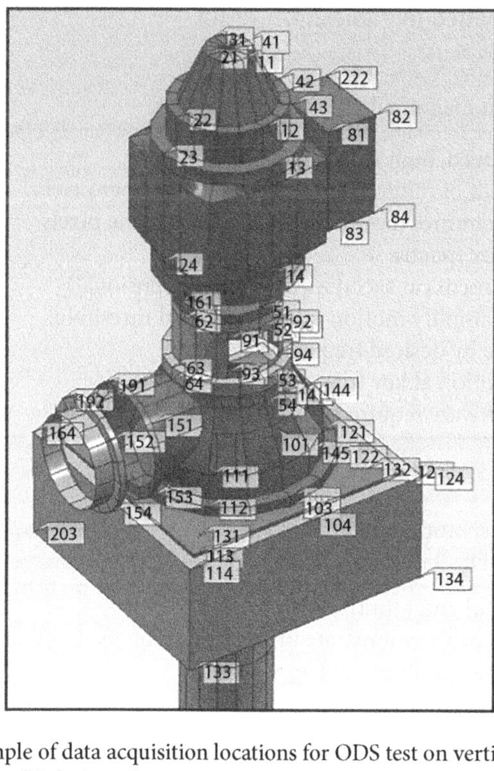

Figure 2.30 Example of data acquisition locations for ODS test on vertical pump. Courtesy Mechanical Solutions, Inc.

makes evident soft foot and even crack locations if the test contains enough vibration pick-off points.

Typical problems such as imbalance, misalignment, and resonances are usually immediately "visible" as well. A typical ODS model, showing test points on a vertical pump as performed by the author's company, is shown in Figure 2.30.

ODS methods typically take one to two days to properly implement in the field, followed by more effort back at the office. This is compared with MMV techniques, which typically take less than an hour for set up, video acquisition, and playback/evaluation. However, sometimes lighting is not adequate for MMV or displacements at important support locations are too small to be detected by MMV. Therefore, there remains a place for both methods in vibration diagnostics and often, today, people will use them in tandem with each other, observing "low hanging fruit" with MMV and if needed, then providing the added low displacement data for possibly important pump-supporting components with ODS. As a refresher of what was discussed earlier, the MMV approach characteristics are listed in Table 2.3. Following this, a list of advantages and disadvantages of each method are provided in Table 2.4.

Table 2.3 Motion magnified video characteristics.

- Uses high-speed, high resolution video
    - Equivalent of millions of accelerometers, 1 per pixel
- Analyzes/quantifies motion based on video scene pixels
    - Frequency spectra
    - Displacements at each frequency (2-dimensional)
- Algorithms amplify motion to human visual threshold
    - Filterable by desired frequencies
    - Slows motion at key frequencies
- Specialty software required

Table 2.4 ODS vs. motion magnified video.
Accelerometer-ODS-Based Pros and Cons

✓ Powerful and intuitive diagnostic tool
Can clearly demonstrate modes and frequencies of vibration
✓ Proven over decades of application
× Time consuming
Data acquisition – hundreds of data points
Post processing – compiling database to match model points

- × Potential for bookkeeping error
  Must match all gathered data points to appropriate place on model
- × Requires proximity – not appropriate for restricted access
  Heat, radiation, accessibility/scaffolding requirements

Motion Magnified Video Pros and Cons

- ✓ Powerful and intuitive diagnostic tool
  Realistically demonstrates modes and frequencies of vibration
  Easy for non-experts to understand (management, etc.)
  Comprehensive set of data points
- ✓ Does not require "contact" – perfect for restricted areas
  Heat, radiation, accessibility/scaffolding requirements
- ✓ Fast
  Millions of data points ready to evaluate in minutes, not days
- ✓ Cost-effective
  Helps focus effort for accelerometer-based ODS
- × New
  Requires training
- × Frequency range
  At first only effective at lower frequencies ($<$100 Hz)
  Even advanced systems display vibrations at ISO levels limited to frequencies 500-1000 Hz
- × 2-dimensional
  Requires additional view angles to get third dimension
  Longer term efforts appear to be successfully developing stereo-vision and 3-dimensional analysis

Conclusions

The purpose of this chapter has been to provide an introduction to state-of-the-art condition monitoring and troubleshooting processes for centrifugal pumps so End Users can either learn to do it themselves or carry on intelligent review of evaluations performed on their behalf by OEMs or qualified consultants.

Nomenclature

BEP = best efficiency operating point of the pump
FRF = Frequency response function

f = frequency (cycles per second, Hz)
f_n = natural frequency (cycles per second, Hz)
G_c = gravitational unit (386 in/s2 or 9800 mm/s2)
k = spring constant (lbf/in or N/mm)
M = mass (lbm or kg)
N = shaft rotational speed (revolutions per min, rpm)
t = time (s)
v = vibration velocity amplitude, peak (in/s or mm/s)
x = vibration displacement, peak (mils or mm)
a = acceleration of vibration (in/s^2 or mm/s^2)

References & Bibliography

Kovats, A., 1962, Vibration of Vertical Pumps, ASME Journal of Engineering for Power.

Dodd, V.R., 1974, Total Alignment, The Petroleum Publishing Co., Tulsa, OK.

Blevins, Robert, 1977, Flow-Induced Vibration, von Nostrand Reinhold.

Blevins, Robert, 1979, Formulas for Natural Frequency and Mode Shape, von Nostrand Reinhold.

Black, H.F., 1979, Effects of Fluid-Filled Clearance Spaces on Centrifugal Pump Vibrations. 8th Turbomachinery Symposium, Texas A&M Univ.

Fraser, W.H., 1985, *Centrifugal Pump Hydraulic Performance and Diagnostics, Pump Handbook 2nd Edition,* McGraw-Hill.

Childs, D., and Moyer, D., 1985, Vibration Characteristics of the HPO Turbopump of the SSME, ASME J. of Engineering for Gas Turbines and Power, v. 107, pp. 152-159.

Spettel, T., 1985 (April), The Dynamics of a Vertical Pump, Orbit Magazine, Bently Nevada Corp.

Marscher, W.D., 1986, The Effect of Fluid Forces at Various Operation Conditions on the Vibrations of Vertical Turbine Pumps, Proc. IMechE, Radial Loads and Axial Thrusts on Pumps.

Marscher, W.D., 1987, The Relationship Between Pump Rotor System Tribology and Appropriate Vibration Specifications for Centrifugal Pumps, Proc IMechE 3rd Euro Congress on Fluid Machinery for the Oil & Petrochemical Industries, Hague, Netherlands.

Bolleter, U., 1988, Blade Passage Tones of Centrifugal Pumps, Vibrations, Vol. 4, No. 3, pp. 8-13.

Marscher, W.D., 1989, Analysis and Test of Multistage Pump Wet Critical Speeds, ASLE/ASME Joint Tribology Conf., Ft. Lauderdale, FL.

Bowman, D., Marscher, W., and Reid, S., 1990, Pump Rotor Critical Speeds: Diagnosis and Solutions, Proc. International Pump Symposium, Texas A&M University.

Marscher, W.D., 1990, The Effect of Variable Frequency Drives on Vibration Problems in Vertical Pumps, Proc. Water & Wastewater Conference, Barcelona, Spain.

Marscher, W D, 1997, Vibration Chapter and Predictive Maintenance Chapter, STLE Tribology & Lubrication Handbook: STLE Park Ridge, IL.

Marscher, W.D., 1999b, The Determination of Rotor Critical Speeds While Machinery Remains Operating through Use of Impact Testing, IMAC Conf. Orlando FL, SEM.

Marscher, W.D., 2002, Avoiding Failures in Centrifugal Pumps, Tutorial in Proc. TAMU International Pump Symposium, Houston, TX.

Marscher, W.D.; Onari, M.M.; Boyadjis, P.A., *et al*, Pump Mechanical Behavior & Vibration, Chapter in Centrifugal Pump Handbook, 4th Edition, McGraw-Hill, c. 2008. Edited by Paul Cooper *et al.*

ISO 10816-7, 2009, Mechanical vibration — Evaluation of machine vibration by measurements on centrifugal pumps, Geneva SWITZ (Note that 10816 and 7919 are in the process of being combined into 20816).

API-RP-684, 2nd Ed., 2010, Amer. Pet. Instit., Washington D.C.

Marscher, W.D., Boyadjis, P.A., Gamarra, J., and Gruener, J., 2013, Effect of Component Interference Fit and Fluid Density on the Lateral and Torsional Natural Frequencies of Pump and Turbomachinery Rotor Systems, Lecture in Proc. TAMU International Pump Symposium, Houston, TX

HI/ANSI Standards 9.6.4, 2017, Rotodynamic Pumps - Vibration, The Hydraulic Institute, Parsippany NJ.

HI/ANSI Standards 9.6.5, 2017, Rotodynamic Pumps - Condition Monitoring, The Hydraulic Institute, Parsippany NJ.

Marscher, W.D., Onari, M.M., Gamarra, J., 2017, Pump Cavitation Evaluation Using Accelerometers, Lecture in Proc. TAMU International Pump Symposium, Houston, TX

HI/ANSI Standards 9.6.8, 2019, Dynamics of Pumping Systems, The Hydraulic Institute, Parsippany NJ.

ISO 5348, 2021, Mechanical mounting of accelerometers, Geneva SWITZ.

API 610, 12th Ed., 2021, Amer. Pet. Instit., Washington D.C.

API STD 610, 12th Edition, January 2021, *Centrifugal Pumps for Petroleum, Petrochemical, and Natural Gas Industries,* API.org.

Marscher, William, D., Mechanical Solutions, *An End-Users Guide to Centrifugal Pump Rotordynamics,* Texas A&M Turbomachinery Symposium, 2007.

Acknowledgements

The author would like to thank Mechanical Solutions, Inc. for its support of the author in the construction of this chapter. In particular, Maki Onari, Dan Owens, and Erik Fillebrown who contributed portions of the test methods and results. Special thanks to Mike Marscher and Brian Fletcher who authored much the graphics. Many thanks also to the manufacturers and users of pumps worldwide who have employed MSI in the pursuit of the many interesting design challenges, analyses, and troubleshooting testing that provided valuable background and experiences. In addition, the author would like to express his gratitude to the many key individuals in the turbomachinery industry (both OEM's and End Users) with whom he has had the pleasure to work with over the years on various projects around the world.

3

Proximity Probes are a Good Choice for Monitoring Critical Machinery with Fluid Film Bearings

By Robert X. Perez

Proximity Probe Benefits

Eddy current transducers, also called proximity or "prox" probes, are frequently used to monitor rotating machinery equipped with fluid film bearings like the compressor seen in Figure 3.1. Proximity probes are especially well suited to measure the relative motion between the shaft and the bearing housing (see Figure 3.2). When properly installed, proximity probes can be used to analyze bearing and rotordynamic phenomena, which would be difficult or impossible to detect with case mounted vibration probes. Proximity probes can be used to:

1. Measure shaft dynamic motion relative to the bearing housing
2. Identify rotor issues related to imbalance, misalignment, looseness, oil whirl, etc.
3. Assist in identifying rotor critical speeds
4. Determine the relative phase relationships of rotor mode shapes by comparing phase measurements across a machine train
5. Determine the relative axial position of rotor, i.e., thrust bearing monitoring, relative to the casing

Today, most critical process machines have internally mounted proximity probes at each bearing to ensure they have a clear view of bearing journals. However, in some instances, probes can be temporarily mounted close to bearings for testing purposes.

Robert X. Perez (ed.) Condition Monitoring, Troubleshooting and Reliability in Rotating Machinery, (77–86) © 2023 Scrivener Publishing LLC

Figure 3.1 The barrel-type, centrifugal gas compressor shown here is a typical example of a high-speed process machine employing proximity probes to monitor rotor vibrations and axial shaft position.

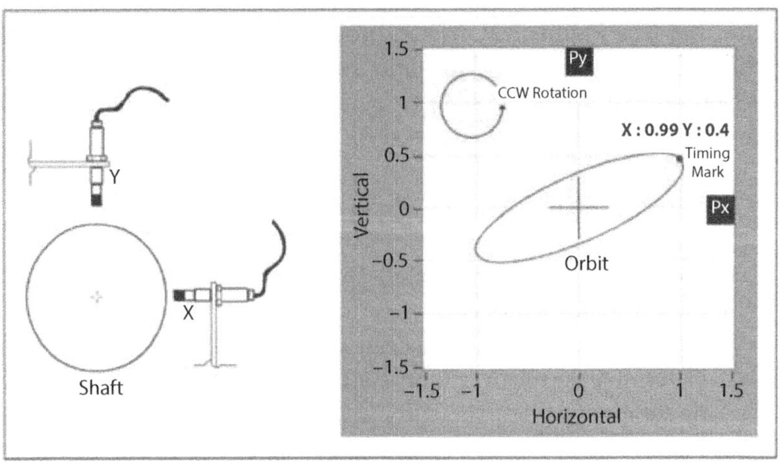

Figure 3.2 A vertical (x) and horizontal (y) probe, oriented 90 degrees apart, as seen on the left, can be used to sense shaft static and dynamic motion at each bearing plane. These dynamic x-y signals can be combined to display the path of the shaft, referred to as the shaft orbit, as it moves within the bearing clearance.

Theory of Operation

Eddy current transducers work on the proximity theory of operation. An eddy current sensing system consists of a matched component system made up of a probe, an extension cable, and an oscillator/demodulator. A high frequency RF signal @2 MHz is generated by the oscillator/demodulator, which is then sent through the extension cable and radiated from the probe tip. When a conductor is placed near the probe's electromagnetic field, some of the energy of oscillation is transferred into the metal of the target in the form of circulating electric currents, called eddy currents. The changes in the probe's impedance due to nearby eddy currents are detected and used to determine the distance between the sensing probe and the shaft. The oscillator/demodulator demodulates the proximity probe signal and provides a 1) modulated DC voltage, which is directly proportional to the probe gap (distance) and 2) an AC (alternating current) portion, which is directly proportional to dynamic shaft motion. This dual signal capability allows eddy current probes to be used for both radial vibration and axial position measurements.

The output signal from the oscillator/demodulator is designed to be proportional to the physical gap between the probe and the shaft, as long as the measurement gap is held within the system's linear range (see Figure 3.3). Notice that in Figure 3.3, the probe's linear range is between a gap of about 22 to 102 thousandths of an inch, or 0.020" to 0.102". The repeatable, linear relationship between the probe's output voltage and the gap between the probe and target is a major reason eddy current probes are widely used for

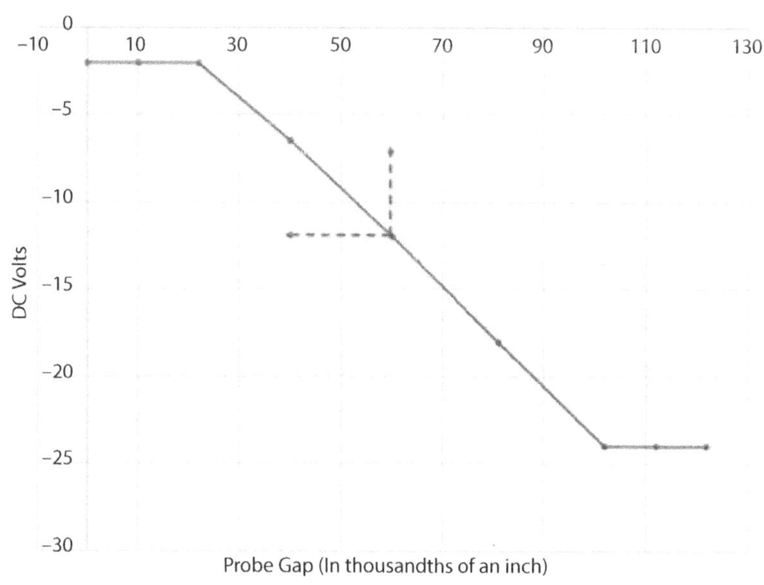

Figure 3.3 Proximity probe calibration curve.

monitoring turbomachinery. It is recommended that all proximity probe systems be 1) checked before their installation and during extended outages to ensure they are working properly and 2) properly gapped to ensure all probes are operating close to the center of their linear ranges.

On this plot, we can see that a gap of 60 mils between the probe and target results in a DC gap voltage of about -12 volts.

Runout Concerns

Because proximity probes are sensitive to the shaft smoothness, shaft imperfections, scratches, pits, and indentions will show up as false vibration, which is commonly referred to as runout. The American Petroleum Institute (API) Standard 670 recommends that *"the combined total electrical and mechanical runout does not exceed 25 percent of the maximum allowed peak to peak vibration amplitude or 0.25 mil (6 micrometers), whichever is greater."* A smooth (64 micro inch) area approximately 3 times the diameter of the probe should be provided for a viewing area to minimize probe measurement errors. The prepared journal area on most shafts should be wider than the bearing itself to allow for probe installations immediately adjacent to the bearing.

Proximity probes are also sensitive to the magnetic permeability and resistivity of the target material. Since the magnetic field of the transducer extends into the surface area of the shaft to a depth of approximately 15 mils (0.015"), care must be taken to avoid the use of coating shafts near probe target areas. Residual magnetic fields in the shaft are another form of electrical runout that can also cause measurement problems. Shaft degaussing may be required to remove any significant residual magnetic fields around probe target areas.

Grounding and Noise

Electrical noise can be a serious problem when using proximity probe systems, so special care needs to be taken to reduce electrical noise levels in the output signal to a minimum. Since most plant electrical noise contains a 60 HZ component and many machines' running speed is also 60 HZ, it can be difficult to separate noise from actual vibration signal. When encountering a 60 HZ signal in the field, first test to see if the signal is related to a mechanical or electrical issue. A high resolution spectral analysis of the peak will allow you to determine if the problem is exactly 60 HZ, which would confirm an electrical issue, or some other frequency.

Shaft Orbits [1]

A shaft rotating within a horizonal rotor system will vibrate within its bearing clearances in a characteristic pattern that is typically elliptical in nature rather than perfectly circular. The actual shape of an orbit depends on bearing stiffness and damping properties and rotor loads, such as gravity, which tends to keep the rotor riding in the bottom half of the bearing and hydraulic forces that may be present. Machinery alignment conditions, balance conditions, lubricant conditions, rotative speed, rotor critical speeds, and other rotordynamic aspects will all affect the two-dimensional motion of the rotor within a bearing under both normal and abnormal operating conditions.

As shown in Figure 3.2, when XY probes are available, the complete 2-dimensional motion, or orbit, of the shaft can be sensed, allowing the maximum peak-to-peak motion of the shaft to be measured. The orbit is the path you would see if a pencil lead at the shaft's centerline traced the shaft's path on a piece of paper and then magnified so it could be seen with the naked eye. In Figure 3.1, the shaft motion labeled as "Orbit" depicts a healthy elliptical shaft orbit, orientated at about 24 degrees from the horizontal axis. This orbit's single timing mark, labeled as "Timing Mark" on the right side of Figure 3.2, indicates that the predominant frequency of the shaft vibration is one time (1x) running speed. By using an orthogonal probe pair, the shape, magnitude, frequency content, and location of the orbit within the bearing can be ascertained and displayed. In contrast, a single proximity probe per bearing arrangement cannot provide a shaft orbit or be used to determine the maximum peak-to-peak value of shaft vibration.

The shape of vibration orbits can be used to identify a broad range of machinery maladies. The orientation and ellipticity of the journal filtered orbits (1X, 2X) can provide significant information about the dynamic stiffness of the bearings and how the oil film is performing. Here are a few examples of how orbits can be used to identify mechanical issues in the field:

1. A deformed elliptical orbit may be indicative of an unwanted bearing preload related to internal or external machine misalignment. An excessive preload can collapse the orbit into a "banana" shape. Extremely large preloads can result in an orbit resembling a figure eight shape.
2. An orbit with a single timing mark, as seen in Figure 3.1, is indicative of the presence of 1x shaft vibration.

3. Orbits with multiple loops with a single timing mark are an indication of vibrational content higher than 1x vibration, i.e., 2x, 3x, 4x, etc.
4. Non-synchronous rotor whirling may be an indication of a rotor or bearing stability problem. For monitoring systems with phase reference capabilities, a normal "blank-bright" convention within the orbit indicates forward precession of the rotor. Conversely, "bright-blank" phase marks within the orbit indicates reverse precession of the orbit.
5. Vibration orbits can also be used to detect runout issues. For example, runout due to a single deep scratch on the shaft will look like an orbit with two arms radiating outwardly, which are oriented 90 degrees apart. Always check slow roll vibration data after overhauls to ensure that runout levels are below API guidelines.

A proximity probe monitoring system with phase reference capabilities can also be used to generate operating defection shape (ODS) plots for a complete machine train. ODS plots can be used identify and address rotor-dynamic issues.

I will now review the various machine classes and offer some advice about whether users should install proximity probes or seismic sensors for condition monitoring. When it comes to machinery monitoring sensors, one size does not fit all.

General Machinery Monitoring Recommendations

Proximity probes and seismic sensors, such as accelerometers, piezovelocity, and moving coil sensors, make up most of the vibration sensor population used to monitor industrial machinery. Each type has their own strengths and weaknesses. One major drawback of proximity probes is that they must be embedded in bearing housings to sense shaft motion, while seismic sensors must be mounted on bearing housings so that bearing housing motion and rolling element bearing defects can be monitored.

Some considerations in the sensor selection process are [2]: 1) the casing-to-rotor mass ratio of your machine, 2) bearing support stiffness ratio of your machine, 3) the expected frequency of expected internal defects, and 4) operating temperature:

- If your machine has a heavy casing and a light rotor, similar to the barrel compressor seen in Figure 3.4, then only a small

Figure 3.4 A centrifugal barrel compressor like the one shown here is an example of a machine with a high casing mass to rotor mass ratio and high support stiffnesses, where proximity probes make the most sense for monitoring rotor vibration.

fraction of the rotor vibration is likely to be transmitted to the casing. Therefore, machines with casing-to-rotor weight ratios greater than 10:1 should probably be monitored with proximity probes.

- Machines with high support and casing stiffnesses usually require the use of proximity probes to accurately sense rotor motion, since only a minimal amount of motion will be transmitted to the casing. A barrel compressor is an example of a process machine with high support and casing stiffnesses.
- If your machine has a light casing and heavy rotor, it is possible for the rotor to generate sufficient force to cause the casing to vibrate sufficiently for measurement. Machines with a less than 5:1 casing to rotor weight ratio can usually be monitored with seismic sensors.
- If a machine's bearing supports are flexible enough to transmit most of the shaft vibration to the bearing housing, then seismic probes make sense.
- In some cases, it might make sense to use both proximity probes and seismic sensors on a machine train. For example, a machine train with a large gear box might need case mounted accelerometers to detect gear defects and also require embedded proximity probes to monitor relative rotor to bearing motion.

Table 3.1 [3] Where proximity probes and seismic sensors are best used.

Machine description	Vibration probe type	Comments
Centrifugal Compressors	Proximity	Machines in this class tend to have very heavy casings and rigid bearing housings.
Large Steam Turbines	Proximity	Similar to centrifugal compressors, machines in this class tend to have very heavy casings and rigid bearing housings.
Large Motors/ Generators	Proximity/ Seismic	Rotors are generally light compared to casing, but vibration related to the casing motion (@120 Hz) is generally better measured with seismic sensors. Talk to the OEM to see what they recommend for your specific machine.
FD/ID Fans	Seismic	Rotors are heavy and supports/pedestals are flexible. Proximity probes can be used but may be insensitive for some designs.
Gas Turbines	Proximity/ Seismic	Casings are typically relatively flexible, even for large industrial gas turbines. Often, both sensor types are recommended by OEMs. Talk to the OEM to see what they recommend for your gas turbine.
Centrifugal Pumps	Proximity/ seismic	Large, heavy pumps with journal bearings are better monitored with proximity probes. Small to medium size pumps with anti-friction bearings are better monitored with seismic sensors.

(*Continued*)

Table 3.1 [3] Where proximity probes and seismic sensors are best used. (*Continued*)

Machine description	Vibration probe type	Comments
General Purpose Steam Turbines	Seismic	Smaller steam turbines tend to have relatively small casing to rotor weight ratios and flexible bearing housings. Talk to the OEM to see what they recommend for your specific steam turbine design.
Anti-friction Bearing Machines, such as Small Pumps and Electric Motors	Seismic	It is not recommended that proximity probes be used to monitor smaller machines with anti-friction bearings.

- The seismic sensors are often the preferred choice in harsh environment applications, such as at the hot end of gas turbines where the temperature is far too high for traditional proximity probes.

If in doubt about which type of sensors to use to monitor your machinery, discuss that matter with the original equipment manufacturer (OEM).

Table 3.1 summarizes where I believe proximity probes and seismic sensors are best used. In general, it is recommended that proximity probes should be used in machines with fluid film bearings and rigid bearing housings. In contrast, seismic sensors are typically used to detect casing motion in rolling element bearing applications or where casings are relatively flexible. Accelerometers are the most common seismic sensors used on industrial machinery today due to their cost and reliability. Keep in mind that some machine trains may require both proximity probes and seismic sensors due to the nature of their construction and criticality.

Final Thoughts

The decision of whether to install XY (orthogonal) probe pairs versus a single probe at each radial bearing of a given rotating machine train or not should be based solely on economics. If an economic justification is required, the user must first quantify the lifetime production losses related to extended downtimes that could have been avoided if sufficient vibration data was available to quickly resolve a given vibration issue. Then, a detailed cost-benefit analysis can be done to determine if multiple (XY) probes at every bearing is justified. The reader should keep in mind that in some applications there may be only enough room for one probe.

While it may be appealing to instrument the machine with only a single probe at each bearing to control initial project costs, the resulting monitoring system may not be able to perform all the machinery protection and machinery diagnostic functions required for the application. In my experience, when all installation costs are accounted for, adding a second radial probe will cost a fraction of the cost of the initial proximity probe.

Advice: If you are considering using a single proximity probe monitoring arrangement, I recommend that you have the equipment packager quote a single probe and XY probe system. Having both quotes will allow you to compare the actual cost difference between the two systems. You may be surprised at the actual difference between the cost of single versus XY probe monitoring system.

The cost differential between single and dual probe systems will be justified the first time a major machinery problem occurs and a comprehensive picture of machinery behavior is required. As a rule, I recommend that dual probes be specified on all critical machines.

References

1. Kingham, John and Winterton, John, Orbit magazine, Volume 30, Number 1, 2010, pages 5–9.
2. Hady Abdel Hady, *Why is vibration monitoring important?*, Vibration and Dynamic Measurements On-line Blog.
3. Kelm, Ray, *Introduction to Journal Bearings & Orbit Analysis*, Vibration Institute: Piedmont Chapter #14, May 2018.

4

Optimizing Lubrication and Lubricant Analysis

By Jim Fitch and Bennett Fitch

Introduction

Lubrication is an unmistakably integral part of machine reliability. Rotating machines are dependent on lubrication decisions made, such as which lubricant to use, how the lubricant needs to be applied to the tribological zones, and what is done during operations to monitor and control the integrity of these frictional zones.

At the same time, lubrication is too often not top of mind when considering the critical aspects of rotating machines, partly because there is a lack of general understanding of the crucial role the lubricant plays in reliability. But even for those who may understand this, it still is not intuitive to manage these factors carefully. Rather, there are incorrect assumptions that lubrication is straight forward; in other words, simply "just having oil or grease in the machine is largely all that is necessary" is a perspective of many. This, coupled with the fact that lubrication is messy and not as exciting as the many other maintenance tasks, often challenge workforce culture.

As a result, the industry suffers from stagnant practices and lethargic attitudes. Although, the dismal state of an old and generally unexciting field is a huge opportunity in disguise. For plant maintenance personnel that see this opportunity, improvements in lubrication not only help avoid unnecessary costs in repairs and downtime, but also have a huge impact in improving the maintenance culture and creating a foundation for sustainable growth. But what should be the focus for improvement and achieving lubrication excellence?

Lubrication Excellence is a term used to describe the holistic achievement of maintaining all lubrication practices from start to finish. One way

Robert X. Perez (ed.) Condition Monitoring, Troubleshooting and Reliability in Rotating Machinery, (87–134) © 2023 Scrivener Publishing LLC

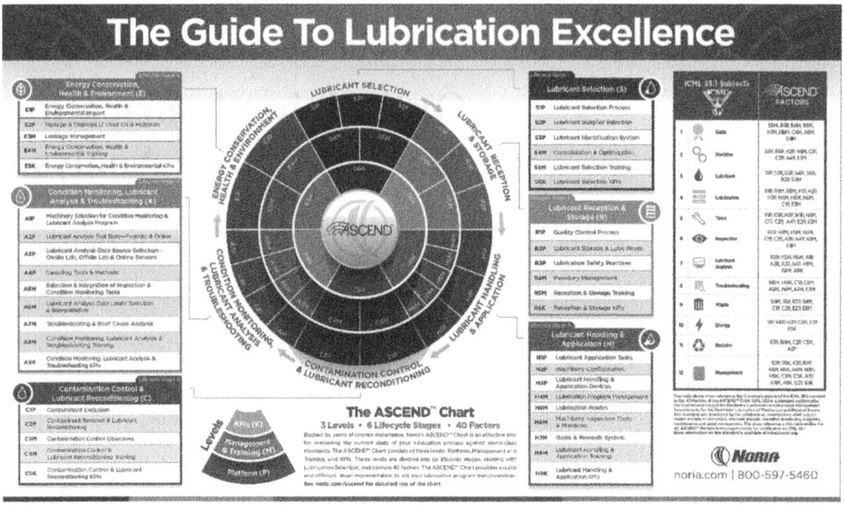

Figure 4.1 Ascend™ chart.
Note: The 40 factors in this chart will be explained in more detail later in this chapter.

to observe this is considering the lifecycle of the lubricant by way of the Ascend™ Methodology (developed by Noria Corporation) [10], as represented in the Ascend chart (Figure 4.1, above). It has been a powerful tool and metric to evaluate, design, codify, implement, and control the essence of the lubrication excellence transformation.

Ascend is based on the notion that identifying the starting point on the journey to lubrication excellence always requires evaluating current practices in contrast with lubrication best practices, also referred to as the Optimum Reference State (ORS). (Note to reader: Table 4.6 defines all acronyms used in this chapter.) Once the gap is identified, it is necessary to establish an optimal road map that ensures resources, effort, and time are allocated to targeted initiatives that produce the highest return on investment. Before reviewing the Ascend methodology, let's first understand how the ORS is applied in real lubrication examples.

Optimum Reference State [12]

When it comes to making lubrication decisions, there are many mutually influencing factors that must be considered. The Optimum Reference State (ORS) serves as a guiding principle on how decisions are best made. This principle is critical in the journey to world-class lubrication and enhanced machine reliability. In short, it is defined as *the prescribed state of machine configuration, operating conditions and maintenance activities required to achieve and sustain specific reliability objectives.* Lubrication excellence is achieved when the current state of lubrication approaches that of the Optimum Reference State.

There are many different critical attributes of the ORS. These attributes relate to people preparedness, machine preparedness, precision lubricants, precision lubrication, and oil analysis. Achieving the ORS almost always involves change or modifications, including:

- Modifying Machines
- Modifying Behaviors (People)
- Modifying Lubrication Procedures, Methods, and Frequency
- Modifying Preventive Maintenance (PMs) and Inspections
- Deploying Vigilant Lubrication Monitoring (Oil, Grease, and Filter Analysis)

Each attribute must be:

- Precise and definable (e.g., a specific lubricant sump level),
- Measurable (e.g., a specific viscosity) or verifiable (e.g., a sample port location),
- Controllable (by modification) and sustainable (by program continuity),
- Able to achieve the desired reliability objectives related to the financial benefit, safety, and machine readiness.

Throughout this chapter, lubrication practices and oil analysis will be viewed through the lens of the ORS. The state of machine and plant reliability should not be maximized, but instead optimized. When considering lubrication decisions at specific machines, take for example the modifications seen in Figure 4.2 comparing two identical centrifugal pumps. Each have specific and unique needs with respect to the ORS. For example, they require different modifications influenced by their respective criticality to plant-wide reliability (downtime risk) and cost of repair.

This is a very visual example of how the ORS can be applied, although the same lens can be given to decisions related to sampling frequency, filter selection, inspection frequency, training plans, and oil analysis test selection. Table 4.1 shows how these activities can pose a risk to the organization if the ORS is not properly applied (optimized). Too far to either extreme results in higher costs in the form of waste or a state of excess.

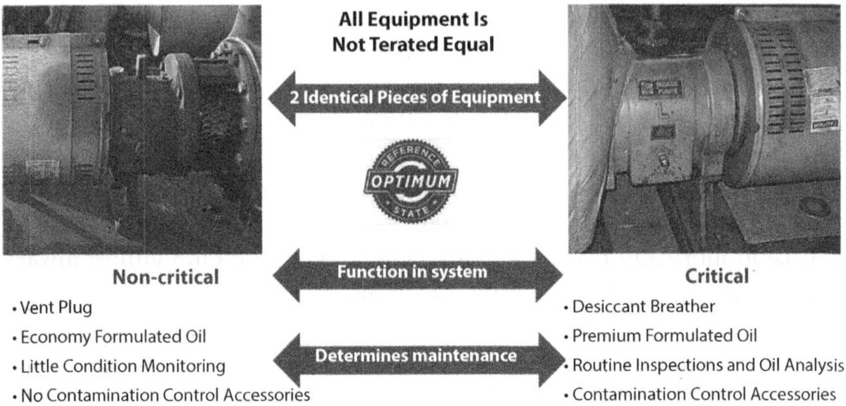

Figure 4.2 Example of optimum reference state on two identical centrifugal pumps.

Table 4.1 The optimum reference state avoids waste and excess.

Sub-optimum	ORS = Precision Lubrication	State of Excess
	Activity	
Price-selected base oils	Machine and application-specific base oil selection	Across-the-board synthetics
Bi-annual sampling	Setting sampling frequency based on criticality and failure probability	Continuous sampling
Factory filters only	Selecting filters based on criticality, machine sensitivity and surge ingression risk	Across-the-board 3-micron filters
Weekly oil-level checks	Engineered daily PM inspection checklists	Sensor-based real-time monitoring
Old-timer apprentice training	Professional skills training using subject-matter specialists and peer-reviewed curriculum	Degreed engineers and chemists only
Economy-based one-size-fits-all test slates	Application-specific oil analysis test slate selection with exception tests	Premium test slates for all machines

When making plant-wide lubrication decisions, such as selecting which machines are appropriate for oil analysis or how to consolidate lubricants to a manageable selection across the plant, the ORS applies here as well. From a macroscopic perspective of decision making, managing risk and costs can override (or complement) the decisions made for each machine. In fact, when the sum of all lubrication ORS attributes in the plant are considered from both micro- and macroscopic perspectives, this creates an engineering specification for Lubrication Excellence.

Fundamentally, there is a need for users to understand how their current lubrication practices may be sharply different than what might be considered the optimized state. This is further underscored by showing the value to the organization and to maintenance workers individually (financially and career development). And ultimately, it needs to be specified, through good engineering practices, in such a way that deployment can be sustainable with manageable risk and cost. The Ascend methodology structures, codifies, and prioritizes this comprehensively based on the ORS in all essential areas of lubrication-enabled reliability.

Lubrication Excellence and the Ascend Chart

The Ascend Methodology deploys six lifecycle stages of a lubricant that follow in chronological order starting at the top going clockwise represented by angular wedges on the Ascend Chart. Each stage plays an important part in lubrication excellence, machine reliability, and asset management.

- Lubricant selection (S)
- Lubricant reception and storage (R)
- Lubricant handling and application (H)
- Contamination control and lubricant reconditioning (C)
- Condition monitoring, lubricant analysis, and troubleshooting (A)
- Energy conservation, health, and the environment (E)

Following the lubricant life cycle allows users to correlate their day-to-day activities at each stage with fundamental business objectives and how each stage functions individually as a part of the whole in executing correct principles of lubrication.

Ascend punctuates the importance of selecting the right lubricant, with performance specific to the requirements of the machine and its operational context. But it doesn't stop there. It also needs to specify the correct condition, storage, and handling practices, application methods, and frequency. Precision lubrication also includes application of the correct amount, with the right tools, by a person with the right knowledge and skills. Lastly, there is a fundamental need for sustainable contamination control of lubricant in storage and use and, through lubricant analysis, a metric on the state of lubrication and overall machine health.

Lubrication activities across the lifecycle of the lubrication process are divided into 40 Factors (see Figure 4.3). Each of these factors is interrelated to one or more of the 12 areas of the lubrication plan as specified in clause 5.0 of the standard ICML 55.1[1] Asset Management Requirements for the Optimized Lubrication of Mechanical Physical Assets.

Figure 4.3 40 Ascend™ factors.

However, it should be noted that these 40 factors do not all share the same weight (priority) or urgency of implementation. Therefore, to emphasize and clarify these differences, the factors are arranged into three levels (see Figure 4.4):

- Platform (P)
- Management and Training (M)
- Performance Indicators (K)

The Ascend™ methodology uses Six Sigma's DMAIC (Define, Measure, Analyze, Improve, and Control) tool to assess the maturity of the lubrication program from an established starting point through the transformation plan. The Optimum Reference State is defined based on the requirements of the ICML 55.1 standard. ORS compliance is assessed through inspections, measurement, verification, and interviews.

Differences between required ORS performance and current performance are analyzed by the Ascend methodology to prioritize the implementation plan based on impact on business objectives and reliability. When an assessment is performed using the Ascend methodology, a maturity of each factor based on the degree of compliance (completeness) is defined along with a list of specific improvements aimed at complying with the ORS in each of the 40 factors of the lubrication process. It identifies the critical actions needed to ensure that the process is sustainably implemented with suitable controls.

Similarly, a maturity score can be identified for each of the six lifecycle stages and the three levels. Finally, a holistic indicator of maturity of the lubrication program, called the Ascend Compliance Level (ACL), is a sophisticated calculation based on a proprietary mathematical algorithm that includes a Balanced Score Card (BSC) of the lubrication process and the maturity of each of the factors. The active ACL value can be displayed in the center of the Ascend Chart and as a guiding key performance indicator for lubrication excellence.

Figure 4.4 Three levels of ascend™.

Bringing Awareness to Lubrication, Contamination, and Oil Analysis

The following sections in this chapter are not intended to be a comprehensive overview in lubrication, contamination, and oil analysis, as this would be too extensive and there many other published references, including ICML 55.1, that very adequately suffice. Rather, what is written here is aimed at uncovering key aspects in lubrication and contamination that are often less understood yet are crucial in establishing foundational knowledge on the impact it has on machine reliability.

Then, oil analysis is brought into focus to carefully identify what is necessary for end-users to get right in order to achieve the expected return on investment. While it has long been a common practice for monitoring the conditions in the lubricant, thus the machine's conditions, there is much that can unknowingly go wrong and jeopardize the integrity of the efforts. Furthermore, it is important to connect this to newer (and often simpler) techniques to monitoring lubricant conditions in improving lubrication and controlling contamination

What You Might Not Know About Lubrication

Machine Surface Interaction

It is crucial to understand how two metal surfaces within a machine interact with each other. Regardless of how smooth a surface may appear, each metal surface has high points, known as surface asperities, and low valleys. When the two surfaces move past each other, the asperities on each surface come into contact (collide), as illustrated in Figure 4.5.

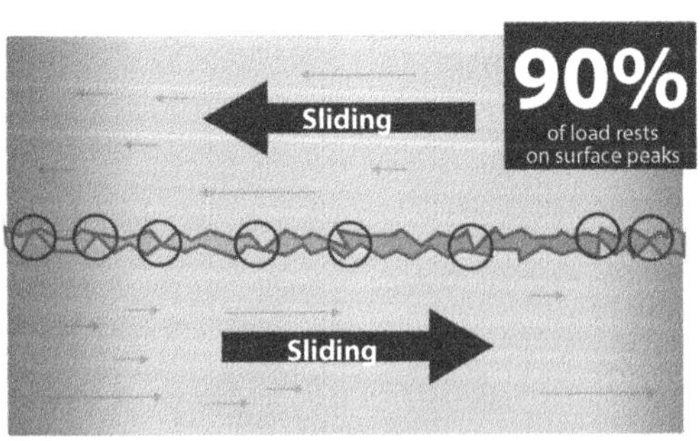

Figure 4.5 Circles represent asperity contact during sliding conditions.

The height and amount of the asperities dramatically affect the friction between the surfaces. It is the lubricant's job to keep asperities apart and prevent or mitigate destructive contact and mechanical friction. Metal-to-metal friction causes two surfaces in relative motion to generate heat, consume energy, and often weld together, causing adhesion (galling) and abrasion (cutting). In severe cases, the metal may completely weld together.

Reducing friction can be accomplished with:

- Proper lubricants and lubrication.
- Replacing sliding friction with rolling friction.
- Improved materials selection and surface finish – softer babbitt materials are often used to reduce friction during start-up and shutdown.

The Lubricant Film

When thinking of lubrication, one should consider how the base oil creates a film to separate metal surfaces to minimize mechanical contact. For the base oil to provide separation, there must be a balance of three contributing factors: relative surface velocity, base oil viscosity, and applied load. These three factors are also influenced by other elements such as temperature, contamination, and surface topography. The film thickness that is achieved is called hydrodynamic lubrication.

In applications involving rolling contact, film thickness can still occur, even with greater localized pressure (unit load). In fact, these pressure points play an important role in achieving the film. The base oil's pressure-viscosity relationship results in the viscosity to increase temporarily and significantly. This is called elastohydrodynamic lubrication. It is best to keep machine surfaces separated and the film thickness provides the best opportunity to reduce friction and wear. But what happens if these film thickness conditions are not met, such as when there is insufficient surface velocity, inadequate viscosity, and/or excessive load?

When the hydrodynamic or elastohydrodynamic lubrication prerequisites are not met, the base oil will require support from boundary contact conditions. This support involves using wear and friction-control additives. The base oil and additives are mixed together to produce a lubricant, either oil or grease. This lubricant is then formulated to mitigate the anticipated boundary conditions. This process ensures the lubricant will have the correct amount of film strength and boundary lubrication properties.

Film Strength

Film strength is the lubricant's ability to lessen the effects of friction and control wear by means other than viscosity-induced film thickness. When the base oil viscosity is insufficient to overcome metal-to-metal surface contact, the base oil and additive chemistry work together to create a surface protection mechanism. Even when loads and temperatures are higher and relative surface velocities are lower, functional film strength can be achieved.

Unlubricated Surface Interactions

When observing frictional surfaces on a microscopic level, they are rough, form a topography of undulating peaks and valleys, even while they may look smooth or even polished to the unaided eye. This is similar to how spherical the Earth looks from space but in fact is brimming with mountains and valleys on the surface. Because the surface has these mountains and valleys, when unlubricated metal surfaces come into contact, the actual contact area is actually less than the apparent (perceived) contact area. The surfaces will only touch (bearing load) where the higher asperities reach the opposing surface. However, these asperities can elastically deform under pressure, making the contact area increase with an increased load.

Friction and Wear Generation

Many consider surface roughness the primary contributing factor of friction. However, when considering that the real contact area may be less than 1 percent of the apparent contact area, the actual roughness becomes much less relevant. The significant process contributing to friction is a result of the adhesive bonds occurring at the atomic level of asperity contact.

In conditions where there is inadequate lubricant film thickness, the asperity contact points can lead to cold welding, which leads to adhesive wear. The adhesion undergoes a work-hardening process, which strengthens it. Thus, the shear point happens layers below the asperity contact point where the metal has not been strengthened. As the metal shears, the asperity tip is either transferred to the other surface or broken off, becoming an abrasive particle.

Mitigating Surface Interactions

Friction and wear-control additives are formulated in small quantities within the base oil and have polar properties that foster metal surface attraction. These attractions are encouraged to chemically react with the surface. When machine surfaces interact with higher pressures and temperatures, the additives mitigate the typical effects of wear by creating initial flexible molecular layers on the machine surface. These friction-control layers become sacrificial and directly reduce shear strength during contact, as illustrated in Figure 4.6.

The initial layers can mitigate friction by allowing the lubricant's weaker molecular bonds to release with less force compared to that of the strong bonds that result from the metal-to-metal asperity boundary conditions. The formation of low-shear-strength films is also influenced by the base stock type and the metallurgy of the mechanical surfaces.

There are three types of lubricant additives that help reduce this friction and control wear formation:

- Friction modifiers
- Anti-wear additives
- Extreme-pressure additives

Physics and Chemistry

The interactions of asperities at the actual contact pressure points are the main concern when poorly lubricated machine surfaces come into sliding

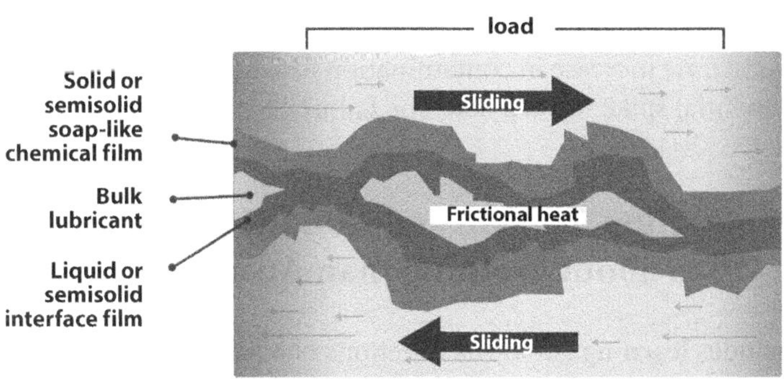

Figure 4.6 Sliding and frictional heat.

contact. At this molecular scale of the machine surfaces, boundary conditions are subject to numerous principles of physics and chemistry.

The role of oxidation, corrosion, chemisorption, and other chemical reactions on the machine surfaces must be carefully balanced when additive compounds are selected for film strength protection [7]. These friction and wear-control additive films reduce the shear strength at the contact points. The low-shear-strength films are sacrificial during physical interactions and protect the surface from the effects of adhesive, abrasive, and fatigue wear.

While the base oil's viscosity is always preferred to protect the machine surfaces with hydrodynamic and elastohydrodynamic lubrication, boundary conditions will still exist. Therefore, to protect against boundary conditions, a properly formulated lubricant with friction and wear-control additives should be used to provide a film strength that is proportional to the exhibiting mechanical interactions.

Contamination: The Antagonist to Lubrication

Contamination is often a silent lurking danger and can alter or destroy critical mechanical zones of your machinery. All lubricants have some level of solid contamination. Even with new lubricants, complete cleanliness is unattainable. The key is to ensure that there are more contaminants being filtered out than contaminants being ingested or generated within. Problems arise when too many contaminants are allowed to exist in the workings of a machine.

Since contaminants can silently accumulate and generate within machines, this cyclical act proliferates to a point where the filter cannot maintain a healthy balance in contamination levels. So, while oil analysis may show that contaminant levels remain at an acceptable rate, it may only take a moderate increase in contamination to set this off balance and create an exponential spike toward machine failure [1].

Contamination Control and Condition Monitoring is More Often about Training than Advanced Technology

This includes learning about the functions of a lubricant, how to select the right lubricant for each application, how to manage them in storage and properly apply them to machines, and of course, it is all about monitoring lubricant and machine conditions through inspections and oil analysis. There are two specific areas of lubrication that must be communicated by bringing awareness to those who work with and around plant equipment. The first one is the importance of contamination control, which is discussed here. The second is inspections, which will be discussed later in this chapter.

Contamination Control

Contamination is defined as "any foreign or unwanted substance that can have a negative effect on system operation, life or reliability." This is more than just solid particulates from the environment, it includes other factors such as water, air, glycol, soot, and fuel. Even the wrong lubricant being mixed into the current lubricant is considered a form of contamination referred to as cross-contamination.

Contamination control for lubrication includes the "planning, organizing, managing, and implementing all activities required to determine, achieve and maintain a specified contamination level." Notice that neither the word "eliminate" nor "remove" is used in this definition.

Don't Leave It to Instinct

A lot of what is important with contamination control is not intuitive. For example, take the physical size of solid contaminants that could damage a rolling or sliding contacting component. Oil films are usually 5-20 microns for sliding contact (turbine bearings, gears, pistons), all the way down to less than one micron for rolling contact (rolling element bearing, gears, cams). Typical airborne particulates that ingress into machines are usually much smaller than 40 microns, which is the visibility limit of the unaided eye.

This makes it common for there to be a misperception of the degree of cleanliness needed in and around lubricated machines. It is not intuitive for us to understand the importance of these virtually invisible contaminants with everyday practices. This is just one example of what must be learned through careful training. It is important to provide a discussion and explanation why contamination control is important, rather than just telling them what to do and what not to do.

Creating a Balance Between Exclusion and Removal

It is often incorrectly assumed that contamination control is just about filtration. While it is a big part of contamination control, filtration is only necessary because contamination is first allowed to get into the oil and the machine. The actions that must be taken to control contamination include both exclusion (seals, breathers, clean new oil, etc.) and removal (filtration). In fact, it will always be much cheaper (at least one-tenth the cost) to exclude a gram of dirt from getting into a machine than it is to remove it through filtration.

Nevertheless, neither exclusion nor removal is perfect; they must be considered together as a contamination control solution and where appropriate based on the Optimum Reference State. Contamination control requires a balance of a two-part approach, just like our bodily caloric control, where we strive to burn more calories than we consume. For machines, we can monitor contamination levels, such as with oil analysis, to verify that this is staying in balance. If more contaminants are accumulating in the oil than are being removed, it is only a matter of time that the contamination will induce a failure. Component contaminant sensitivity, contamination likelihood, criticality, safety risks, and downtime are examples of variables that go into justifying the ORS contamination control requirements. It is important that those who make decisions about breathers, seals, filtration, and other everyday oil sump management have learned about contamination control and optimizing these decision variables to ensure enough is being done to keep this in balance.

Scenario Analysis: Many maintenance teams don't realize the benefits of contamination control. For decades, countless industry studies by OEMs and end-user groups have identified that contamination is the number one cause of wear on rolling element bearings, gears, and the majority of lubricated components. Additionally, it is well established that the cost of controlling contamination through optimized best practices will be considerably less than the cost savings from mechanical wear-related failures decreasing over that period.

Then why is this not often realized? This is where bringing awareness is necessary and must influence the maintenance culture. As mechanical wear occurs from moderate levels of contamination, it propagates a gradual Failure Development Period (FDP) that appears largely uneventful to the untrained person. As the wear gets worse, eventually predictive maintenance (PdM) may trigger a corrective action through vibration analysis, inspections, or other means. If this becomes a common occurrence, then a preventative maintenance (PM) task may get scheduled to replace these components on a fixed interval that is significantly less than the intended design life and unfortunately, this is very common.

These PdM catches and scheduled PMs are rewarded, but these habits form an unhealthy maintenance culture focused on reacting to failure rather than establishing proactive measures to recognize the root cause (contamination) and improve proactive maintenance (via contamination control).

When a root cause failure analysis is performed, it is usually difficult to pinpoint one single cause. Rather, the root cause is viewed as a collection of decisions and practices that impact contamination. Good practices include everyday activities or decisions such as:

- Managing new oils and keeping them clean and dry before use
- Transferring new oils in clean, sealable-and-refillable containers or filter carts
- Managing machines' headspaces by using quality desiccant breathers
- Monitoring contamination levels with particle counting on critical machines
- Establishing filtration needs effectively, either through continuous stationary filtration or through periodic filtration with a filter cart
- And many more daily activities like careful machine washdowns, keeping machine areas tidy and clean, walk-by inspections, etc.

The actions and decisions that influence contamination control are part of a collective effort involving nearly everyone working around the machines including maintenance, operators, lube techs, reliability engineers, supervisors, etc. Similarly, when these teams go through contamination control trainings as a team, everyone builds a collective awareness and a better understanding of what each of their roles entail. The benefit of contamination control training multiplies as the importance is bought-in together, especially when the training takes place in person as one group.

Ultimately, contamination control is everyone's responsibility. When contamination control is analyzed as part of the overall lifecycle of a lubricant in the Ascend Methodology, it tends to dominate the next step requirements that follow. This is especially true through metrics that identify the impact of contamination on lubrication. It becomes comparatively important to selecting the right lubricant or monitoring the conditions, such as through oil analysis. Lubrication awareness training sets the tone from the top down that lubrication is not a trivial part of maintenance but instead requires carefully made decisions, quality daily actions, and, most importantly, it impacts the bottom line.

Why Perform Oil Analysis [3]

Oil analysis is performed to understand the condition of the oil in an effort to bring awareness to the condition of the machine from which the oil sample was taken. It is a practice that has been around for a long time, yet it is a challenge to get right for many and its viability is too often disregarded. Why? Laboratory-based oil analysis is often monitoring data that can be difficult to comprehend because it is not as straight forward as many other condition monitoring technologies such as vibration, thermography, or even the more recent addition of motion amplification. All of these technologies are viable and can provide value when performed correctly.

However, oil analysis finds itself in a category that arguably provides the most value as a condition monitoring strategy when performed correctly. This is because it can do an excellent job of detecting problems ahead of machine failure by looking for changes in fluid properties or contamination levels in addition to machine faults in the form of wear debris. It targets issues largely in the proactive maintenance domain and closer to the root cause. This is why it is so important to figure out. These three aspects of oil analysis are often clearly defined as categories of oil analysis.

Fluid Properties Analysis

This type of oil analysis focuses on identifying the oil's current physical and chemical state as well as on defining its remaining useful life (RUL). It answers questions such as:

- Does the sample match the specified oil identification?
- Is it the correct oil to use?
- Are the right additives active?
- Have additives depleted?
- Has the viscosity shifted from the expected viscosity? If so, why?
- What is the oil's RUL?

Contamination Analysis

By detecting the presence of destructive contaminants and narrowing down their probable sources (internal or external), oil analysis can help answer questions such as:

- Is the oil clean?
- What types of contaminants are in the oil?
- Where are contaminants originating?
- Are there signs of other types of lubricants?
- Is there any sign of internal leakage?

Wear Debris Analysis

This form of oil analysis is about determining the presence and identification of particles produced as a result of mechanical wear, corrosion, or other machine surface degradation. It answers questions relating to wear including:

- Is the machine degrading abnormally?
- Is wear debris produced?
- From which internal component is the wear likely originating?
- What is the wear mode and cause?
- How severe or threatening is the wear condition?

Achieving Oil Analysis Success by Looking Holistically [5]

To get the answers to the questions mentioned above, a number of practices must be well established in the oil analysis program (see Figure 4.7), such as: How often should oil samples be taken? Which machines should be sampled? What test should be performed? In fact, there are several crucial factors that must be carefully considered when building out an oil analysis program. If one of these factors are overlooked, then the entire oil analysis program may not live up to its return on investment. The next few sections review some of these crucial factors, grouped into three categories:

- Obtaining a Representative Oil Sample
- Ensuring Reliable Testing
- Determining the Optimum Course of Action

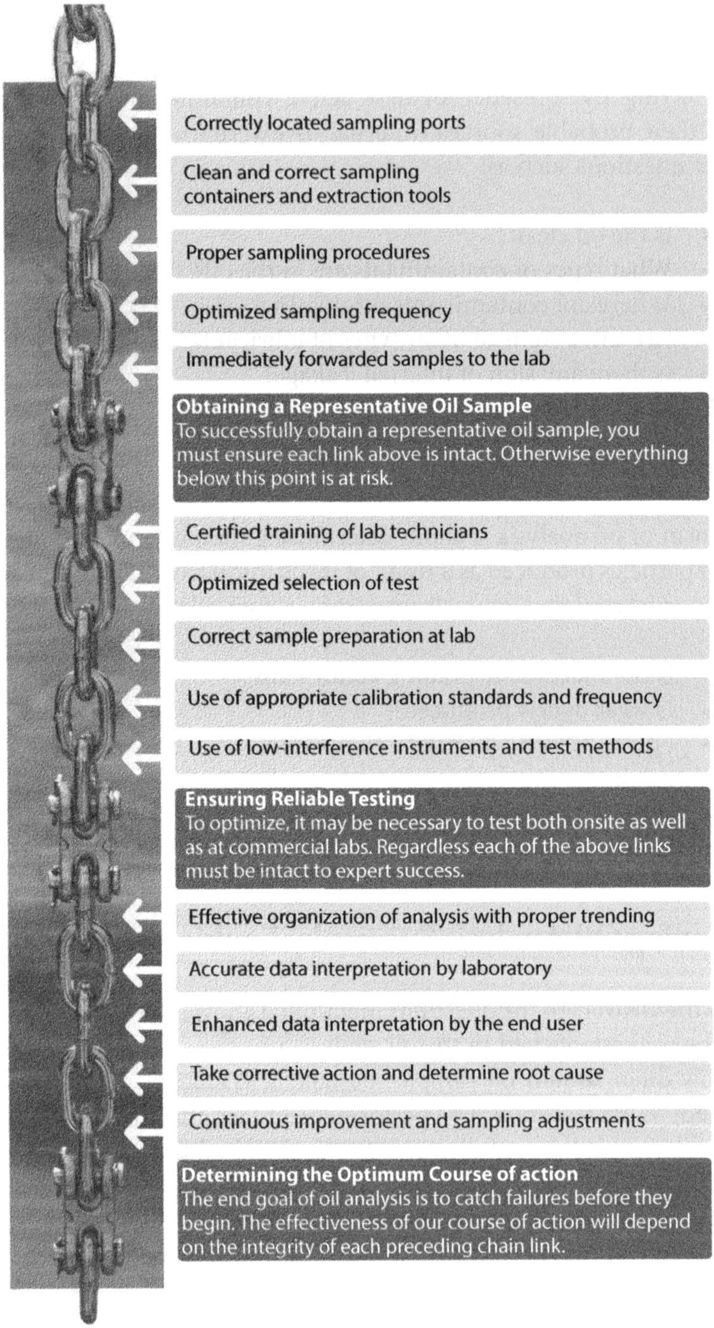

Figure 4.7 Holistic approach steps.

The first and third category are weighed greatly on the end-user to get right. Therefore, these will be focused on more below. Following this, we will discuss how oil analysis should be considered beyond just what is laboratory-based. After all, any time an analysis of the oil is performed this is considered oil analysis and often times, what is perceived as more basic and unsophisticated, such as a visual inspection, is exactly the opposite. These visual inspections should occur even while taking a sample. Although, let's start with what it takes to obtain a representative oil sample and why that is important.

Obtaining a Representative Oil Sample

Select the Right Machines for Oil Analysis

Each sample obtained for oil analysis can be costly, so sampling every machine in a facility is not feasible. The best method is to determine the Overall Machine Criticality (OMC). The Overall Machine Criticality is a risk-profile assessment that can be calculated to a single numerical value. The lower the OMC, the lower the risk. The OMC is the multiplied product of two factors: the Machine Criticality Factor (MCF) and the Failure Occurrence Factor (FOF).

A risk-profile assessment directly justifies which machines need oil analysis (or other condition monitoring techniques) in order to help lower the risk. For example, let's consider a machine with an MCF of 5 which may be associated with high repair costs if a failure were to occur and has moderately highly impact on production. With the implementation of better oil analysis practices (aided by carefully collected representative oil samples), this can effectively reduce the machine criticality factor to a 2.5 with early detection achieved through oil analysis. (See Figure 4.8 below.)

Machines across a facility should be viewed in a similar manner to choose which machines would benefit from early detection. Other factors, such as accessibility or a similar risk-profile assessment on the lubricant to calculate the Overall Lubricant Criticality (OLC) should be considered.

Clean and Correct Sampling Containers and Extraction Tools [2]

One of the main objectives of oil sampling is minimizing data disturbance. Using the right sampling tools and ensuring their cleanliness will be vital.

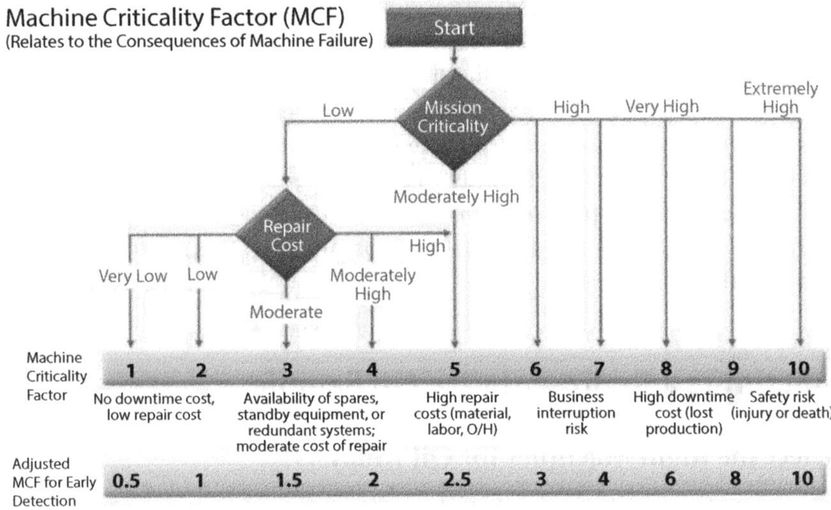

Figure 4.8 Machine Criticality Factor (MCF) (Relates to the consequences of machine failure).

Oil sampling bottles should be certified as one of the three cleanliness levels: clean, super clean, or ultraclean (per ISO 3722[1]).

For extraction tools, nothing in the fluid's pathway from the machine to the bottle should further contaminate the sample and disturb the data. In addition to the bottle, this takes into consideration the sample valve, the sample tube, and vacuum devices. It only takes a small amount of contamination to trigger premature cautionary or critical alarms (false positives).

Correctly Located Sampling Ports

Along with using the right sample valves, the precise location where an oil sample is extracted must be carefully chosen so the analysis results will be representative of the oil in the machine's wear protection zones. Two samples taken from the same machine but in different locations will likely have different results for tests such as particle counts, elemental analysis, and Fourier transform infrared (FTIR) spectroscopy.

Therefore, it is best to take a sample in a live (active) zone of the machine, which typically requires a sample valve with a pilot tube for splash/bath lubricated machines, as seen in the Figure 4.9 below. In a wet sump circulating system, the best location is usually after the pump and before the filter and for dry sump circulating systems, this would be on the return line

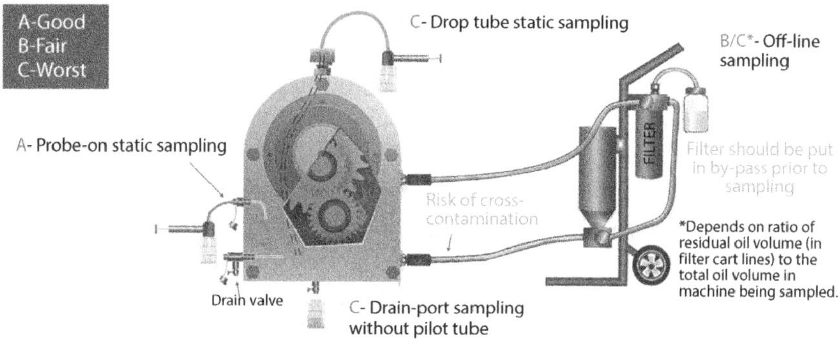

Figure 4.9 Sample locations for splash/bath lubricated machines.

before a filter and before the oil returns to the reservoir. Once chosen, the sample location should remain consistent to ensure the trends of data can be reliable.

Proper Sampling Frequency

The sampling frequency should be frequent enough to detect dangerous spikes in unfavorable results but not so often that time and money are wasted. Typically, this may range anywhere from two weeks to three months for routine oil sampling in the context of the Optimum Reference State for each machine and plant-wide reliability objectives. Periodic adjustments to these frequencies should be addressed based on the trends of the data and known maintenance practices, such as sampling more frequently in the early stages after installation or as machines are actively experiencing possible failure modes.

Proper and Consistent Sampling Procedures

During the sampling process, an oil sample can be altered in a variety of manners, many of which are not apparent. As discussed with sampling hardware cleanliness being imperative to avoid false positive, the procedure to take the sample can have significant influence.

Scenario Analysis: Imagine two lube techs independently take a sample from the same machine on the same day using the same type of tools and same location. It may surprise you the difference in data that may result.

Just the amount of oil to flush through the sampling hardware prior to the sample collection can have an influence, which is recommended to be 10 times the estimated volume of the fluid pathway from the originating location to the sample bottle.

Using the correct procedure consistency is the name of the game. Otherwise, data trends will become unreliable and the sampling procedure will become a point of blame when data is showing concerns. Following a proven sampling procedure (such as the best practice procedure often referred to as the "Clean Oil Sampling Procedure") is vital for obtaining a representative sample.

Forward Samples Immediately to the Laboratory

This one is simple but often overlooked: Do not wait any longer than necessary to send your samples off to the laboratory after collection and assess the state of your machines.

Scenario Analysis: This is from a real example where a crusher gearbox at a copper mine failed, losing ~$10,000 every hour that the machine was down. Oil samples were taken days before the failure but left on a desk to finish collecting samples from other machines before shipping off. The sample was eventually tested and sure enough the indications of a concern were clearly seen in the changes in fluid properties and rise in wear debris data. The mine learned the hard way how important it is to get quick results.

Ensuring Reliable Testing

Certified Training of Laboratory Technicians

Many people assume that their laboratory will have a staff of properly trained and certified technicians who know how to operate all the lab's instruments, but this may not always be true. A quality-control program should be in place with written procedures for consistently providing the best instruments for your oil analysis. The tests performed on your oil samples deserve uniformity. Without it, accurately comparing one sample result to another will be impossible.

Optimized Selection of Tests

A typical laboratory will have an assortment of oil analysis tests to choose from. Don't allow the lab to select just the standard test slate for all the samples. Unless the appropriate tests are chosen for each sample point, early warning signals may be missed. More often than not, opportunities for maintenance cost savings are overlooked.

Scenario Analysis: Testing for particle counts are sometimes left off of the "basic package" by commercial laboratories because of the hard costs to the lab associated to this, although this is one of the most important tests. The vigilant practice of monitoring particles in oil effectively tracks the number one cause of wear (particles) and effect of wear (wear particles) using a single technology.

Test slate optimization requires a selection of cost-effective tests for routine analysis and exception tests that are triggered based on the results of the routine tests. For example, analytical ferrography is a more expensive test but is critically important to have triggered anytime particle counts and ferrous density reach critical alarms. This methodology should be used with all test packages for each machine type and criticality [13]. For more information on the most common oil analysis tests, review the section on this topic.

Onsite Oil Analysis

An excellent way to enhance any oil analysis program, regardless of whether a commercial (offsite) laboratory has been selected, is to incorporate onsite oil analysis. Certain types of tests are standard and should be conducted on every sample, such as particle counts and moisture content. But, there is often a need to get same-day data. If this is the case, consider investing in onsite lab equipment to handle some of the oil analysis needs or at least some "field tests" that can provide a cost effective and time sensitive validation of data. This may include viscosity comparators, blotter spot tests (for analyzing insolubles and other contaminants), or the crackle test (a pass/fail test for water). Visual inspections as an important part of oil analysis will be discussed later [4].

Determining the Optimum Course of Action

Effective Organization of Analysis with Proper Trending

Although laboratories generally attempt to display oil analysis results in a user-friendly manner, they usually are not straightforward. Comprehending how results are organized will be necessary for proper interpretation. The organization of oil analysis results is initially the laboratory's responsibility, although the data can be transferred into other software to allow for more user preferences. The more effective the organization is for interpretation, the less likely a preventable machine failure will be missed. This is elaborated further in the following sections on Interpreting an Oil Analysis Report and Following the Trends.

Accurate Data Interpretation by the Laboratory

End-users often do not have enough time to interpret oil analysis results and must rely on the laboratory for interpretation. Oil analysis results typically come with a summary paragraph that suggests actions the end-user should take. Laboratory personnel interprets the data to the best of their ability, but without a representative sample taken and crucial details about the machine provided, a diagnosis or prognosis can be inaccurate. Some of these important details include:

- The machine's environmental conditions (extreme temperatures, high humidity, high vibration, etc.)
- The originating component (steam turbine, pump, etc.), make, model, and oil type currently in use
- The permanent component ID and exact sample port location
- Proper sampling procedures to confirm a consistently representative sample
- Occurrences of oil changes or makeup oil added, as well as the quantity of makeup oil since the last oil change
- Whether filter carts have been in use between oil samples
- Total operating time on the sampled component since it was purchased or overhauled
- Total runtime on the oil since the last change
- Any other unusual or noteworthy activity involving the machine that could influence changes to the lubricant

Enhanced Data Interpretation by the End-User

Following the laboratory's interpretation, the end-user must make the final decision. No one knows the machine's history and application better than those who see the machine daily. The oil analysis program can greatly benefit from the involvement of these personnel if they are also trained to interpret an oil analysis report. For those individuals, here are a few things to look for when reviewing an oil analysis report

1. Read and check the data on the oil type and machine type for accuracy.
2. Verify that reference data is shown for new oil conditions and that trend data is at an understood frequency (preferably consistent).
3. Check the measured viscosity.
4. Verify elemental wear data and compare to reference and trended data. Use a wear debris atlas to match elements to their possible source.
5. Check the elemental additive data and compare to reference and trended data. Use a wear debris atlas to match elements to their possible source.
6. Verify elemental contamination data along with particle counts and compare with reference and trended data. Use a wear debris atlas to match elements to their possible source.
7. Check moisture/water levels and compare to reference and trended data.
8. Verify the acid number and base number and compare to reference and trended data.
9. Check other analyzed data such as FTIR oxidation levels, flash point, demulsibility, analytical ferrography, etc.
10. Compare any groups of data that are trending toward unacceptable levels and make justifications based on these trends.
11. Compare written results and recommendations with known information on the oil and machine, such as recent changes in environmental or operational conditions or recent oil changes/filtration.
12. Review alarm limits and make adjustments based on the new information.

Take Corrective Action and Determine the Root Cause

Every crucial factor mentioned that goes into building a successful oil analysis program is all providing the confidence to the end user of what step to take next. This confidence must be achieved, otherwise all the efforts to obtain a representative oil sample and use a capable laboratory are pointless. You should not expect a return on the time and money spent unless action is taken as a result of the oil analysis. Be sure to always follow up on the results and be intentional with a conclusion on what the corrective action should be, if any. Remember, it takes just one catch, such as the discovery of elevated wear debris levels on a critical machine to justify the cost of the oil analysis program. Always focus on eliminating the root cause with a proactive maintenance strategy.

Continuous Improvement and Key Performance Indicator (KPI)

Whenever oil analysis is effective in providing proactive maintenance, machine failures are significantly reduced. As time passes, it may eventually be forgotten that the oil analysis is the reason for failures, especially for new personnel. Maintaining this awareness through proper training and key performance indicators (KPIs) will be the key to the sustainability of successful oil analysis.

Oil analysis performance indicators are essential for realizing the overall health of a plant over time. While oil analysis KPIs may be beneficial for your machines, they can also prove to be valuable indicators of the progress made in improving your plant's overall health as well as establishing the viability of oil analysis.

Oil Analysis Tests [8]

For a standard piece of equipment undergoing a recommended oil analysis, the test slate would consist of "routine" tests. If more testing is needed to answer advanced questions, these would be considered "exception" tests.

Routine tests vary based on the originating component and environmental conditions but should almost always include tests for viscosity, elemental (spectrometric) analysis, moisture levels, particle counts, Fourier transform infrared (FTIR) spectroscopy, and acid number. Other tests that are based on the originating equipment include analytical ferrography, ferrous density, demulsibility, and base number testing.

Viscosity [14]

Viscosity is a lubricant's most important characteristic. Monitoring the oil's viscosity is critical because any changes can lead to a host of other problems such as oxidation, glycol ingression, or thermal stressors. Several methods are used to measure viscosity, which is reported in terms of kinematic or absolute viscosity.

Too high or too low viscosity readings may be due to the presence of an incorrect lubricant, mechanical shearing of the oil, and/or the viscosity index improver, oil oxidation, antifreeze contamination, or an influence from fuel, refrigerant, or solvent contamination.

Limits for changes in the viscosity depend on the type of lubricant being analyzed but most often have a marginal limit of approximately 10 percent and a critical limit of approximately 20 percent higher or lower than the intended viscosity.

Acid Number and Base Number

Acid number and base number tests are similar but are used to interpret different lubricant and contaminant-related questions. In an oil analysis test, the acid number is the concentration of acid in the oil, while the base number is the reserve of alkalinity in the oil. Results are expressed in terms of the volume of potassium hydroxide in milligrams required to neutralize the acids in one gram of oil. Acid number testing is performed on non-crankcase oils, while base number testing is for over-based crankcase oils.

An acid number that is too high or too low may be the result of oil oxidation, the presence of an incorrect lubricant, or additive depletion. A base number that is too low can indicate high engine blow-by conditions such as soot, the presence of an incorrect lubricant, internal leakage contamination (glycol), or oil oxidation from extended oil drain intervals and/or extreme heat.

FTIR

FTIR is a quick and sophisticated method for determining several oil parameters including contamination, oil degradation products like oxides, nitrates, and sulfates, as well as the presence of additives such as zinc dialkyldithiophosphate (ZDDP) and phenols.

The FTIR instrument recognizes each of these characteristics by monitoring the shift in infrared absorbance at either a specific or a range of wavenumbers. Many of the observed parameters may not be conclusive, so these results are often coupled with other tests and used as supporting evidence.

Elemental Analysis

Elemental analysis works on the principles of atomic emission spectroscopy (AES), also called wear metal analysis. This technology detects the concentration of wear metals, contaminants, and additive elements within the oil. The two most common types of atomic emission spectroscopy are rotating disc electrodes (RDE) and inductively coupled plasma (ICP).

Both methods have limitations in analyzing particle sizes, with RDE limited to particles less than 8 to 10 microns and ICP limited to particles less than 3 micron, but they are still useful for providing trend data [6]. Table 4.2 shows possible sources for common elements.

The best way to monitor this type of data is to first determine what is expected to be in the oil. An effective oil analysis report will provide reference data for the new oil so any amounts of additive elements can be easily distinguished from those of contaminants. Because many types of elements should be expected at some level, it is better to analyze trends rather than focus on any specific measurement of elemental analysis data.

Particle Counting

Particle counting measures the size and quantity of particles in the oil. Many techniques can be used to assess this data, which is reported based on ISO 4406:21. This standard designates a range number that correlates to the particle counts of particles greater than 4, 6, and 14 microns. Table 4.3 provides an illustration of how different particle counts are assigned specific ISO codes.

Table 4.2 Possible sources for common elements.

ELEMENT	POSSIBLE SOURCES
Aluminum	Pistons, bearings, pumps, thrust washers
Antimony	Bearings, grease
Barium	Rust and oxidation inhibitor additives, grease
Boron	Anti-corrosion additives in coolant, dust, water
Calcium	Detergent/dispersant additives
Chromium	Piston rings in internal combustion engines
Copper	Bearings, brass/bronze alloys, bushings, thrust washers
Iron	Shafts, rolling-element bearings, cylinders, gears, piston rings
Lead	Bearings, fuel additives, anti-wear additives
Lithium	Grease, additives
Magnesium	Transmissions, detergent additives
Molybdenum	Piston rings, electric motors, extreme-pressure additives
Nickel	Bearings, valve train, turbine blades
Phosphorus	Anti-wear additives, extreme-pressure gear additives
Potassium	Coolant additives
Silver	Bearing cages (plating), gear teeth, shafts
Silicon	Dust/dirt, defoamant additives
Sodium	Detergent or coolant additives
Tin	Journal bearings, bearing cages, solder
Titanium	Bearing hub, compressor blades
Zinc	Neoprene seals, grease, anti-wear additives

Moisture Analysis

Moisture content within an oil sample is often measured with the Karl Fischer titration test. This test reports results in parts per million (ppm), although data is often shown in percentages. It can find water in all three forms – dissolved, emulsified, and free. The crackle test and hot-plate test are non-instrument moisture tests for screening before the Karl Fischer method is used. Possible reasons for a moisture reading being too high or too low would include water ingression from open hatches or breathers, internal condensation during temperature swings, or seal leaks.

Table 4.3 Assigning particles ISO codes.

EXAMPLE PARTICLE COUNT	
SIZE IN MICRONS (C)	COUNT LARGER THAN SIZE PER ML
4	1,752
6	517
10	144
14	55
20	25
50	1.3
75	0.27
100	0.08

Renard Series Table

$R_4/R_6/R_{14}$
ISO 18/16/13

1,752 particles > 4 μm/ml
517 particles > 6 μm/ml
55 particles > 14 μm/ml

NUMBER OF PARTICLES PER ML		
MORE THAN	UP TO AND INCLUDING	RANGE NUMBER (R)
5,000,000	10,000,000	30
2,500,000	5,000,000	29
1,300,000	2,500,000	28
640,000	1,300,000	27
320,000	640,000	26
160,000	320,000	25
80,000	160,000	24
40,000	80,000	23
20,000	40,000	22
10,000	20,000	21
5,000	10,000	20
2,500	5,000	19
1,300	2,500	18
640	1,300	17
320	640	16
160	320	15
80	160	14
40	80	13
20	40	12
10	20	11
5	10	10
2.5	5	9
1.3	2.5	8
0.64	1.3	7
0.32	0.64	6
0.16	0.32	5
0.08	0.16	4
0.04	0.08	3
0.02	0.04	2
0.01	0.02	1

4 μm	6 μm	14 μm	ISO CODE
1,301	321	41	18/16/13
2,500	640	80	18/16/13
2,501	641	81	19/17/14
5,000	1,300	160	19/17/14

one more particle
4 times as many particles

If only two range numbers are used: ISO */16/13 or ISO 16/13

Interpreting Oil Analysis Reports

The first thing to check on an oil analysis report is the information about the customer, originating piece of equipment, and lubricant (see Section A in Table 4.4: Example Oil Analysis Report). Including these details is the customer's responsibility. Without this information, the effectiveness of the report will be diminished.

Knowing which piece of equipment the oil was sampled from affects the ability to identify potential sources of the measured parameters, especially wear particles. For example, the originating piece of equipment can help associate reported wear particles with certain internal components. The lubricant information can provide a baseline for several parameters such as the expected viscosity grade, active additives, and acid and base number levels.

Section B in Table 4.4 shows the elemental analysis findings. This data can help identify contamination, wear metals, and additives present within the oil and these parameters are reported in parts per million (ppm). Still, this does not mean a contamination particle, for example, can only be indicated by sodium, potassium, or silicon spikes.

Table 4.4 Example oil analysis report[1].

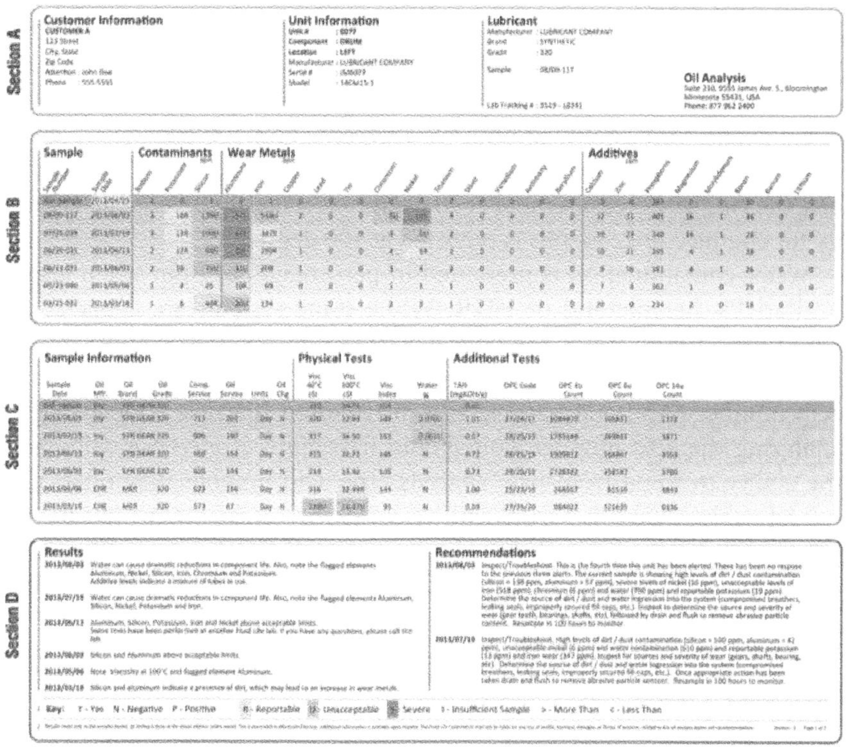
Section A
Customer Information
Unit Information
Lubricant
Oil Analysis
Section B
Sample
Contaminants
Wear Metals
Additives
Section C
Sample Information
Physical Tests
Additional Tests
Section D
Results
Recommendations

In the example above, the rise in silicon and aluminum could indicate dust or dirt contamination as the root cause. One likely explanation for these spikes is that as dirt, or silicon, enters the oil from an external source, followed by three-body abrasion occurring within the machine causing wear debris including aluminum, iron, and nickel to increase.

With a better understanding of the metallurgy within the system's components, any spikes in wear metals can be better associated, allowing a proper conclusion for which internal components are experiencing wear. Keep in mind that this trend analysis benefits from representative samples taken at an appropriate and uninterrupted frequency.

With elemental data related to contaminants and wear metals, alarms are set for upward trends in the data. For elemental data about additives, alarms are set for downward trends. Having a baseline of new lubricant reference data is critical in assessing which additives are expected and at what levels. These baselines are then established to help determine any significant reduction in specific additives.

Section C in Table 4.4 presents previously identified sample information from the customer such as oil manufacturer, brand, viscosity grade, and in-service time, as well as if an oil change has been performed. This is important data that can provide an explanation for what could be false positives in alarming data changes. The "physical tests" section of a report offers details on viscosity at both 40 degrees C and 100 degrees C, along with the viscosity index and percentage of water. For common industrial oils, the viscosity measurement at 40 degrees C is usually given. The viscosity for engine crankcase oils is reported at 100 degrees C.

Water contamination is presented in percentages or ppm when tested by Karl Fischer method or FTIR. While some systems are expected to have high levels of water, the typical alarm limits for most equipment are between 50 to 300 ppm. The "additional tests" section shows two final tests: acid number (AN) and particle size distribution (particle count). When analyzing the acid number, you should have both a reference value and the ability to trend from past analysis. The acid number may often rise considerably when this occurs with viscosity increases or may indicate the oil is oxidizing rapidly and should be changed. If FTIR data or other testing is performed, water content may also show up in the "additional tests" sections.

Section D in Table 4.4 provides written results for each of the final few test samples along with recommendations for required actions. As mentioned previously, these recommendations are entered manually by laboratory personnel and based on information provided by the customer and the data collected in the lab. If there is an explanation for the data that stems from something not explicitly stated by the customer, this is where the end user must further the interpretation of the data to consider factors associated to the machine's history of environmental and operating conditions. Understanding the information given here is critical. Remember, there is always an explanation for each exceeded limit and the root cause should be the focus of an investigation.

Following the Data Trends

Monitoring cautionary and critical limits for oil analysis results is essential and irreplaceable as groundwork in an oil analysis program. Nevertheless, the data fluctuations observed, even if they are within the established limits, can still prove to be valuable. In these conditions, trending oil analysis data is where hidden value can be realized.

Simply obtaining a snapshot of data from an oil sample is essentially worthless without something with which to compare it. Therefore, trending data in oil analysis reports is beneficial. It not only allows you to determine if the current oil properties are unfavorable, but also if they will become unfavorable soon. Quality trending provides a powerful means of recognizing when an oil property is moving in an unhealthy or threatening direction (see Figure 4.10).

The most effective way to follow a trend is by plotting the results on a property-versus-time graph. The "property" can be anything from the remaining additives within the oil to the base oil's changing properties or the number and types of particles. Trending can quickly reveal the rate-of-change over time (slope on the plot) associated with a series of monotonic

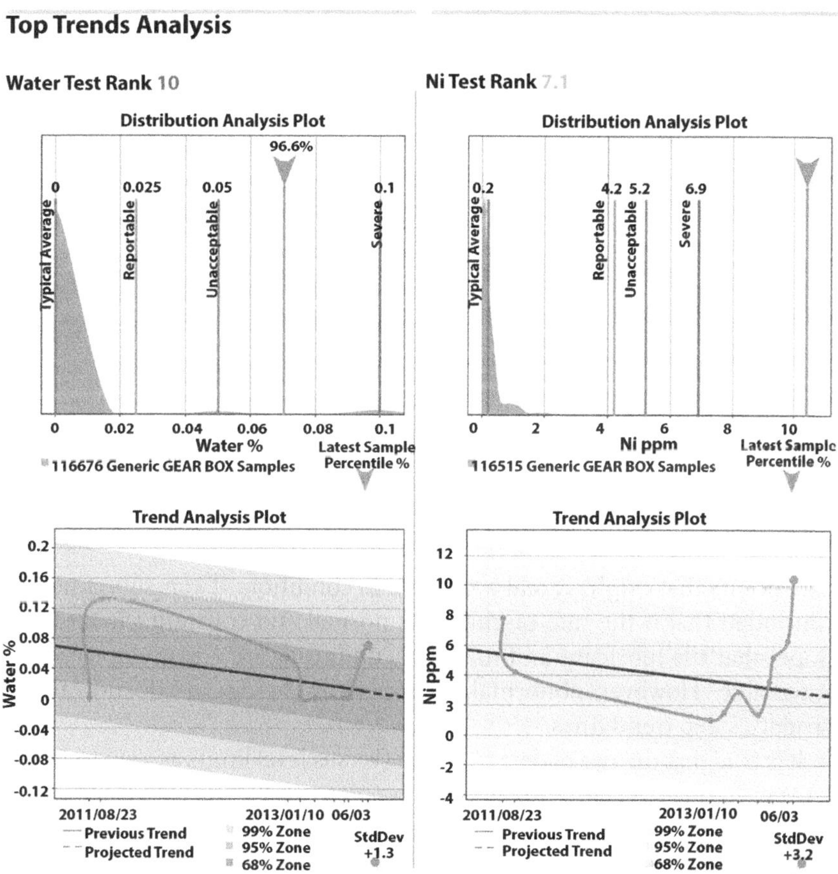

Figure 4.10 Graphs in an oil analysis report can help illustrate notable trends in the data[1].

(Continued)

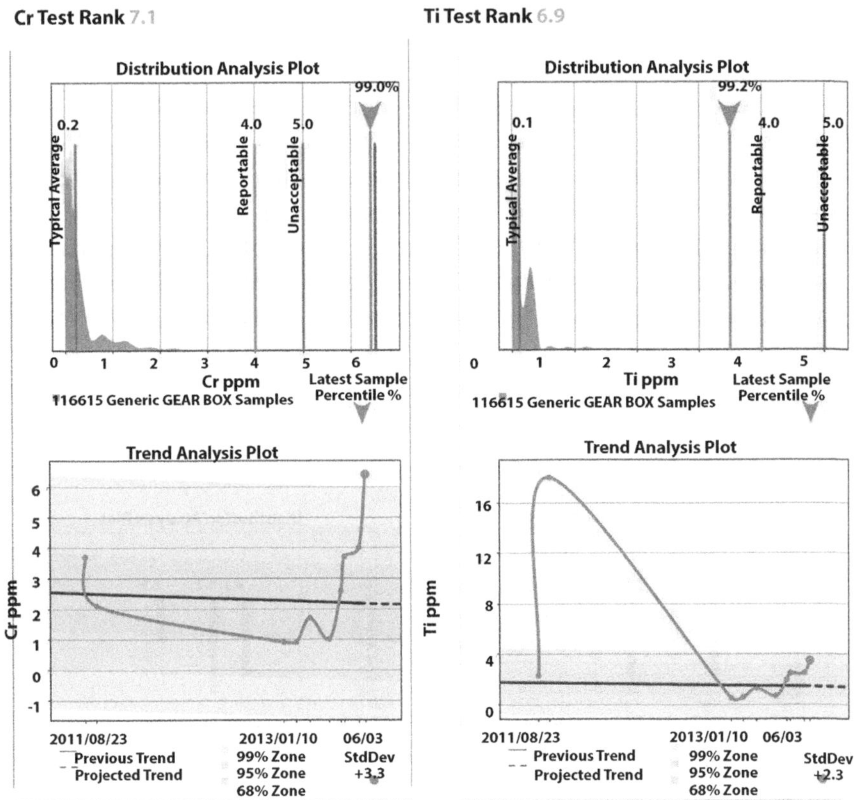

Figure 4.10 (Continued) Graphs in an oil analysis report can help illustrate notable trends in the data[1].

data points that might reveal a reportable condition. It can sometimes be concluded that if the rate-of-change is normal and constant (linear trend slope) that the lubricant and machine conditions are equally normal and acceptable. However, abnormal or unhealthy conditions do not always produce steep trend lines.

It is true that the use of data trending (versus level limits as seen in Figure 4.11) overcomes certain problems or complexities that have plagued the oil analysis field for years. This works best when all other factors in oil analysis (as mentioned throughout previous sections of this chapter) are locked down and kept consistent to minimize their influence on the data changes. Other factors important to lock down to improve trend capabilities include:

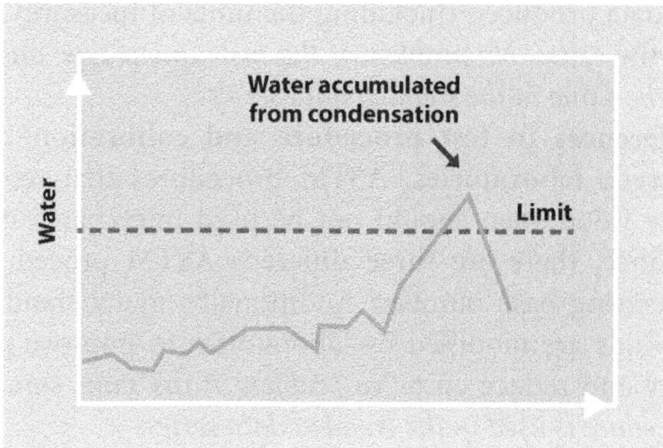

Figure 4.11 Moisture trend approaching a level limit.

1. Samples taken from the same test port using the same method
2. Oil and machine service life (in hours) are known
3. Makeup rates are known
4. When machine operating conditions and environment are constant
5. Oil type and formulation remain fixed
6. Exact same laboratory test instrument and procedure are used

When this regiment is followed, trending can correct for differences that are often outside of the control of the laboratory. For instance, when you use trend analysis and follow the six points above, the following trend-corrupting conditions would not occur.

- **Data reproducibility discrepancies between identical instruments.** There is often data disagreement between two instruments of the same type used for the same procedure. This trend corruption might be noticed if you send samples randomly to different laboratories using the same instruments. *No problem if the instrument used is always the same in the trended data series.*
- **Differences in instrument types used by different laboratories.** There are many diverse instrument types that target the same exact property. For instance, ferrous density can be measured by several different proprietary instruments, but

the data produced (including the units of measure) is often not the same. *No problem if the instrument type and brand are the same in the trended data series.*
- **Differences in test procedure and calibration method between laboratories.** ASTM procedures that report the same value often should not be used interchangeably. For instance, there are three different ASTM procedures for measuring base number. Additionally, many standardized methods are modified by laboratories to improve productivity and reduce cost. *No problem if the same suitable test procedure is used in the trended data series.*
- **Machine wear rate and operating environment differences.** Identical machine types often produce sharply different wear rates, preventing the data between machines from having any mutual statistical significance in setting alarms. Some machines are "high readers" while others are "low readers" compared to statistical averages. Nominal readers, however, follow statistical averages for a machine group. *No problem if the trended data is compared only to historical data from the same machine in the same operating environment.*

Trend analysis requires diligence with the entire oil analysis program. When successful, it is an intrinsic part of data interpretation of the lubricant. When combined with other alarming tactics, it can recognize such things as bad samples, an oil filter going into bypass, additive depletion, a new forcing function (abnormal wear), and wrong oil in use. While computer software is extremely helpful, it is hard to improve upon the ability to detect discernable oil analysis trends by simply plotting the data graphically and using our eyes.

It is imperative that oil samples are carefully collected and that all variables are minimized or at least addressed. Factors that can influence the results include:

- Sample location consistency
- Service life of the machine and oil
- Makeup oil rates
- Changes in environmental or operating conditions
- Oil formulation changes
- Testing procedure consistency

The key to success with trending is to learn from the past. This includes others' past failures, not just those of your machines. Start by identifying when certain oil properties have typically been healthy and use this as the standard. Also, take note of when a change in an oil property has previously led to a machine issue or failure. You must develop the awareness to recognize when a change in a particular property could eventually lead to a problem with the machine.

Looking Back at the Past

The world's population growth offers a good example of the types of trends that can exist within machinery. The earth's population has been growing for thousands of years, but it wasn't until around 1800 that it reached 1 billion people. While this was a major milestone, it only took approximately 120 more years to double to 2 billion. Less than 100 years later, the population is rapidly approaching 8 billion people. Many factors have influenced this recent trend, such as the Industrial Revolution and advanced medicine. Figure 4.12 shows how this rise in population would appear on a graph.

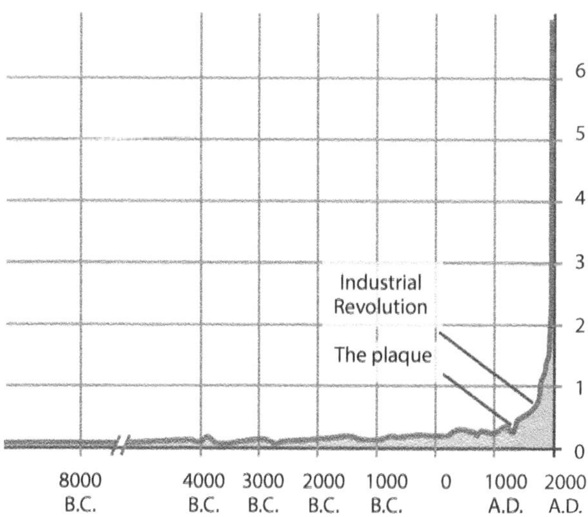

Figure 4.12 World population growth.

This trend can be compared to the growth of particle contamination in machinery. Particles produce particles. In fact, one particle can generate as many as 20 new particles within a machine. Of course, this will depend on many variables such as particle ingression rates, the filtration rate, and the likelihood of wear generation. Regardless, when particles are the instigator of new particles being created, the contamination can quickly escalate.

Inspection 2.0: Advances in Early Fault Detection Strategy [15]

Arguably one of the most important oil analysis functions doesn't require anything done at the laboratory. What it requires instead are skillful inspections that are rapid, comprehensive, and frequent. In its most basic form, they have been around forever. However, Inspection 2.0 is a radical reinvention of the whole concept of machine inspection. It has little to do with conventional practices of doing daily machine rounds.

With Inspection 2.0 you don't just "look" at a bearing, seal, coupling, or pump. Instead, you "examine" these components with a keen and probing eye. Inspection 2.0 is intense and purposeful. It seeks to penetrate and extract information from what's been referred to as machine sign language. Inspection 2.0 requires polished linguistic skills to translate this sign language into prescribed activities and instructions that stabilize reliability.

The technologies of machine condition monitoring have been advancing at a near break-neck pace for years. These innovations will continue for decades to come. Still, for most machines, there is currently no fault-detecting technology more effective than the razor-sharp and relentless focus of a human being. The potential of a human being as a condition monitoring instrument depends on transformation, specifically from the going-through-the-motions inspections of the past to mission-intensive detective work inspections of the future. That is the essence of Inspection 2.0.

Low-Hanging Fruit

Often, the simplest solution is the best solution and the right solution. How do you get the optimum level of reliability at the lowest possible cost? How do you achieve a synergistic blend of condition monitoring activities that unifies Inspection 2.0 with the range of other options being advanced and currently available?

Inspection presents some benefits and advantages that are difficult, if not impossible, to duplicate with other condition monitoring options. These include:

- Inexpensive, simple, lasting deployment
- Operator-driven (total productive maintenance emphasis)
- More emphasis on examination skills, less on technology [11]
- The power of frequency and the one-minute daily inspection
- Root-cause-oriented to avoid developing fault bubbles; more proactive, less reactive
- Early fault detection; more predictive, fewer misses and "just-in-time" saves

We all seek more for less and no one likes the pain and frustration that often come with exceedingly complex solutions to simple problems. KISS (keep it simple stupid!) solutions should always be your priority. Their application is at the core of Inspection 2.0. No array of sensors and computer intelligence can outperform a human inspector at a large number of condition monitoring tasks.

Inspection Frequency Trumps High Science

Why not perform oil analysis every day on just about every machine? Yes it sounds expensive, but it doesn't have to be. Oil analysis can be done with your senses, aided by inspection windows. Visual oil analysis is real oil analysis. Who said a laboratory is a requirement for oil analysis anyway?

Many are familiar with the P-F interval from the teachings of reliability-centered maintenance (RCM). As shown in Figure 4.13, "P" is the point-of-failure first detection, while "F" is the endpoint of functional inoperability. Although the P-F interval is a theoretical concept that has useful application, it is rarely applied in real-world machines. This is because the real world comes with many variable events. These events distort the predictability of the P-F interval. They include:

- Multiple components on a single machine or drive train, each with its own P-F tendencies
- Multiple failure modes for any single component
- Variable duty cycle (speeds, loads, shock, temperature, etc.)

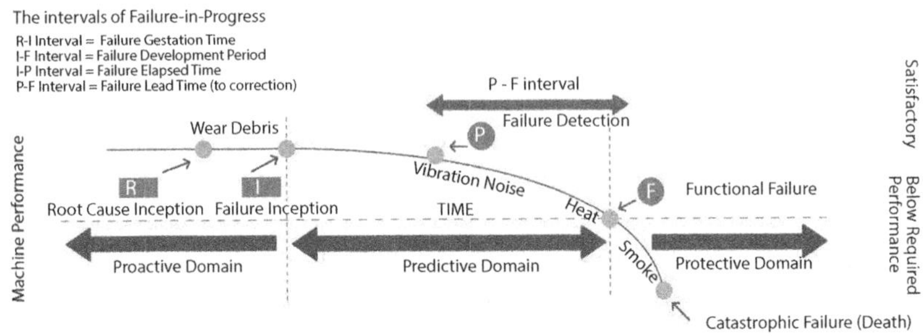

Figure 4.13 Condition monitoring and time domains of machine failure.

- Remaining useful life (RUL) varies with age. For any given fault mode, the P-F interval shrinks as the machine ages.
- Failure detection methodology and effectiveness vary (ability to detect faults early)

The best countermeasure for uncertain P-F intervals is frequency. For certain machines, real-time monitoring using imbedded sensors is justified, especially high-speed, high-risk machines. However, for nearly all other machines, the simple solution for early detection is daily inspection aided by inspection windows and tools.

Even the world's best oil analysis laboratories can't see faults in the "non-sample." Inspection 2.0 asks you to deploy your senses intensely every time you walk by the machine. The "oil sample" is examined carefully in the sight glass, but it never leaves the machine.

The power of frequency is illustrated in Figure 4.14. In this example, the failure development period (from inception to functional failure) is one month. If your condition monitoring interval is quarterly or bi-monthly, you won't catch the fault prior to functional failure (this is a condition monitoring "miss"). If you use a monthly monitoring interval, you catch

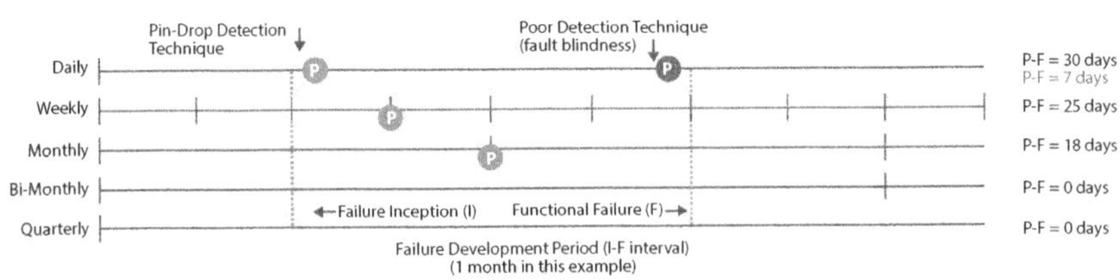

Figure 4.14 Failure detectability technique and inspection periodicity influence the P-F interval.

the fault with an 18-day P-F interval (lead time to corrective measure). Note: the longer the P-F interval, the better.

If your condition monitoring interval is weekly, your P-F interval jumps to 25 days (better). However, if you can inspect this machine daily using detection-sensitive Inspection 2.0 methods, your P-F interval is 30 days, which is better than weekly testing with the best condition monitoring technology (vibration, oil analysis, thermography, etc.). By comparison, a poor daily inspection technique yields a P-F interval of just seven days.

Beware of Short P-F and Sudden-Death Failures [9]

As mentioned, the P-F interval is almost impossible to predict for a variety of machine-specific reasons. In fact, the interval can vary from seconds to decades. Maintenance departments like long reaction times to schedule needed corrections. Still, Murphy's Law always looms to ruin an otherwise perfectly good day.

The best strategy to mitigate sudden-death failures is to focus on the early detection of root-cause fault bubbles. This is a fundamental proactive maintenance strategy (see the left side of Figure 4.13). Fault bubbles are escalating conditions that threaten the onset of an active failure event. As much focus should be spent on preventing the inception of failure as on detecting a failure in progress. Every failure mode has one or more root causes. Ensure good root-cause alignment with your inspection strategy.

Following are a few examples of short P-F and sudden-death failure modes and fault bubbles. What intervention strategy focused on root causes would you apply to detect and neutralize these threatening conditions?

- Oil filter rupture
- Wrong oil or severely degraded oil
- Fish-bowl conditions (disturbed and mobilized bottom sediment)
- Severe shaft misalignment
- Stiction/silt lock of hydraulic valve (motion impediment)
- Grease "cake lock" starvation of automatic lubrication system
- Impaired oil supply of a splash-lubed gearbox
- Heavy fuel dilution of a diesel generator
- Heavy chemical contamination of a compressor oil
- Gross seawater contamination of a shipboard hydraulic fluid
- Shock loading of a large thrust bearing

Inspection Windows and Zones

Inspection 2.0 is searching, i.e., detective work. It puts the machine under the microscope day by day. To do this, the machine's exoskeleton must be penetrated. You have to find ways to see through steel plate and cast iron. You must also "ready" machines for world-class inspection. New products, including modernized sight glasses, are being developed to bring vision to critical zones within the machine.

Figure 4.15 shows the use of windows for convenient zone inspections and a list of example root causes and faults that can be visually detected using these windows. The third zone, the bottom sediment and water (BS&W), is one of the best examples of where a modernized sight glass, in this case a BS&W bowl (Figure 4.16) can provide value for fault detection for common oil reservoirs and sumps. Each zone has a story to tell about the oil and machine.

If the BS&W bowl is properly positioned, anything that is heavier than the oil will accumulate there for quick inspection with a good light. This includes sediment, water, sludge, wear debris, coolant, dead additives, and dirt. If your BS&W bowl is clear, bright, and without sediment, there are many things that could be going wrong with your oil and machine that are not going wrong simply because this sight glass passes inspection.

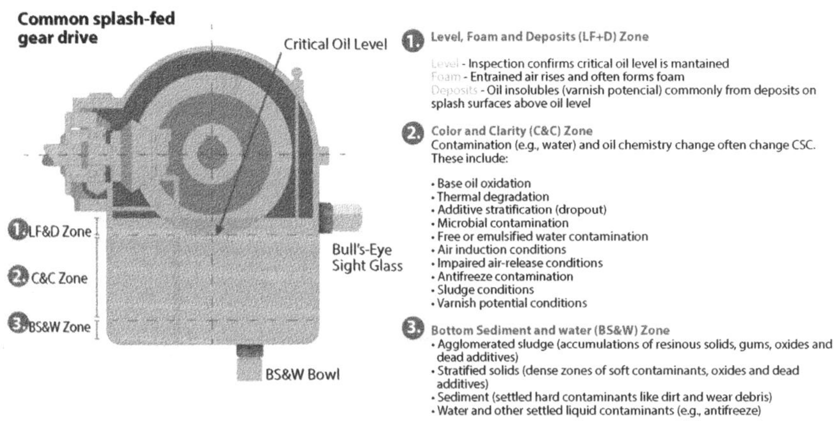

Figure 4.15 Zone inspections for early problem detection.

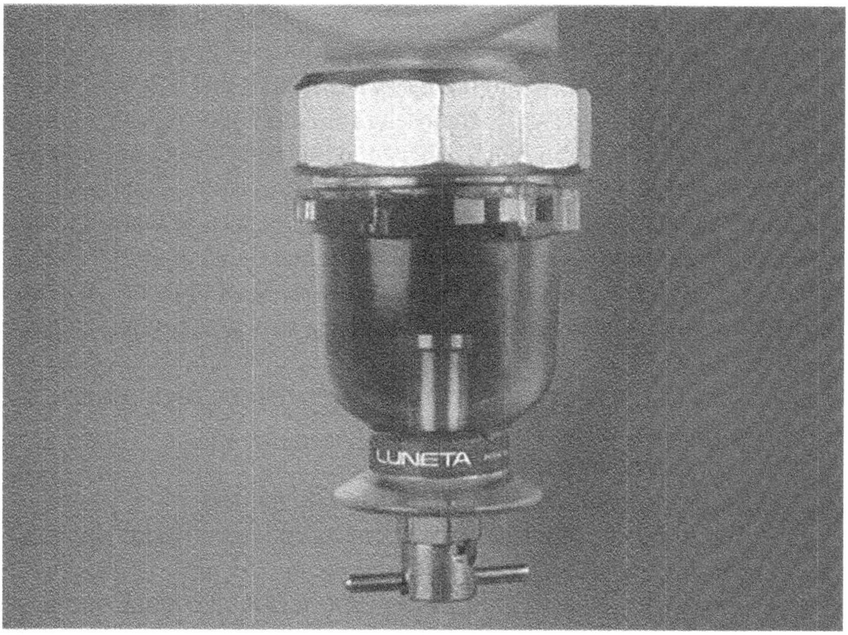

Figure 4.16 Modern bottom sediment and water (BS&W) Bowl Sight Glass.

Inspection 2.0 is a Nurturing Strategy

Applied tribology is a behavioral science. This means that in most plants the practice of tribology and lubrication is people-intensive. Lubricants are what we buy and lubrication is what we do. The main reason machines fail prematurely is the result of what people do or don't do.

Inspection is a subset of tribology (and lubrication) and likewise is very much a behavioral science. People have to passionately want to find faults and reportable conditions. The people I'm referring to are operators, technicians, lube techs, millwrights, mechanics, etc. It's no longer just "looking" during the inspection route and then checking the box on the report. Instead, it's about intense examining, probing, digging, and searching. The people factor will make or break any good reliability and maintenance undertaking and is perhaps truer with inspection.

People respond to work tasks and challenges differently. Much of it has to do with the leadership and maintenance culture. A positive, nurturing maintenance culture is a critical plant asset. Consider that when people do good work, they feel good about themselves and their job. When people do bad work, they feel bad about themselves and their job. The simple solution is to enable people to do good work that is recognized and celebrated.

Culture drives behavior. Behavior influences quality of work. Quality work is fundamental to plant reliability and the cost of reliability. Of course, this most definitely includes inspection activities. The following list delineates the minimum requirements for building a strong inspection culture:

- Training and inspection skill competencies (optimizing inspection skills readiness)
- Celebrating inspection "saves"
- Inspection KPIs and other performance metrics
- Installing penetrating inspection windows (optimizing machine inspection readiness)
- Availability and use of inspection aids and tools (optimizing tool inspection readiness)
- Promptly responding to inspection-generating alerts and red flags

Side-by-Side Comparison

What differentiates Inspection 2.0 from conventional inspection practices? It's mostly about execution. Think about how to make it 10 times more effective with very little extra cost. Table 4.5 details several of the main differentiators that distinguish and empower Inspection 2.0 to this higher level of performance. This is an exceptional low-hanging-fruit opportunity in machine reliability as well as a foundational element for lubrication excellence.

Final Tips to Help Error-Proof Your Lubrication Program [4]

Developing and managing a Lubrication Program can quickly become overwhelming without the right guidance. The Ascend methodology is one effective way to organize this and provide a guiding North Star on what the best practice is. This has been the methodology used for the hundreds of lubrication assessments performed as part of Noria's Lubrication Program Development. The results of a comprehensive questionnaire across all 40 factors provides a holistic analysis on a plant's current lubrication practices and stages of improvement

Table 4.5 Main differentiators between convention inspections and inspection 2.0.

Distinction	Conventional inspection	Inspection 2.0
Emphasis on daily inspections	Sometimes	Always
Emphasis on inspection location	Rarely	Always
Installed inspection windows	Rarely	Always
Inspection alignment to failure mode ranking	Sometimes	Always
Inspection designed to preempt fault bubbles	Rarely	Always
Emphasis on early "weak-signal" detection	Rarely	Always
Use of advanced inspection aids and tools	Rarely	Always
Inspectors who are highly skilled and motivated	Sometimes	Always

during reassessments. Using a tool like this to monitor progress and establish a firm understanding on what the current state is, is a great way to error-proof the lubrication program from the very beginning.

The Optimum Reference State (ORS) is another important way to help make the right decisions in machinery lubrication and oil analysis and keep them aligned to overall plant reliability objectives. Without the right tools, established practices, and team training, it can be common for lubrication programs to struggle. Often, it comes down to a lack of ownership or champion who maintains the focus, keeps key performance indicators (KPIs) tracked, and aligns them with maintenance and reliability goals.

As a final thought, the following is a list of a few tips to help error-proof your lubrication program:

- *Spread the Ownership but Maintain One Champion:* Everyone needs to be incentivized by the success of a lubrication program. It's easy for the executives to want this, as they gain from the profitability, but ownership must be realized and rewarded by the individuals making the daily efforts in lubrication activities. Overall, there should be one champion who has complete focus on the program's quality control and ensures nothing is falling behind.

- *Train Everyone Together and Frequently*: Training is a vital part of any lubrication program, but when it's done together as a team, there is a bit of comradery created that rallies everyone to make the needed efforts together. Buy-in is effective. Especially with Inspection 2.0, there is the value of learning from each other's experience and how to search for the subtle clues of lubricant issues or machine faults.
- *Prioritize and Follow up Routinely*: Every task must be prioritized because sometimes not everything can get done. Follow up is one of the ways to have a tracking metric. What gets measured gets done.
- *Keep Procedures Easy to See and Understand*: It is considered a best practice to keep procedures easily accessible and posted nearby and on the machines. Posting every routine procedure directly on the equipment provides clear instruction every time.
- *Track Compliance and Be Intentional*: Visually post your progress and lubrication activities with metrics on walls that are seen constantly. This shows everyone why these activities are done and why they must continue.
- *Consider Ergonomics with Machine Inspections*: The best inspections are those that are recognized by the untrained eye. For example, this would include using sight glasses that are installed next to a common walking path with the maximum and minimum levels clearly marked. Even better would be to post the contact information of the person to inform if these levels are breached.
- *Keep It Simple (KISS Principle)*: If your lubrication tasks are prepared to be easy to use and understand, they will be completed more often and more effectively.

Noria Corporation

This chapter is comprised of the knowledge and experiences of Noria Corporation's Jim Fitch and Bennett Fitch. Founded in 1998, Noria's single focus is helping user organizations improve machine reliability through the deployment of best practices in lubrication and oil analysis. For over 20 years, their world-class approach has changed how organizations manage and monitor lubricants to maintain optimum reliability and safety. The experts at Noria have become trusted advisors to many of the world's leading organizations.

Table 4.6 Acronym list.

ACL	**Ascend compliance level**
AES	Atomic Emission Spectroscopy
AN	Acid Number
BSC	Balanced Score Card
BS&W	Bottom Sediment and Water
DMAIC	Define, Measure, Analyze, Improve, Control
FDP	Failure Development Period
FOF	Failure Occurrence Factor
FTIR	Fourier Transform Infrared
ICLM	International Council for Machinery Lubrication
ICP	Inductively Coupled Plasma
ISO	International Organization for Standards
KPI	Key Performance Indicator
MCF	Machine Criticality Factor
OEM	Original Equipment Manufacturer
OLC	Overall Lubricant Criticality
OMC	Overall Machine Criticality
ORS	Optimum Reference State
PdM	Predictive Maintenance
PM	Preventative Maintenance
PPM	Parts Per Million
RCM	Reliability-Centered Maintenance
RDE	Rotating Disc Electrodes
RUL	Remaining Useful Life
ZDDP	Zinc Dialkyldithiophosphate

References

1. Applied Industrial Technologies. "Lubricant Failure = Bearing Failure." *Machinery Lubrication*, Jan. 2009, www.machinerylubrication.com/Read/1863/lubricant-failure.
2. Fitch, Bennett. "Anatomy of a Representative Oil Sample: Part 2 – Sample Extraction Tools." *Machinery Lubrication*, Apr. 2014, www.machinerylubrication.com/Read/29705/oil-sample-extraction.
3. Fitch, Bennett. "The Benefits of Contamination Control Training." *Machinery Lubrication*, Feb. 2022, www.machinerylubrication.com/Read/32119/the-benefits-of-contamination-control-training.
4. Fitch, Bennett. "Error-Proof Your Lube Program with Visual Lubrication Best Practices." *Machinery Lubrication*, 2018, www.machinerylubrication.com/Read/31535/lubrication-best-practices.
5. Fitch, Bennett. "Follow the Chain of Success for Oil Analysis Optimization." *Machinery Lubrication*, Aug. 2014, www.machinerylubrication.com/Read/29817/oil-analysis-optimization.
6. Fitch, Bennett. "Follow the Trend for Successful Oil Analysis." *Machinery Lubrication*, Feb. 2015, www.machinerylubrication.com/Read/30079/oil-analysis-trend.
7. Fitch, Bennett. "The Importance of Lubricant Film Strength." *Machinery Lubrication*, www.machinerylubrication.com/Read/30835/lubricant-film-strength.
8. Fitch, Bennett. "Oil Analysis Explained." *Machinery Lubrication*, Dec. 2013, www.machinerylubrication.com/Read/29598/oil-analysis-report.
9. Fitch, Bennett. "The Silent Danger of Abrupt Lubricant Failure." *Machinery Lubrication*, Oct. 2014. www.machinerylubrication.com/Read/29894/abrupt-lubricant-failure.
10. Fitch, Jim, and Gerardo Trujillo. "Ascend to Lubrication Excellence." *Machinery Lubrication*, Dec. 2020, www.machinerylubrication.com/Read/31946/ascned-lubrication-program-excellence.
11. Fitch, Jim. "The Daily One-Minute Inspection." *Machinery Lubrication*, Jan. 2007, www.machinerylubrication.com/Read/966/daily-inspection.
12. Fitch, Jim. "How to Optimize the State of Lubrication." *Machinery Lubrication*, Feb. 2013, www.machinerylubrication.com/Read/29278/optimize-lubrication-state.
13. Fitch, Jim. "A New Look at Criticality Analysis for Machinery Lubrication." *Machinery Lubrication*, Apr. 2014, www.machinerylubrication.com/Read/29346/machinery-criticality-analysis.
14. Fitch, Jim, *et al. The* Practical *Handbook of Machinery Lubrication.* Fourth ed., Tulsa, Oklahoma, Noria Corporation, 2017.
15. Fitch, Jim. "Why Inspection 2.0 Is Your Best Strategy for Early Fault Detection." *Machinery Lubrication*, Aug. 2016, www.machinerylubrication.com/Read/30562/inspection-fault-detection.

5

Troubleshooting Temperature Problems

By Robert X. Perez

One of the most common types of problem seen around process machinery are those related to high temperature indications. A high temperature can be indicative of a bearing failure, electric motor overload, a reciprocating compressor valve failure, internal leakage problem, gas turbine firing issue, etc. Temperature problems can appear as a temperature alarm, paint discoloration, an elevated reading taken with an infrared gun or contact thermometer, or an abnormal temperature sensed by hand. When a temperature issue is identified, the first questions are: Is the temperature normal? What has it been running? It is too high? Should I shut down now or can I continue monitoring the temperature?

Every process machine is unique and poses unique challenges based on its design and construction. Some machines have rolling element bearings and some have fluid film bearings, which have different temperature limits. There are also centrifugal compressors and reciprocating compressors that offer unique challenges based on process conditions and the gases they compress. Because the designs of process machines are so broad, so is the topic of temperature monitoring. We will break down the discussion in this chapter into two parts: a) bearing temperature assessments and b) compressor discharge temperature assessments.

Temperature Assessments

Temperature is a practical and straightforward indicator of internal machine problems [1]. It can be used to assess the condition of a bearing, identify a cooling system problem, detect leakage on reciprocating compressor valves, and many other problems. Anyone interested in the condition of their machines should employ some type of temperature indicating equipment. Critical machines typically have imbedded thermocouples or thermistors that continuously report temperature, while less critical

Robert X. Perez (ed.) Condition Monitoring, Troubleshooting and Reliability in Rotating Machinery, (135–166) © 2023 Scrivener Publishing LLC

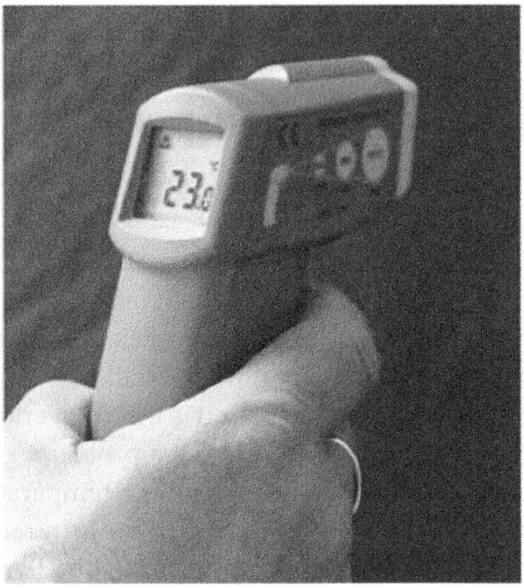

Figure 5.1 Infrared temperature gun.

machines tend to be monitored with portable devices. One of the handiest temperature measuring devices is the infrared temperature gun like the one seen in Figure 5.1.

How do Infrared Thermometers Work?

The most basic infrared thermometers (Figure 5.1) consist of a lens to focus the infrared (IR) energy on to a detector, which converts the energy to an electrical signal that can be displayed in units of temperature after being compensated for ambient temperature variation. This configuration enables temperature measurement from a distance without contact with the object to be measured. For these reasons, the infrared thermometer is useful for measuring temperature under circumstances where thermocouples or other probe type sensors cannot be used or do not produce accurate data for a variety of reasons, such as when the surface to be measured is moving or inaccessible.

It is important to remember that the measurement area sensed by the gun increases as the distance from the gun to the measurement point increases very similar to the area covered by the light emitted from a flashlight. In other words, if you want the temperature of a tight area, the temperature gun must be held near the target. Consequently, if the gun is held far away from the surface of interest, it will report the average temperature of a wide spot of the machine.

Takeaway: If you wish to know the temperature of a particular spot on a machine you need to get the gun close to the target area.

Bearing Temperature Trending

Temperature trending is a powerful, yet easy, data analysis method that allows you to assess your bearing's relative condition by regularly monitoring a representative temperature on or near the bearing of interest. It is prudent that you begin your data series by recording and saving representative baseline bearing temperature readings.

To record baseline conditions, it is best to wait until the machine is under full load and at steady state conditions. Additionally, to reduce variability in successive readings, you should select locations on the bearing housings that are near the maximum load regions. Be sure to carefully label or tag them so that the same locations are measured during each visit and be sure to include seasonal changes too. These steps will enable you to better detect changes in their condition or operating environment.

Once you have begun collecting temperature data, you can enter them into a spreadsheet program in order to generate a trend of temperature as a function of the date, similar to Figure 5.2 below. This will enable you to easily spot dramatic changes in recorded temperatures. (Of course, if you have installed permanent bearing temperature sensors, your process computers can be configured to automatically trend temperatures.)

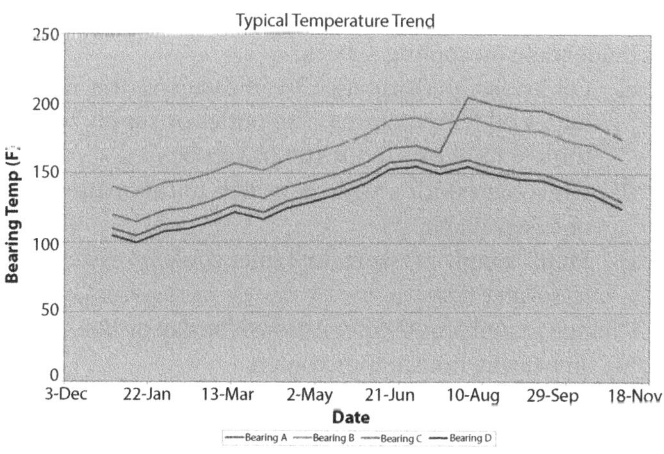

Figure 5.2 Typical temperature trend.

Notice the temperature of "Bearing C" jumps in August, while the other bearings appear to have normal temperatures.

All events where a bearing experiences an unexplained jump (10 to 20 °F) in temperature or exceeds a defined alarm level should be investigated. Remember to take seasonal temperature changes into account, since ambient temperatures have marked effect on running temperatures. As you can see in the figure above, all the temperatures seem to general follow seasonal changes until a problem is encountered.

Here are some possible causes of an upward bearing temperature trend:

1. Failing bearing—Check oil analysis for signs of bearing metals
2. Internal clearances too tight. Check for:
   a. Wrong bearing fabrication or selection (this only applies to a recently installed bearing)
   b. Distortion that may be closing internal clearances; can be caused by uneven cooling or excessive bearing crush
3. Inadequate oil supply. Check for:
   a. Wrong or plugged oil supply orifice
   b. Oil pressure too low
   c. Oil level too low
   d. Oil level too high
   e. Oil ring issues
   f. Oil mist pinched off
4. Wrong oil—Check the most recent oil analysis and consider retesting to see if possibly the wrong oil was added by mistake.
5. Inadequate oil cooling. Check:
   a. Oil cooler performance by measuring the inlet and outlet oil temperatures. The outlet or supply temperatures should run about 100 to 120 °F.
   b. Oil cooler(s) for plugging; switch to the spare cooler if one is available
   c. Water supply pressure; low inlet cooler pressure means less water flow
6. Change in ambient temperature—Hot day or hot air blowing on bearing housings or coolers
7. Plugged air filters (electric motors)—Hot stators can lead to hot bearings; trend motor stator temperatures to see if they are trending up

8. Change in axial bearing load (thrust bearings). Possible causes:
   a. Plugged balance holes (impellers)
   b. Fouling of rotating blades in steam turbines
   c. Significant change in differential pressure (centrifugal pumps and compressors)
9. Change in radial bearing load (radial bearings). Possible causes:
   a. Off design flow conditions (centrifugal pumps)
   b. Machine to machine misalignment
   c. Piping strain
   d. Poor internal alignment due to machining problems
   e. Excessive belt drive tension
10. Possible grease issues
   a. Over-greasing
   b. Mixing incompatible greases
   c. Low temperature grease is not adequate at actual operating temperature

Rolling Element and Sleeve Bearing Temperature Guidelines

Bearing temperature monitoring may be accomplished by measuring either the

1. Oil sump or grease temperature
2. Bearing metal temperature
3. Bearing housing temperature

The bearing metal temperature may be detected with an embedded or bayonet type of sensor. The embedded detector is preferred because it can usually be placed closer to the loaded bearing surface than the bayonet detector. Of all these methods, the bearing housing temperature measurement offers the most variability, due to inaccessibility to the entire bearing, changes in ambient conditions, heat transfer path, metallurgy, etc. This variability is best handled by long-term trending in order to better understand what is normal for your machine.

Tables 5.1a and 5.b can help you better manage machine bearing temperatures by providing general operating limits. Table 5.2 contains the dropping point for various greases. The dropping point of a soap-thickened lubricating grease is the temperature where the grease passes from the semi-solid state to a liquid state under specific test conditions. Try to limit grease temperatures to less than 50 degrees Celsius (122 °F) of the stated dropping point.

Table 5.1a Bearing temperatures guideline (*Celsius*).

Bearing Temperatures Guidelines (see notes below)		
Anti-Friction Bearing—Oil Lubricated	**Bearing Temp.**	**Oil Temp.**
Typical Running Temperature	<71°C	<51°C
Alarm Temperature	82°C	62°C
Trip Temperature	93°C	73°C
Anti-Friction Bearing—Grease Lubricated	**Bearing Temp.**	**Grease Temp.**
Typical Running Temperature	<90°C	<80°C
Alarm Temperature	110°C	100°C
Trip Temperature	120°C	110°C
Sleeve Bearing— Oil Lubricated	**Bearing Temp.**	**Oil Temp.**
Typical Running Temperature	<80°C	<60°C
Alarm Temperature	90°C	70°C
Trip Temperature	95°C	75°C

Notes: For large and critical machines consult manufacturers recommendations.

Table 5.1b Bearing temperatures guidelines, (*Fahrenheit*).

Bearing Temperatures Guidelines (see notes below)		
Anti-Friction Bearing—Oil Lubricated	**Bearing Temp.**	**Oil Temp.**
Typical Running Temperature	<160°F	<124°F
Alarm Temperature	180°F	144°F
Trip Temperature	200°F	163°F
Anti-Friction Bearing—Grease Lubricated	**Bearing Temp.**	**Grease Temp.**
Typical Running Temperature	<194°F	<176°F
Alarm Temperature	230°F	212°F
Trip Temperature	248°F	230°F
Sleeve Bearing— Oil Lubricated	**Bearing Temp.**	**Oil Temp.**
Typical Running Temperature	<176°F	<140°F
Alarm Temperature	194°F	158°F
Trip Temperature	203°F	167°F

Notes: For large and critical machines consult manufacturers recommendations.

Bearing Temperature Notes:

1) When monitoring bearing temperatures, it is best to measure metal temperatures as close to the loaded portion of the bearing as possible.
2) The temperature guidelines in the tables above assume you have access to a location near the maximum metal temperature of bearing under assessment. These temperatures should be reduced by 5 °C to 10 °C (9 °F to 18 °F) if this is not possible.
3) Bearing detectors can be either thermocouples or resistance temperature devices (RTD's).
4) Ensure the grease being used has a maximum service temperature that exceeds all the measured grease temperature guidelines shown in Table 5.2. If not, lower these guidelines accordingly. See the table below for the maximum service temperatures of some selected greases.

Table 5.2 Dropping points for various greases.

The dropping point of the grease is not the same point as the useful temperature limit of the grease. A good rule of thumb is to consider the useful temperature limit to be 50° Celsius less than the stated dropping point.		
Type	**Dropping point, °C**	**Maximum service temperature, °C**
Simple Lithium	175	120 to 135
Lithium Complex	>250	150 to 180
Polyurea	>245	150 to 180
Lime Soap	90	60 to 70
Anhydrous Calcium	140	90 to 110
Calcium Complex	>232	130 to 150
Modified Clay	>280	190 to 220
Sodium	165	190 to 120
Sodium Complex	>250	150 to 180
Aluminum Complex	>250	130 to 150
Calcium sulfonate Complex	>260	190 to 200

Rule of Thumb for Rolling Element Bearings:

Limit the maximum bearing temperature rise to 45 °C (113 °F) for machines running 1800 rpm and lower and to 50 °C (122 °F) for machines running 3600 rpm and higher. This is based on an assumed ambient of 40°C (104 °F).

Bearing Temperature Guidelines for Instrumented Hydrodynamic Bearings [2]

Most high-speed centrifugal compressors rely on fluid film bearings-such as tilting pad journal (TPJ) bearings and equalized (Kingsbury style) thrust bearings for rotor support and positioning. A compressor's rotor radial position must be maintained within a few thousandths of an inch to operate efficiently and reliably. A bearing failure will not only mean a compressor shutdown; it could also result in secondary or peripheral damage to costly internal parts (impellers, vanes, etc.). Early detection of a bearing failure may only result in an outage lasting a few days, but an undetected failure causing major internal damage can force an outage for weeks or months.

Condition monitoring has proven important in the installation and maintenance of all types of turbomachinery throughout the past 30 to 40 years. Babbitt bearing temperature is primarily important because bearings are the critical links between the rotating and stationary components in a machine. If temperatures are taken in the correct location with respect to the direction of bearing load, to direction of shaft rotation, and to the distance of the tip of the sensor from the bearing running surface, then Babbitt metal temperatures can be the best indicators of a bearing's operating condition.

For consistency and repeatability, the 4th Edition of API 670 (2000) defined the recommended measuring points in babbitted bearings. These locations closely approximate the zones of peak temperature as they relate to the peak hydrodynamic oil film pressure zones.

Recommended Guidelines for Babbitt Bearings*

Alarm temperature = 230 °F
Shutdown temperature = 250 °F

*Note that these temperatures assume that the temperature sensor has been positioned in the bearing as per API 670 guidelines. Only temperature measurements with embedded temperature sensors can be considered reliable.

Bearing Temperature Sensor Placement

As a practical consideration, the choice of temperature sensors has been narrowed to three basic types: Thermocouple (TC), Resistance Temperature Detector (RTD), and Thermistor. TCs and RTDs have metal-to-metal sensing junctions and are the most popular. They can be fashioned into miniature sensing tips for ease of installation. Thermistors are ceramic semiconductors of limited use, mainly used for establishing close monitoring over a limited temperature range.

For consistency and repeatability, the 4th Edition of API 670 (2000) defined the recommended measuring points in Babbitted bearings. These locations closely approximate the zones of peak temperature as they relate to the peak hydrodynamic oil film pressure zones. These are worth repeating here for sleeve bearings, tilt pad journal bearings (Figure 5.3), and tilt pad thrust bearings.

Sleeve Bearings

The probe or probes should be located 30 degrees ± 10 degrees beyond the point of maximum bearing load in the direction of shaft rotation. More precisely, the best location is in line with the point of minimum oil film. Use the following guidelines to determine the axial placement of the probes: 1) If the length to diameter L/D ratio is less than or equal to 0.5, a single sensor shall be axially located in the center of the bearing or 2) If the length to diameter L/D ratio is greater than or equal to 0.5, two sensors shall be located at a distance of 0.25L from the end of the bearing running face.

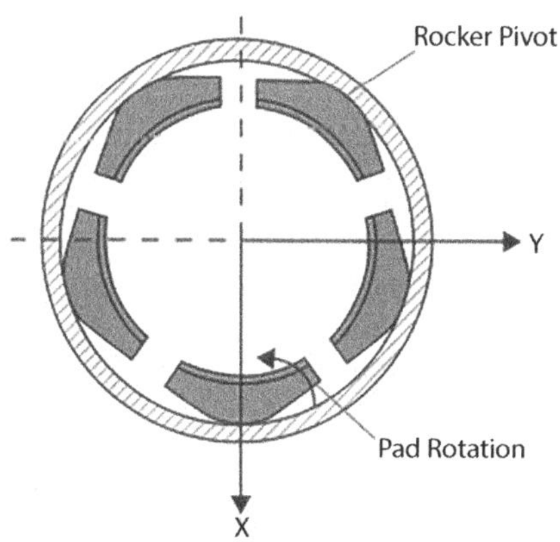

Figure 5.3 Tilting pad bearing.

Embed temperature sensors between 0.060" and 0.100" below the babbitt/shell interface.

Tilting Pad Journal (TPJ) Bearings-Load on Pad

The probe or probes should be located in the pad carrying the greatest share of load, usually the bottom centerline pad. Probes should be embedded and located 75 percent of the circumferential pad length from the leading edge in the direction of rotation. For pads with self-aligning pivots, only one probe on the axial centerline is required regardless of pad length. For line contact pivots, use the following guidelines to determine the axial placement of the probes: 1) If the length to diameter L/D ratio is less than or equal to 0.5, a single sensor shall be axially located in the center of the bearing or 2) If the length to diameter L/D ratio is greater than or equal to 0.5, two sensors shall be located at a distance of 0.25L from the end of the bearing running face. Embed temperature sensors between 0.060" and 0.100" below the babbitt/shell interface.

Tilting Pad Journal Bearings-Load between Pads

The probe or probes should be located in the trailing of the two loaded pads in the direction of rotation. In this pad, the probe or probes should be located 75 percent of the circumferential distance from the leading edge in the direction of rotation. For pads with self-aligning pivots, only one probe is required regardless of pad length. For line contact pivots, use the following guidelines to determine the axial placement of the probes: 1) If the length to diameter L/D ratio is less than or equal to 0.5, a single sensor shall be axially located in the center of the bearing or 2) If the length to diameter L/D ratio is greater than or equal to 0.5, two sensors shall be located at a distance of 0.25L from the end of the bearing running face. Embed temperature sensors between 0.060" and 0.100" below the babbitt/shell interface.

Thrust Bearings-Tilting Pad

Probes or sensors should be located in at least two pads in the designated active thrust bearing, at the 75%/75% location (see Figure 5.4) to ensure sensing close to the maximum pad temperature. Regardless of the number

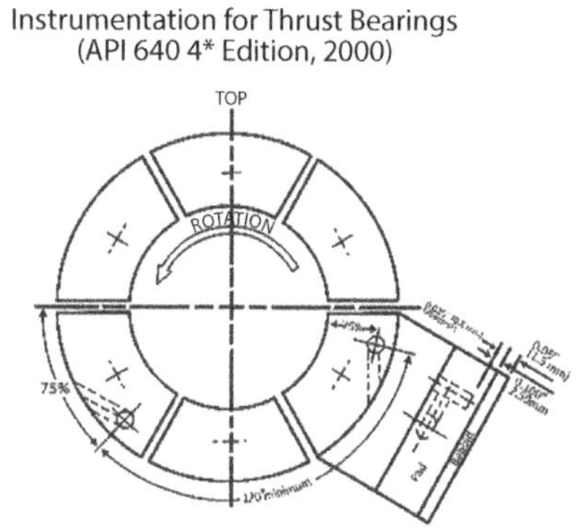

Figure 5.4 Recommended location for temperature sensors in titling pad thrust bearing.

of pads in a bearing, attempt to have sensors in two pads located in the bottom half of the bearing assembly at least 120-deg apart. When installing probes, the tilting action of individual pads should not be compromised. Embed temperature sensors between 0.060" and 0.100" below the babbitt/shell interface.

General Temperature Probe Installation Guidelines

As a general guideline, all bearings-sleeve, tilt pad journal, and thrust-probes or sensors can be installed from any direction that is most convenient from a machinery assembly standpoint as long as bearing function is not compromised and lead wires are not subjected to excessive stress, movement, or fretting.

The distance of the sensor from the running surface of the Babbitt is more critical in steel and bronze than in the backing material of copper because of copper's better thermal conductivity. Regardless of backing material, there is a drop in temperature from the running surface to the sensor tip as heat travels through the Babbitt layer, the backing material and the contact zone between the backing and the sensor tip.

Standardizing on sensor location will develop confidence in bearing temperature readings. Even though the Babbitt thickness and embedded depths are usually out of our control, we can request that the Babbitt/shell interface is not penetrated and that the sensors are below the bond line as prescribed in these guidelines.

Compressor Discharge Temperature Assessments [3]

Heat of Compression

If you compress a gas adiabatically (without allowing heat to enter or leave) you are doing work on the gas, which increases its internal energy. Another way to look at the heat of compression:

compression occurs➔ gas volume decreases ➔ the frequency of gas particle collisions increases➔the gas temperature increases.

Remember that temperature is a measure of the internal energy (kinetic energy of the molecules for an ideal gas). Thus, compression increases gas temperature.

Another way to explain the heat of compression is by remembering Boyle's Law: Temperature increases because temperature and volume of gas are inversely proportional to each other according to gas laws. Therefore, when a gas is compressed rapidly, its volume decreases leading to an increase in temperature.

The heat of compression effect is why we sometimes need cooling between compression stages or on the discharge of a compressor. Air fin coolers are the most common discharge gas coolers found in the industry (see Figures 5.5 and 5.6). Horizontal air fin coolers, like the one shown below, force air across finned tubes filled with hot compressor discharge gas. Typically, the

Figure 5.5 Air fin cooler, which is located on the far right of the skid, provides interstage cooling for the reciprocating compressor (far left).

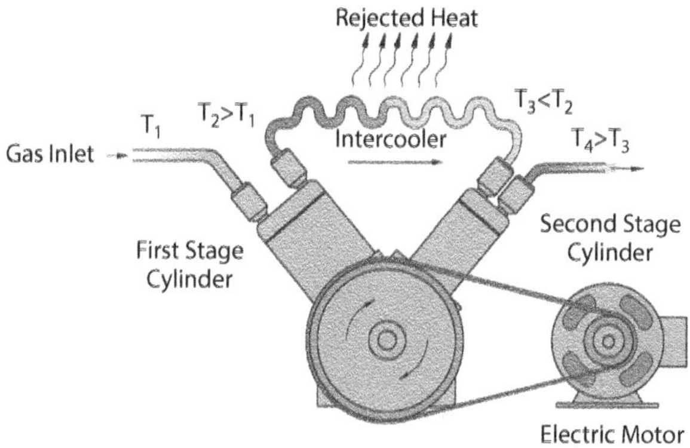

Figure 5.6 Schematic of a two-stage reciprocating compressor with an intercooler between stages. Notice that gas entering the second stage of compression has been cooled (T3 < T2) with an air-cooled heat exchanger before entering the second stage of compression.

fan is driven by the driver shaft. If designed properly, the cooler will drop the gas temperature to a manageable level before it enters the next stage of compression or the gas distribution piping.

Types of Compression Processes

It takes power to compress a gas from a lower pressure to a higher pressure. Power is defined as force applied over a distance. In reciprocating compressor terminology, the distance traveled by a piston is called the piston stroke and the force required to move the piston is the difference in the pressure across the piston times the piston area (P x Piston Area). During the compression process, the volume of the gas decreases and its temperature increases as the total internal energy of the system increases. In other words, during the compression cycle, energy is being converted from mechanical energy into gas energy in the form of increased temperature and pressure.

Recall that when we studied Boyle's Law (P_1V_1 = constant = P_2V_2), it was stated that for this law to be valid the temperature (T) of the gas must be kept constant. This means that for Boyle's Law to be accurate, the compression and expansion process must occur very slowly so that the gas remains at a constant temperature. In fact, all the gas laws we discussed earlier in this book only apply during isothermal compression, i.e., compression occurring at a constant temperature, inside lab devices that move slowly and are poorly insulated. In reality, these conditions never occur in

Figure 5.7 Reciprocating compressor performance can be approximated using the formula for adiabatic compression.

compressors due to their speeds and construction. It would be impractical to build compressors that moved slowly and compressed isothermally. Therefore, to properly analyze the performance of gas compressors, we must understand the various types of gas compression processes found in the industry. The two most common types are the adiabatic compression process (see Figure 5.7) and polytropic compression process.

Adiabatic Compression

A thermodynamic process in which there is no heat transfer into or out of a system is called an adiabatic process. To perform an ideal adiabatic process, a system would have to be surrounded by a perfect heat insulator. If a compression or expansion of a gas takes place in a short time, it can be considered nearly adiabatic, for example the compression stroke of a gasoline or a diesel engine.

In an adiabatic process, change is occurring within a system because of transfer of energy to or from the system in the form of work only, i.e., no heat is transferred. The adiabatic process obeys the relation:

$$pV^k = C = \text{constant}$$

We can put this into words by saying that pressure (p) times the volume (V) raised to the power of k, which is the ratio of specific heat, is equal to a constant, C.

To get a clearer idea of what adiabatic compression is, let's consider a simple example: Consider a piston cylinder in a reciprocating compressor. As we push the piston from point B to C, the volume of the cylinder decreases, causing the pressure to increase. This is illustrated in the pressure vs volume graph below (Figure 5.8). Since the piston is moving at a high rate of speed, there is likely very little heat added or removed from the compressor cylinder during the compression process. In this example, we can conclude this situation approximates an adiabatic process. For similar reasons, we usually assume positive displacing compressors, such as reciprocating and screw compressors, can be reasonably described as an adiabatic process.

One useful calculation for predicting the discharge temperature of a reciprocating compressor is the theoretical discharge temperature equation for adiabatic processes. The equation is written as shown here:

$$T_f = T_i \times \frac{P_f}{P_i}^{\left(\left(k-1\right)/k\right)}$$

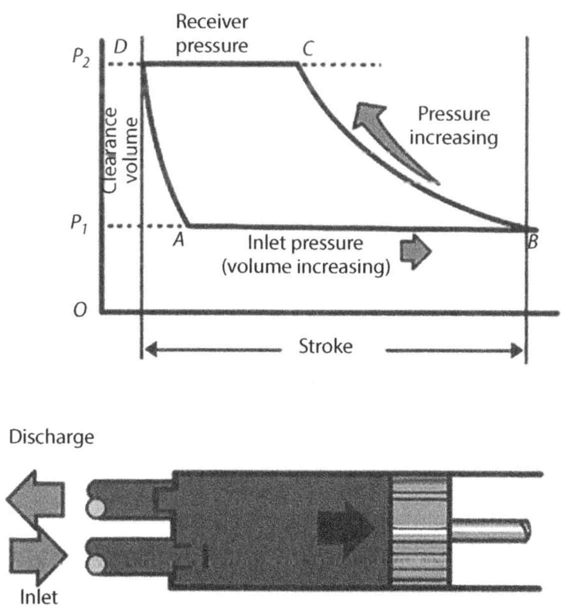

Figure 5.8 Pressure versus stroke diagram for reciprocating compressor.

Where

T_f is the final absolute temperature (T in degrees Rankine = T in degrees Fahrenheit + 459.67)

T_i is the initial temperature (T in degrees Rankine = T in degrees Fahrenheit + 459.67)

P_f is the final absolute pressure (psia = psig + 14.7)

P_i is the initial absolute temperature (psia = psig + 14.7)

k is the ratio of specific heat

See Table 5.3 below for a list of ratios of specific heat (k) for common gases.

The ratio of specific heats is a thermodynamic parameter used in many compressor performance calculations, usually denoted by k. The ratio of specific heats is defined as $k = c_p/c_v$, *where* c_p *is the specific heat for a gas in a constant pressure process and* c_v *is the specific heat for a gas in a constant volume process.*

Table 5.3 The ratio of specific heats.

Gas	Ratio of specific heats
Carbon Dioxide	1.3
Helium	1.66
Hydrogen	1.41
Methane or Natural Gas	1.31
Nitrogen	1.4
Oxygen	1.4
Standard Air	1.4

Before the reader has a math anxiety attack, let me explain how this equation can be analyzed in steps using a common scientific calculator.

Step 1: Calculate the compression ratio (Pf/Pi) using the absolute pressure values. Remember that absolute pressure is the pressure in psi plus 14.7.
Step 2: Look up the k value for your gas (see Table 3.1) and calculate the exponent (k-1)/k. Write the value down.

Step 3: Using a scientific calculator, raise the compression ratio (Pf/Pi) to the (k-1)/k power. Use the x^y function on your calculator to perform this mathematical operation.
Step 4: Multiply the number obtained in Step 3 with your absolute inlet temperature, i.e., Ti in degrees Fahrenheit + 459.67.
Step 5: Subtract 459.67 from the number obtained in Step 4.
The final number represents the final or discharge temperature in degrees Fahrenheit.

Adiabatic Example 1:
Let's assume we are compressing methane at 60 °F (k = 1.31 @ 60 °F) from 100 psig to 250 psig. First, we calculate the compression ratio: (250+14.7) x (100+14.7) = 2.308. Next, we plug all the values into the theoretical discharge temperature formula:

$$T_f = T_i \times \left(\frac{P_f}{P_i}\right)^{\left((k-1)/k\right)} = (60 + 459.67) \times (2.308)^{\left((1.31-1)/1.31\right)} = 173.74\ F$$

This result suggests that we can expect a discharge temperature of 173.74 °F. In reality, the actual temperature may vary from this value depending on the actual condition and construction of the compressor. As a reciprocating compressor wears and internal leakage becomes more and more significant, the difference between the actual discharge temperature and the theoretical adiabatic discharge temperature will increase. Keep the basic equation in mind when you want to assess the condition of a reciprocating compressor under field conditions.

Adiabatic Example 2:
Now, let's assume we are compressing hydrogen at 60 °F (1.41 @ 60 °F) from 100 psig to 250 psig. The compression ratio is the same: (250+14.7) x (100+14.7) = 2.308. Next, we plug all the values into the theoretical discharge temperature formula:

$$T_f = T_i \times \left(\frac{P_f}{P_i}\right)^{\left((k-1)/k\right)} = (60 + 459.67) \times (2.308)^{\left((1.41-1)/1.41\right)} = 202.65\ F$$

Notice by simply changing the type of gas being compressed that the theoretical discharge temperature increased from 173.74 °F to 202.65 °F.

This simple example illustrates why it's important when troubleshooting to know if your gas conditions have changed from your original design conditions. Changes in the composition of the compressed gas can affect your discharge temperature as well as the horsepower required for compression.

Polytropic Compression

A polytropic process is a thermodynamic process that obeys the following relationship:

$$pV^n = C = \text{constant}$$

We can put this into words by saying that pressure (p) times the volume (V) raised to the power of n, which is the polytropic index (defined below), is equal to a constant, C. The polytropic process equation can describe multiple expansion and compression processes which include heat transfer.

If the ideal gas law applies, a process is polytropic if and only if the ratio (K) of energy transfer as heat to energy transfer as work at each infinitesimal step of the process is kept constant.

An ideal adiabatic process must occur very rapidly without any flow of energy in or out of the system. In practice, most expansion and compression processes are somewhere in between or said to be polytropic. Centrifugal compressors (Figure 5.9) are typically said to behave like a polytropic process.

Figure 5.9 Centrifugal compressor performance can be approximated using the polytropic formula.

One useful calculation for predicting the discharge temperature of centrifugal compressors is the theoretical discharge temperature equation for polytropic processes. The equation is written as shown here:

$$T_f = T_i \times \left(\frac{P_f}{P_i}\right)^{\left((n-1)/n\right)}$$

Where
T_f is the final absolute temperature (T in degrees Rankine = T in degrees Fahrenheit + 459.67)
T_i is the initial temperature (T in degrees Rankine = T in degrees Fahrenheit + 459.67)
P_f is the final absolute pressure (Psia = Psig + 14.7)
P_i is the initial absolute temperature (Psia = Psig + 14.7)
n is the polytropic index

$$\frac{n-1}{n} = \frac{k-1}{k} \times \frac{1}{\varepsilon_p}$$

Where
n is the polytropic index
k is the ratio of specific heats
e_p is the polytropic compressor efficiency. (Typical polytropic efficiencies range from 70 percent to 85 percent.)

As I did before, let me explain how this equation can be analyzed in steps using a common scientific calculator:

Step 1: Calculate the compression ratio (P_f/P_i) using the absolute pressure values. Remember that absolute pressure is the pressure in psi plus 14.7.
Step 2: Look up the k value for your gas (see Table 3.1) and calculate the exponent (k-1)/k. Write the value down.
Step 3: Calculate the polytropic index exponent by dividing your (k-1)/k by the polytropic efficiency, e_p. (polytropic index = (k-1)/(k*e_p)). Write the value down.
Step 4: Using a scientific calculator, raise the compression ratio (P_f/P_i) to the (k-1)/(k*e_p) power. Use the x^y button on your calculator to perform this mathematical operation.

Step 5: Multiply the number obtained in Step 4 with your absolute inlet temperature, i.e., T_i in degrees Fahrenheit + 459.67.
Step 6: Subtract 459.67 from the number obtained in Step 5. The final number represents the final or discharge temperature in degrees Fahrenheit.

Polytropic Example 1:

Let's assume we are compressing methane at 60 °F (k = 1.31 @ 60 °F) from 100 psig to 250 psig with a centrifugal compressor that has a polytropic efficiency of 80%. First, we calculate the compression ratio: (250+14.7) x (100+14.7) = 2.308. Next, we calculate (n-1)/n: (k-1)/k*e_p = (1.31-1)/1.31*0.80= 0.2958. Now, we can plug all the values into the theoretical discharge temperature formula:

$$T_f = T_i \times \left(\frac{P_f}{P_i}\right)^{\left((n-1)/n\right)} = (60 + 459.67) \times (2.308)^{(0.2958)} = 205.86\,F$$

This result suggests that we can expect a discharge temperature of 205.86 °F. Remember that the actual temperature will vary from this value depending on the actual condition and construction of the compressor. As a centrifugal compressor wears and internal leakage becomes more and more significant, the difference between the actual discharge temperature and the theoretical polytropic discharge temperature will increase. Keep the basic equation in mind when you want to assess the condition of a centrifugal compressor under field conditions.

Polytropic Example 2:

Now, let's assume we are compressing hydrogen at 60 °F (1.41 @ 60 °F) from 100 psig to 250 psig with a centrifugal compressor that has a polytropic efficiency of 80%. The compression ratio is the same: (250+14.7) x (100+14.7) = 2.308. Next, we calculate the (n-1)/n: (k-1)/k*e_p = (1.41-1)/1.41*0.80= 0.3635.

Next, we plug all the values into the theoretical discharge temperature formula:

$$T_f = T_i \times \left(\frac{P_f}{P_i} \right)^{\left({}^{(n-1)}/_{n} \right)} = (60 + 459.67) \times (2.308)^{(0.3635)} = 244.64\ F$$

Notice by simply changing the type of gas being compressed that the theoretical discharge temperature increased from 205.86 °F to 244.64 °F. This simple example illustrates why it's important when troubleshooting to know if your gas conditions have changed from your original design conditions. Changes in the composition of the compressed gas can affect your discharge temperature as well as the horsepower required for compression.

Why Compression Ratio Matters

What Role It Plays in Compressor Design and Selection

One of the critical parameters in reciprocating compressor (Figure 5.10) design and selection is the compression ratio, often denoted as "*r*", required for each stage of compression. The compression ratio is simply the ratio of the absolute stage discharge pressure to the absolute stage suction pressure. Because most gases increase in temperature when they are compressed, the final compressor outlet temperature is always a concern. A high discharge temperature can lead to the failure of internal components due to material degradation or excessive thermal expansion. Compression ratio is also important in the determination of the required horsepower, i.e., the higher the ratio the greater the required horsepower for that stage.

Compression Ratio versus Discharge Temperature

Here is a simple example of how to calculate your compression ratio. Let's say you are compressing a gas with a ratio of specific heat of 1.3 (see Table 5.3: R*atio of Specific Heat* box.) from a suction pressure -0.5 psig to a discharge pressure of 35 psig. To calculate the compression ratio, you first must convert both pressures to absolute pressures by adding 14.7 to them and then divide the absolute discharge pressure by the absolute suction pressure, as shown in Equation 5.1.

$$R = P_d/P_s - (35+14.2)/14.7=3.5 \tag{5.1}$$

Figure 5.10 Reciprocating compressor.

Once we know the compression ratio, we can determine the theoretical discharge temperature, which assumes there are no internal losses and adiabatic compression, using Equation 5.2.

$$T_d = T_s \times r^{\frac{k-1}{k}} \tag{5.2}$$

In this equation, T is in degrees Rankin (R=F+460), k is the ratio of specific heat, and r is the compression ratio calculated by Equation 5.1. Assuming a suction temperature of 60 °F, r=3.5, and k=1.3, we arrive at a theoretical discharge temperature (T_d) of 234 °F.

Let's take this exercise a step further by increasing the compressor discharge pressure in 5 psi increments to see what happens to the discharge temperature. Table 5.4 summarizes the results. You can clearly see that as the discharge pressure increases, the compression ratio rises and correspondingly, the discharge temperature (T_d) increases. In this example, T_d increases from 234.3 F for a compression ratio of 3.5 to 335.7 F for a compression ratio of 5.61.

Table 5.4 Effect of discharge pressure on theoretical discharge temperature.

Pd (psig)	Compression ratio	Theoretical discharge temperature	Design limit
35	3.50	234.3	275
40	3.85	249.8	275
45	4.20	264.3	275
50	4.56	277.9	275
55	4.91	290.7	275
60	5.26	302.8	275
65	5.61	314.3	275
70	5.96	325.2	275
75	6.32	335.7	275
Assumptions	Ps	-0.5	
	k=	1.3	
	Ts (F)=	60	
	(k-1)/k=	0.231	
	Efficiency	100%	

Design Temperature Margin

If we compare our results to a hypothetical design limit of 275 °F, we see that at a compressor discharge pressure of 50 psig, we begin to exceed our design limit temperature. This relationship between the theoretical discharge temperature and design limit temperature can clearly be seen in Figure 5.11. I recommend that you select a conservative design temperature limit during the selection phase of a project to ensure you will have a safe operating margin to handle any unexpected process conditions or unexpected internal cylinder losses due to wear or fouling.

Let's assume the compressor you are considering has a recommended discharge temperature alarm limit of 325 °F and an automatic shutdown at 350 °F. If your actual discharge pressure is 60 psig, you should expect a minimum T_d of about 303 °F, as listed in Table 5.4. (Remember that the discharge temperature values in Table 5.1 are theoretical values. In reality, your discharge temperature will be higher due to internal losses as your compressor experiences normal degradation.) If your actual T_d is more like 318 °F, you will only have a margin of 7 °F. This is going to lead to countless alarms and midnight phone calls. The way to avoid this situation is to use a conservative design discharge temperature specification and use more compression stages to ensure smaller compression ratios per stage. Inspection of Table 5.1 will lead you to the conclusion that, for this example, design compression ratios should not exceed 4.5 per stage if you

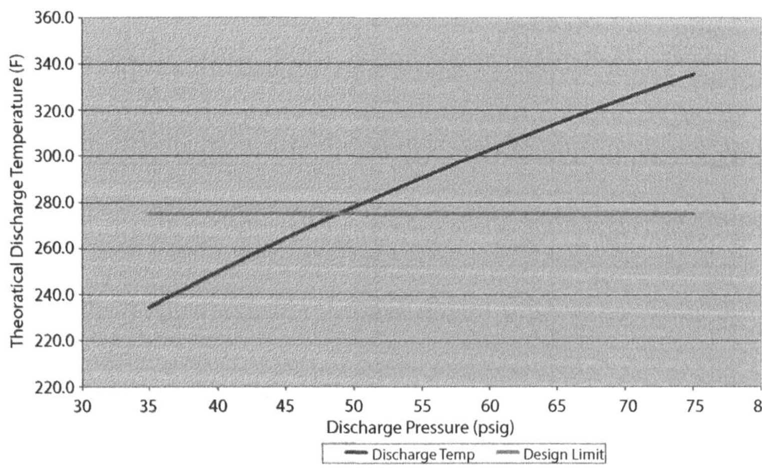

Figure 5.11 How discharge pressure affects compressor's discharge temperature. This plot assumes Ps=0.5 psig, k-1.3, and the efficiency is 100%.

wish to maintain a healthy margin between the operating temperature and alarm limit.

Cautionary Note: The examples shown here are based on theoretical adiabatic compression calculations. Adiabatic compression formulas are used to approximate the performance of reciprocating compressors. Readers should work with compressor manufacturers to determine actual compression ratio and discharge temperature limits for your specific compressor design. The theoretical calculations used in this chapter were used to demonstrate the general trend expected from high and higher compression ratios.

For centrifugal compressors, the following theoretical temperature rise equation based on polytropic compression, Equation 5.3, is commonly used:

$$T_d = T_s \times r^{\frac{n-1}{n}} \tag{5.3}$$

Here, (n-1)/n is defined by the relationship:

$\frac{(n-1)}{n} = \frac{(k-1))}{k} \times \frac{1}{\eta_p}$, where h_p is the compressor's polytropic efficiency.

Equation 5.3 should only be used to estimate the theoretical discharge temperature in dynamic compressors.

Design Tradeoffs

The reader can see there is always a trade-off between number of compression stages and the compressor discharge temperature at each stage. The more compression stages you have, the more costly the compressor installation will be due to its overall complexity, i.e., more piping and cooling required. However, the more compression stages installed, the smaller the temperature rise will be across each stage, allowing the compressor components to operate cooler and more reliably. This is the trade-off all compressor professionals must make as they begin their selection process.

The overall compression ratio is the product of all the individual compression ratios, i.e., $r_0=r_1$ x r_2 x r_3 x, etc. Assuming all stages have the same compression ratio, we can write:

$$r_o = r_s^n$$

In this formula, n is the number of stages and r_s is the compression ratio per stage. From this relationship, we can conclude that the number of stages required to achieve the required overall ratio is:

$$n = \ln(r_o)/\ln(r_s)$$

For example, if your required r_o is 20 and the recommended maximum r_s is 4.5, then we would need two stages of compression since:

$$n = \ln(20)/\ln(4.5) = 1.99$$

In general, we can conclude that the smaller the allowable compression ratio (r_s), the more stages (n) required.

My recommendation is to give your compressors a fighting chance. Try to use the widest economical margin between the calculated discharge temperature and the compressor design temperature limit. In the example above, I recommended a design margin of 50 °F (325-275 °F) to provide a buffer for unforeseen internal compressor degradation. If you expect factors such as dirty gas, you may want to increase your design margin even further.

Remember, your best friend during the compressor selection process will be the compressor manufacturer. They will keep you out of trouble by helping you select the right compressor for your application. For them to be successful, you must provide them the most accurate compression ratio, gas composition, and suction temperature specification available. Together, both of you can select an efficient and trouble free compressor.

Reciprocating Compressor Temperature Monitoring

Temperature is a valuable indication of reciprocating condition monitoring [4]. The primary reciprocating compressor temperature measurements that have proven to be effective under field conditions are cylinder discharge, valve, packing, cross head pin bearings (Item #3 in Figure 5.12), connecting rod big-end bearings* (Item #4 in Figure 5.12), and main bearing temperatures.

*Big End Bearing – Located at the 'Big End' of the Connecting Rod, where it connects to the Crankshaft.

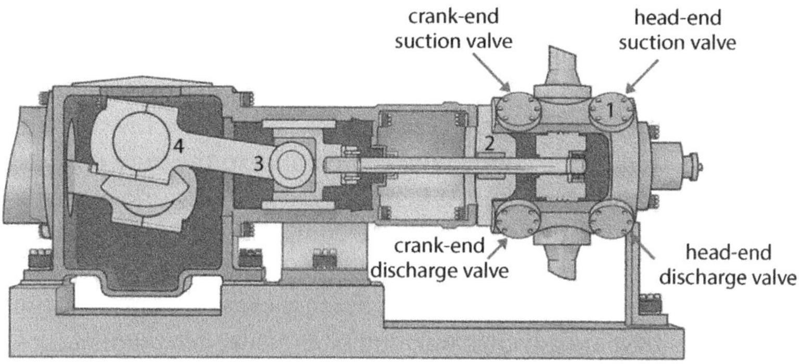

Figure 5.12 Typical reciprocating compressor temperature measurement. Note: Main bearings not shown here.

In addition, the following field measurements have proven helpful in identifying early problems:

1. Cylinder discharge temperature is one of the protection parameters recommended by API 618 since leaks in the rings and valves result in recompression that will raise the discharge temperature.
2. Valve temperature has proven to be valuable in identifying individual valve problem but is most effective if the measurements are made in a thermowell in the valve covers so that the reading being is taken close to the valve plates or poppets rather than the measuring contact temperature of the valve covers. (See Item #1 in Figure 5.12).
3. Packing case temperatures or packing leak-off temperature provide an indication of a packing leak. (See Item #2 in Figure 5.12).
4. Crosshead bearing temperatures can be monitored using eutectic, i.e., turkey popper temperature devices. There are wireless temperature measurement technologies being developed that allow monitoring of these bearings. (See Item #3 in Figure 5.12).
5. Main bearing temperature measurements have proven effective in preventing major damage due to main bearing failures.

Reciprocating compressors can be outfitted with eutectic or "turkey popper" temperature sensors in the connecting rods to prevent catastrophic failures. The sensors use a fusible eutectic material that is designed to fail at a designated temperature, typically about 200 °F. The fuse rod threads into a thermowell in the connecting rod parallel to the bore of the sleeve bearing in the connecting rod. When the fuse rod fails under spring tension it trips a pneumatic switch and the compressor is shutdown. Due to the design, the devices cannot be recalibrated and even testing requires replacing the fuse rod. For this reason, the fuse rods are replaced annually. The operators have no advance warning of an issue; detection occurs only at the point the unit shuts down from a fuse rod failure/trip.

Newer temperature sensors consist of a wireless measurement system based on wireless radar technology without the need of an external power source. A wireless sensor replaces the eutectic device in the connecting rod's thermowell. An antenna replaces the pneumatic switch and receives a signal from the wireless sensor every time it passes, once per revolution. The processing unit software calculates the temperature and transmits it to the supervisory control and data acquisition (SCADA) system.

Valve Temperature Monitoring [5]

The suction and discharge valves are typically the most maintenance-demanding parts of reciprocating compressors. Faulty valves are considered the main cause for unplanned shutdowns of reciprocating compressors. Valve failures can occur as a result of numerous problems, such as particulates and liquids in the gas, misapplication of valve designs, improper valve repairs, etc. However, even in clean gas, valve degradation will occur over time as a result of normal wear and tear. With any amount of valve wear comes a significant decrease in sealing capacity. Whether valve damage occurs as a result of liquid intrusion or is simply caused by normal wear, it is essential that operators maintain an awareness of valve functionality.

The monitoring of valve temperature offers insight into the condition of the valves. If a leak occurs at one of the valves, the valve temperature changes. By monitoring valve temperature, a defect at the valves can be detected quickly. A defective valve can also significantly decrease the efficiency of reciprocating compressors. For these reasons, monitoring valve temperatures is highly recommended.

RTD or thermocouple sensors are normally used to measure valve temperatures (see Item #1 in Figure 5.12). There are two common ways of positioning these sensors:

1. Sensor positioning in the valve through a thermowell
   Valve temperature measurements are more accurate when the temperature sensor is mounted in a thermowell that is embedded into the valve cover than when the measurements are taken on top of the valve cover. There are no environmental concerns when using this type of sensor positioning. The sensors that are installed in the valve cover also provide a faster detection of changing temperatures and valve leaks. However, not all valve designs allow for the installation of thermowells for temperature monitoring.
2. Sensor positioning on the valve cover
   Mounting the temperature sensor directly to the valve cover makes the sensor vulnerable to be influenced by environmental conditions such as sunlight or wind effects. It also has a slight delay on detecting valve leaks compared to using valve cover thermowells.

The signal coming from RTD or thermocouple sensors can then be transmitted to the facility's Distributed Control System (DCS), a PLC, or a machine monitoring system (MMS). In the DCS or MMS, temperatures can be trended to generate long-term information about valve conditions. The temperature data can be monitored in following ways:

- Temperature Spread (valves and packing): Plot all the valve cap and packing temperatures together for comparison. If the difference between the maximum temperature and the minimum temperature in the group is increasing, there is a problem. The point with the highest temperature is the first place you should begin looking for the problem.
- Deviation from the theoretical temp (T_{actual} – $T_{theoretical}$): Plot the difference between the actual compressor discharge temperature and the theoretical adiabatic compressor discharge temperature. If this difference is increasing or higher that the historical value, then there is a problem. Issues such as excessive piston ring or valve leakage can lead to cylinder heating from gas recompression.
- Trending: Plot all temperature values and look for increasing temperature trends. Upward trends are an indication of compressor deterioration.

Temperature Monitoring Example

Table 5.5 is a hypothetical collection of discharge valve temperatures taken with embedded thermocouples. The table shows the individual temperature values along with the maximum, minimum, and spread temperatures for the 1st and the 2nd stage discharge valves. The high spread seen in Table 5.5 on the 1st stage cylinder is a good indication of an issue. Drilling down further, we can quickly see that the #3 valve on the 1st stage is running at the highest temperature (250 °F). This tabular depiction of the data can be helpful in identifying the location of the problem.

Plotting the valve temperature in a bar graph format is another way to locate a problem. Figure 5.13 shows all the valve temperatures from the earlier example in one plot. The #3 valve on the 1st stage cylinder clearly stands out as the hottest valve.

Table 5.5 Discharge valve temperature valves.

Valve	1st Stg valve temps (°F)	2nd Stg valve temps (°F)
1	211	205
2	209	206
3	250	207
4	212	205
5	210	215
6	208	206
7	211	205
8	210	204
Max	250	215
Min	208	204
Spread	42	11

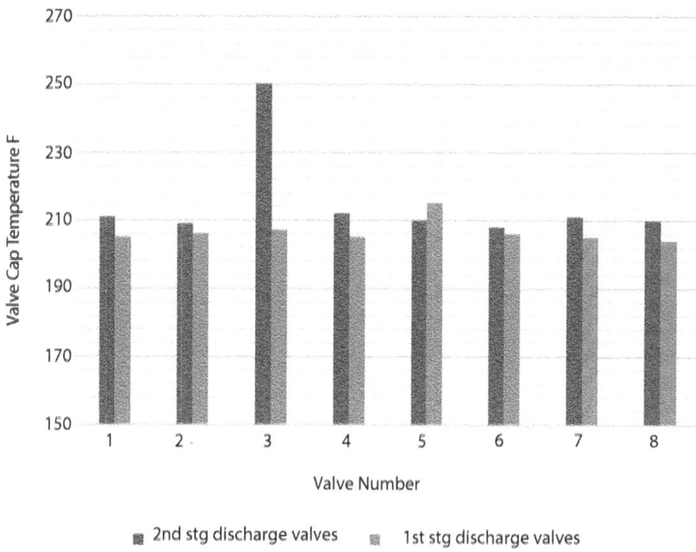

Figure 5.13 Hypothetical compressor valve temperatures.

Summary

Here, we have shown some proven ways to monitor reciprocating compressor temperatures. Temperature monitoring of cylinder discharge, valve, packing, cross head pin bearings, connecting rod big-end bearings, and main bearing temperatures under field conditions can provide the user with valuable indications of the compressor's health. Temperature monitoring should be a part of every reciprocating compressor reliability program.

References

1. Perez, R. X. and Conkey, A.P, "Is My Machine OK?," Industrial Press, 2012, pp 191-202.
2. Whalen, J., K., "Temperature is Best Indication of Bearing Operating Condition," Turbo Components & Engineering Bearing Journal, 4th Quarter, 1998.
3. Perez, R. X., "Operator's Guide to Process Compressors," Scrivener Publishing, 2018, pp31-41.

4. Schultheis, S.M., Lickteig, C.A., and Parchewsky, Reciprocating Compressor Condition Monitoring, Preceedings of the Thirty-Sixth Turbomachinery Symposium, 2007, pp 107-113.
5. Istec International Website, Temperature measurements for reciprocating compressors, Visited on 2/16/2022.

6

Assessing Reciprocating Compressors and Engines

By Robert X. Perez

Rotating Equipment professionals are frequently asked to guess the answer to challenging machinery condition questions such as:

- Can we make it until. . . ?
- When should I plan for a major overhaul?
- Is this compressor safe to run?
- Should we go into the compressor during this outage or can we make it until the next one?

Deciding whether to repair a machine or keep it running can be a time of high anxiety, especially if limited information is available to decide. Fortunately, the right tools and training can now give you the ability to "see" inside your reciprocating compressors to make confident decisions on repairs.

Figure 6.1 Six throw reciprocating gas compressor coupled directly to 8000 Hp, 720 rpm synchronous motor.

Robert X. Perez (ed.) Condition Monitoring, Troubleshooting and Reliability in Rotating Machinery, (167–194) © 2023 Scrivener Publishing LLC

Reciprocating compressors (see Figure 6.1) are unique in that they are the only compressor type I know whose thermodynamic cycle can be accurately analyzed in great detail using commercially available analysis equipment. Advances in portable reciprocating compression analyzers have made them capable of collecting, storing, and analyzing actual P-V plots. P-V plots are pressure-volume diagrams that are generated by measuring dynamic pressure data inside compressor cylinders and plotting them against the displaced volume of the cylinder. The displaced volume is calculated by correlating the crank angle to the position of the piston and multiplying this value by the piston area.

P-V plots, when evaluated along with vibration data, are immensely valuable in assessing performance and detecting internal malfunctions. "Reciprocating Compressor Basics" is a useful overview of reciprocating compressor technology that includes a clear explanation of the thermodynamic cycle and what a typical P-V diagram looks like. Anyone who works with or around reciprocating compressors should read this article.

As some of the most critical and expensive systems at a production facility, reciprocating compressors deserve special attention. Gas transmission pipelines, petrochemical plants, refineries, and many other industries all depend on this type of equipment. Due to many factors including but not limited to the quality of the initial specification/design, adequacy of maintenance practices, and operational factors, industrial facilities can expect widely varying life cycle costs and reliability from their own installations.

Various compressors are found in almost every industrial facility. Types of gases compressed include:

- Air for compressed tool and instrument air systems
- Hydrogen, oxygen, etc. for chemical processing
- Light hydrocarbon fractions in refining
- Various gases for storage or transmission
- Other applications

There are two primary classifications of industrial compressors: intermittent flow (positive displacement), including reciprocating and rotary types, and continuous flow, including centrifugal and axial flow types.

Reciprocating compressors are typically used where high compression ratios (ratio of discharge to suction pressures) are required per stage without high flow rates and the process fluid is relatively dry. Wet gas compressors tend to be centrifugal types. High flow, low compression ratio applications are best served by axial flow compressors. Rotary types are primarily specified in compressed air applications, though other types of compressors are also found in air service.

Overview of Reciprocating Compressors

Reciprocating compressors (see Figure 6.2) represent a class of gas compressors that raises the pressure of a gas by means of a tightly fitting reciprocating piston inside a cylindrical chamber, reducing volume from intake to discharge, in order to generate flow. Check valves within the cylinder are used to control the flow of gas within the cylinder.

Reciprocating compressor components include [1]:

1. A **piston** with rings and rider bands, cross head, and piston rod, which moves back and forth with each rotation of the crankshaft. The rings act like seals to keep gas from bypassing the piston. Rider bands are wide ring shaped devices that support the piston in the cylinder and prevent the two from touching. Some reciprocating compressors are "single acting" and some are "double acting". Single acting reciprocating compressors compress gas only when the piston is moving in one direction. When the piston moves toward the cylinder head, the gas is being compressed but when the piston is moving away from the head, suction is pulling gas into the cylinder.
   Double acting compressor cylinders compress gas on both movements of the piston. When the piston is moving toward the head of the cylinder, it is compressing gas as well as putting it into the process. At the same time, the piston is on the suction stroke of the crankcase half of the cylinder. When the piston begins to move toward the crankshaft, the crankshaft half of the cylinder is on compression and the head

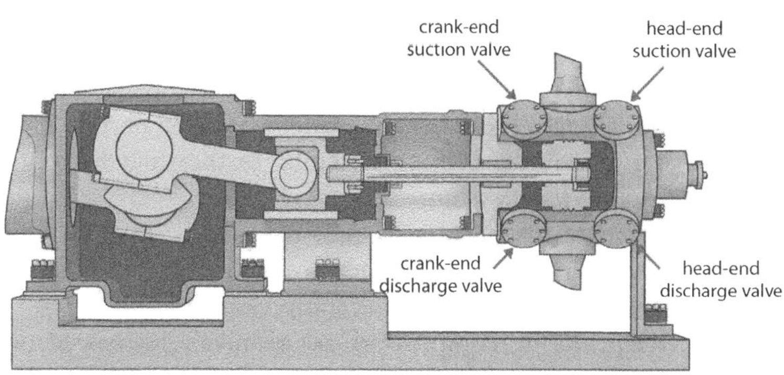

Figure 6.2 Reciprocating compressor cylinder cross section.

end is on suction. The effect is that there is almost twice the output from a double acting cylinder as from a single acting one.

2. A compressor **cylinder** contains the internal pressure and supports the reciprocating piston. The cylinder also has a head on each end with water passages to remove some of the heat of compression. The outer head may also include something called a clearance pocket for adjusting compressor output. The piston rod head has high pressure packing in it to keep the pumped gas inside the cylinder.
3. A set of **suction and discharge valves**. Reciprocating compressor valves are essentially check valves. The suction valves permit flow into the cylinder but not back out and discharge valves permit flow to exit the cylinder but not return back in. Suction and discharge valve failures are the most common component failure in a reciprocating compressor. Unloaders may be part of the suction valve assembly.
4. A **crank shaft, crank case, and connecting rods**. Some reciprocating compressors have two lube systems, one for the crank shaft and cross head and another for the cylinder and packing. The lubricant may be the same in both lubricating systems, but in some cases is different. Be careful not to mix or switch the two oils if they are different.

Suction gases are generally passed through suction strainers and separators to remove entrained particulates, moisture, and liquid phase process fluid that could cause severe damage to the compressor valves and other critical components and even threaten cylinder integrity with disastrous consequences.

Gas may also be preheated to vaporize liquid process gas into the vapor phase. Intercoolers provide an opportunity for heat removal from the process gas between compression stages. (See the following section: The Thermodynamic Cycle.) These heat exchangers may be part of the compressor's oil and/or cylinder cooling system(s) or they may be connected to the plant's cooling water system.

On the compressor discharge side, pressure vessels serve as pulsation dampeners, providing system capacitance to equalize the flow and pressure pulsations corresponding to the piston's compression strokes.

Typically, reciprocating compressors are relatively low-speed devices and are direct- or belt-driven by an electric motor, either with or without a variable speed drive controller.

There are three ways of controlling the output of a reciprocating compressor that is motor driven (constant speed):

i. Reducing suction pressure
ii. Adding a clearance pocket to the cylinder by opening and closing the valve that separates the cylinder from the additional volume in the clearance pocket
iii. Activating unloaders on the intake valves, which will hold them open

If all the compression components are working properly, the flow through a compressor cylinder is controlled by 1) the compressor speed, 2) the displacement of the piston within the cylinder, and 3) the suction pressure. Increasing either the compressor speed, cylinder displacement, or suction pressure will increase the net flow through the cylinder and the horsepower required from the driver. However, the design flow through a compressor cylinder can be reduced by numerous factors, such as internal leakage at the piston rings or valves, preheating of the inlet flow because of high ambient temperatures, a fouled exchanger on the suction gas, a lower suction pressure, etc.

Reciprocating compressors are often equipped with valve unloaders and clearance pockets as a means of controlling compressor throughput. If these flow control components malfunction, compressor flow may decrease dramatically or even drop to zero. Unloaders are control elements designed to hold intake cylinder valves open when they are activated so that no gas can be trapped or compressed inside the cylinder during the compression cycle. Therefore, any compressor cylinders with activated unloaders will not be able to contribute any flow to the compressor's total flow. The more unloaders that are activated, the lower the total flow will be. Unloaders may be operated independently or in groups to produce the desired number of compressor "load steps." In contrast, clearance pockets are used to add or remove internal volume within a compressor cylinder, which in turn adjusts when the intake valves open and vary throughput. Generally, as flow is decreased with the use of unloaders and/or clearance pockets, the required horsepower decreases. Conversely, as flow is increased with the use of unloaders and/or clearance pockets, the required horsepower increases.

Troubleshooting Tips:
If you are tasked with trying to understand why a given reciprocating compressor is not performing as expected, here are some basic relationships that can help identify the cause of your problems:

- If the differential pressure (i.e., the difference between the discharge pressure and the suction pressure) across any compression stage increases, the discharge temperature will increase on that stage. Conversely, if the differential pressure across any compression stage decreases, the discharge temperature will decrease on that stage.
- If the compressor speed increases, the net flow will increase. The horsepower required by a compressor increases at higher speeds, which should be reflected by increasing amps or kilowatts. Conversely, if the compressor speed decreases, the net flow will likely decrease. The horsepower required by a compressor at lower speed will decrease, which should be reflected by decreasing amps or kilowatts. This relationship between compressor speed and horsepower is not seen on motor driven compressors unless the motor is equipped with a variable speed electrical drive.
- If the specific heat ratio of the gas being compressed increases, the required horsepower will also increase for the same compressor ratio. Conversely, if the specific speed ration gas being compressed decreases, the required horsepower will decrease. If you are not sure of the gas composition, get a sample and have it analyzed.
- If the suction pressure increases, the mass flow and required horsepower will also increase due to the increase in the gas density. Conversely, if the suction pressure decreases, the mass flow and required horsepower will decrease. If a higher suction pressure puts more gas into the compressor, more horsepower will be required to achieve the same pressure.
- If a multistage reciprocating compressor begins to lift a relief valve between stages, the stage downstream of the relief device needs to be checked as it is not "keeping up" with the previous stage.
- If throughput falls quickly, check to ensure all unloaders on intake valves are functioning correctly and not allowing the intake valves to close. If clearance pockets are designed to control compressor throughput, ensure the valve that controls their addition or removal from the cylinder volume are functioning properly.

Here are some common causes of field problems to look out for:

- **Abnormal rotational speed.** Compressor speed has a major effect on gas flow, which in turn can affect system pressures and cylinder discharge temperatures. (This is generally not a problem if the compressor is motor driven because motors are constant speed devices.)
- **Abnormal suction pressure.** Low pressure can be caused by a partially opened suction valve or by blockages up-stream of the first stage. High suction pressure can be caused by a control valve problem that regulates the pressure to the compressors first stage.
- **Abnormal discharge pressure.** A high discharge pressure can be caused by a blockage down-stream. A low discharge pressure can be caused by one of the preceding stages not compressing properly or low inlet pressure to the first stage.
- **Abnormal gas density.** This can occur during process upsets, which can change the gas composition and has been known to occur during previous startups. As the ratio of specific heat (k) increases, so do the horsepower requirements for a given compression ratio.
- **Internal gas slippage** due to leaking piston rings. Internal leakage leads to a loss of flow and a corresponding increase in the discharge temperature.
- **Plugged suction strainer** that can result in a drop in flow or a loss of pressure.
- **Discharge restriction** due to downstream plugging or a partially closed valve.
- **Broken valves** will be evident by being much hotter than other valves. This condition can be caused by a bad valve with broken internals, a broken valve gasket, or a bad seat.
- **Fouled compressor valves** due to dirty gas will generally affect the output of the cylinder with the fouled valves.
- **Unusual gas temperatures** can be an indication of a loss of cooling somewhere in the process, usually caused by fouling or obstruction. Check intercoolers and after coolers on both the process side and water side.

Criticality

Compressors of all classes and designs should be monitored closely because they tend to be high-energy, high-horsepower machines. In many cases, they are unspared, which makes them highly critical to the processes they support. When they fail, management takes notice.

Because compressors tend to represent the most critical machines at most sites, it makes sense to watch them closely and keep them maintained and operating efficiently. Here are just a few ways to ensure your compressors perform safely and reliability day in and day out:

1. Perform daily inspections.
2. Check compressor performance on a regular basis. Keep process logs in order to know what is normal. Look at the logs; historical process logs will help you know how your compressors were performing yesterday, last week, and last month.
3. Analyze the lubricating oil regularly by checking levels and color of oil. Increasing oil level, in the crankcase, can mean water building in the crankcase.
4. Watch rod-drop indicators and other indicators of potential problems on a daily basis.
5. If a divider block is used for lubrication, check for any indicator pins that may be popping out, indicating blocked lubrication to a cylinder or packing.
6. If there is a panel for high pressure packing leakage, monitor it for increased flow. Check flow on sweeps of distance pieces.
7. Operators should help perform regular maintenance as prescribed by the manufacturer when appropriate or during turnarounds in order to learn more about a compressor's internal construction. This type of hands-on knowledge is invaluable when troubleshooting and can make you feel more comfortable when inspecting and operating these machines.

General Monitoring Guidelines for Reciprocating Compressors

Vibration Collection and Analysis: Most reciprocating compressors (Figure 6.3) moving flammable gases are equipped with permanently mounted vibration sensors, usually accelerometers or velocity transducers, on crankcases and crosshead guides. These sensors are then tied to local monitors that can be configured to alarms and/or trip the compressor whenever high vibration levels are encountered.

Overall vibration amplitudes can also be trended to see if any changes are occurring. If a significant change in overall amplitude is observed, a frequency analysis is performed in an attempt to identify the nature of the malady.

Figure 6.3 Reciprocating gas compressor.

Temperature Monitoring: Bearing and stage discharge temperatures are both commonly monitored continuously. High bearing temperatures may be an indication of a failing bearing or lack of lubrication. A high cylinder discharge temperature can either signal a cylinder valve failure, cylinder wear issue, or excessive stage compression ratio.

Oil Analysis: Periodic oil analyses are typically performed on reciprocating compressors. These analyses can help detect oil degradation, contamination, or machine wear issues.

Inspection frequency: Monthly and quarterly inspection intervals are common.

Pressure Pulsation Analysis: This type of field analysis is usually conducted whenever damaging compressor/piping interactions are occurring. It is hoped that if a thorough pulsation study and design analysis was performed before machine installation, pulsation will be controlled to low, healthy levels. Clues as to when to conduct a pressure pulsation analysis will be excessive piping vibration, broken bolts, high pulsations observed on a pressure gauge, etc.

Piping Vibration: This type of analysis is normally only conducted if piping failures have occurred or if the piping is deemed to be moving excessively. Piping analysis methods are similar to a vibration analysis performed on rotating machinery except that the assessment criteria are quite different.

Motor Current Analysis: This analysis method only applies to electric motors.

Performance Monitoring: Performance monitoring of reciprocating compressors requires specialized equipment and training. This topic is beyond the scope of this chapter, however, the basic methodology will be presented here. To analyze reciprocating compressor performance, dynamic pressure transducers must be installed in the head-end and crank-end cylinder volumes and a means of determining the crank's angular position is required. By plotting the in-cylinder pressure versus crank position, a pressure-volume (P-V) diagram can be generated similar to the one shown in Figure 6.4 below. PV diagrams help enlighten the analyst as to the condition of critical internal components such as valves, rings, etc. Additionally, vibration versus crank position is collected to determine the condition of critical bushings. (For more information about the reciprocating compression cycle, see "The Reciprocating Compression Cycle" box below.)

Inspection frequency: Quarterly performance evaluations are common.

> The Reciprocating Compression Cycle
>
> Compression occurs within the cylinder as a four-part cycle that occurs with each advance and retreat of the piston (two strokes per cycle). The four parts of the cycle are compression, discharge, expansion, and intake. They are shown graphically with pressure vs. volume plotted in what is known as a P-V diagram (Figure 6.4).
>
> At the conclusion of a prior cycle, the piston is fully retreated within the cylinder at V1, the volume of which is filled with process gas at suction conditions (pressure, P1 and temperature, T1), and the suction and discharge valves are all closed.

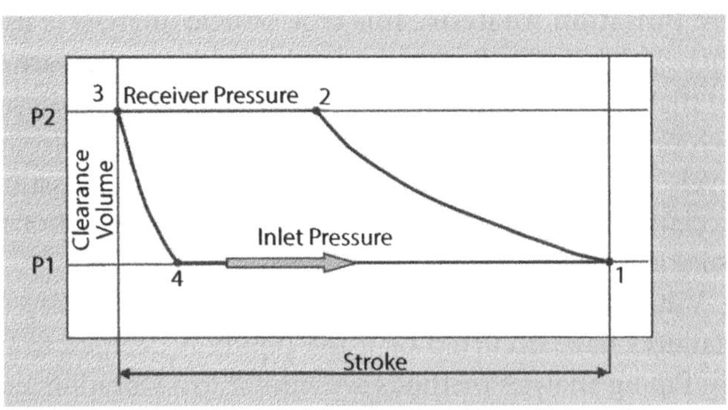

Figure 6.4 Typical pressure volume plot.

This is represented by point 1 (zero) in the P-V diagram. As the piston advances, the volume within the cylinder is reduced. This causes the pressure and temperature of the gas to rise until the pressure within the cylinder reaches the pressure of the discharge header. At this time, the discharge valves begin to open, noted on the diagram by point 2.

With the discharge valves opening, pressure remains fixed at P2 for the remainder of the advancing stroke as volume continues to decrease for the discharge portion of the cycle. The piston comes to a momentary stop at V2 before reversing direction.

Note that some minimal volume remains, known as the clearance volume. It is the space remaining within the cylinder when the piston is at the most advanced position in its travel. Some minimum clearance volume is necessary to prevent piston/head contact and the manipulation of this volume is a major compressor performance parameter. The cycle is now at point 3.

Expansion occurs next as the small volume of gas in the clearance pocket is expanded to slightly below suction pressure, facilitated by the closing of the discharge valves and the retreat of the piston. This is point 4.

When P1 is reached, the intake valves open allowing fresh charge to enter the cylinder for the intake and last stage of the cycle. Once again, pressure is held constant as the volume is changed. This marks the return to point 1.

Comprehending this cycle is key to diagnosing compressor problems and to understanding compressor efficiency, power requirements, valve operation, etc. This knowledge can be gained by trending process information and monitoring the effect these items have on the cycle.

Impact Monitoring

Simply measuring vibration velocity can be unreliable because the increase in velocity from incipient failures is usually small and will be buried in the larger signal due to machine movement. By the time the fault has been detected, major secondary damage may have already occurred. Impact monitoring overcomes this problem by taking raw, i.e., unprocessed, acceleration signal from an accelerometer permanently mounted in the compressor frame and counting the number of excursions that exceed an alarm threshold in a set time. If there are fewer events detected than the internal preset counter value, the count is cleared and begins again. However, a faulty machine will generate more excursions counts per cycle than a healthy machine, so the count is likely to exceed the preset in the specified time and the alarm will annunciate. Judicious setting of the system threshold peak level, allowable counts per measurement interval, and count time interval will permit reliable monitoring without excessive false alarms. This method has been proven to detect various internal faults such as cracks, looseness issues, and excessive clearances.

Rod Drop Monitoring

Rod drop monitoring involves the use of permanently mounted eddy current probes to measure the relative position of compressor rods with respect to the compressor cylinder. For horizontal rods, the probe is mounted vertically above or below the rod so that a change in vertical position of the rod is reported as a change in probe driver output voltage. Ideally, the gap should remain the same over the entire travel of the rod. In practice, the rod position will vary due to running clearances and may even bow when subjected to the operating compression and tension forces. In spite of these complications, it is possible to identify piston and cylinder faults, such as rider band wear, by plotting the probe gap on a continuous basis and looking for changes in relative probe gap.

Using Ultrasonics to Assess Reciprocating Machinery [2]

Ultrasonic energy readings have been recognized to be useful in detecting faulty suction and discharge valves in reciprocating compressors. Figure 6.5 below shows a comparison of healthy crank-end (CE) valve and a leaking CE valve. These types of graphical displays are generated by plotting the ultrasonic sensor's output relative to the compressor crank angle. This allows the analyst to determine when the leakage is occurring relative to the relative crank position.

On an internal combustion engine, ultrasonic readings can show a number of deficiencies such as excessive valve lash, sticky valve guides, necessary bridge adjustments, broken or missing rings, and more. Leakage can be monitored and assessed using a similar graphical presentation of the ultrasonic output versus crank angle. As seen in Figures 6.5 and 6.6, ring deficiencies and valve timing issues are a few of the many capabilities of this analysis method. In the figure below, all of the blue traces indicate a healthy condition, while the red traces show a problem. This technique

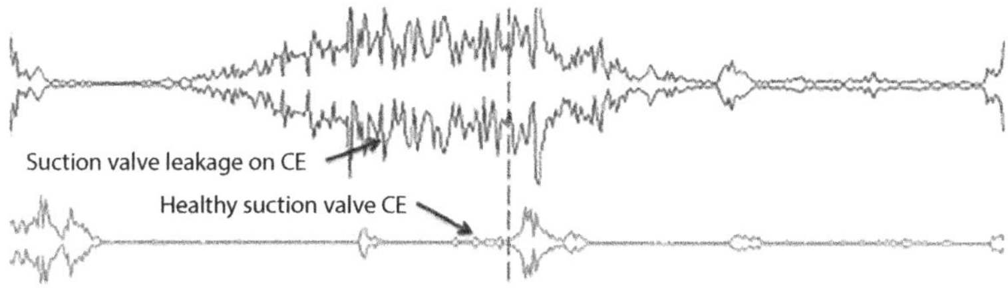

Figure 6.5 Comparison of healthy crank-end (CE) valve and leaking CE valve.

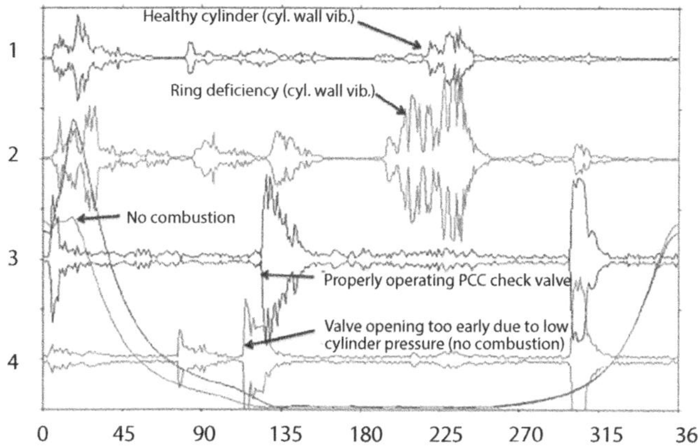

Figure 6.6 Multiple ultrasonic sensor outputs for multiple engine analysis locations (PPC stands for precombustion chamber).

provides a powerful visual means of determining the condition of an engine and where the problem is occurring.

By using a headset instead of the handheld probe, detection of packing, flange, manifold, spark plug, and head gasket leaks is possible, allowing the analyst to pinpoint the origin of a noise. Quick spot checks can also be made at suspect locations around an engine or compressor.

Mystery Reciprocating Compressor Knock

The following case study illustrates how compressor cylinder pressure curves and acceleration traces can be used to understand what's really going on inside of a compressor cylinder (see Figure 6.7).

Former reciprocating compressor analyst, Robert Akins, found that the best way to troubleshoot compressor knocks is to have the head-end and crank-end pressure curves overlaid on top of the accelerometer

Figure 6.7 Reciprocating compressor cylinder.

(vibration) traces. If the knock is occurring due to wear or excessive clearance, there will be an impact indication when the head-end and crank-end pressures equalize. Pressure equalization normally occurs somewhere around mid-stroke of the rod travel. There's a misconception that the rod reversal takes place at the end of the stroke, but in reality the high rod load differential stops the reversal from taking place till the cylinder internal pressures equalize. If the knock occurs at the end of the stroke (head-end or crank-end), this means the piston is making physical contact with either the inboard head or outboard head.

Case Study

A compressor technician asks me to help find a compressor cylinder knock that only occurred after the unit warmed up. The compressor analyst set up his analyzer on the cold unit. He saw that as the unit warmed up and thermal expansion caused the piston rod to grow, he started to see an impact indication on the end of the head-end stroke. The compressor technician said that he had checked the piston spacing and it was correct, so the analyst suggested for him to inspect the piston and outer head for debris.

Findings: The compressor technician found a piece of a metal valve plate imbedded in the piston surface.

If the debris had been on the inner head, the knock would have gone away as the unit warmed up and piston grew away from the inner head.

This same analysis set up can be used to find excessive wear on bearings and wrist pins.

Note: Robert Akins is a former maintenance supervisor for a natural gas pipeline company.

Analyzing the Condition of Natural Gas Engines

Natural Gas Engines [3]

Natural gas engines (NGE) are commonly used to power natural gas compressors, standby electric generators, fire water, and irrigation pumps and are increasingly being used to power primary cogeneration electrical power plants. The main advantages of a natural gas engine over a diesel engine are the lower exhaust emissions of nitrogen oxides (NO_x), carbon monoxide (CO), particulates, and in some cases, lower fuel costs. These stationary gas engines are available in various configurations and sizes. The characteristics include:

- Two- or four-stroke designs
- Less than 100 HP to a maximum of 16,000 HP (800 to 1,500 HP are most common)
- One to 20 power cylinders
- Oil sump capacities of 14 to 6,000 liters (300 to 800 liters (80 to 200 U.S. gallons) are most common)
- Engine speeds ranging from 300 RPM low-speed units to 2,000 RPM high-speed units; most operate at 800 to 1,200 RPM
- Piston bores as large as 22.5 inches (572 mm) in the low-speed units with bores of 3.5 to 9.45 inches (89 to 240 mm) common in high-speed units
- Inlet air may be naturally aspirated or turbocharged (two-thirds of new engines are turbocharged)
- Stoichiometric or lean burn (a relative term for an air-to-fuel ratio higher than stoichiometric) to reduce NO_x emissions
- Engines and compressor units may be either separates, that is, joined end-to-end at the crankshaft by a coupling (as seen in Figures 6.8 and 6.9), or integrals, in which the engine and compressor have a single common crankcase

Figure 6.8 Typical air-cooled, skid mounted natural gas reciprocating compressor.

Figure 6.9 Four throw gas compressors (left) driven by natural gas engine (right).

2 Stoke Engines versus 4 Stoke Engines

In a 2-stroke engine, the air-to-fuel charge is injected with the piston near the bottom of the power stroke. The intake ports are then covered or closed and the piston moves to the top of the cylinder, compressing the charge [4]. Following ignition and combustion, the power stroke starts with the downward movement of the piston. As the piston reaches the bottom of the power stroke, exhaust ports or valves are opened to exhaust, or scavenge, the combustion products and a new air-to-fuel charge is injected. Two-stroke engines may be turbocharged using an exhaust-powered turbine to pressurize the charge for injection into the cylinder and to increase cylinder scavenging. Non-turbocharged engines may be either blower scavenged or piston scavenged to improve removal of combustion products.

Historically, 2-stroke designs have been widely used in pipeline applications. However, current industry practices reflect a decline in the usage of new 2-stroke engines for stationary applications.

Four-stroke engines use a separate engine revolution for the intake/compression cycle and the power/exhaust cycle. These engines may be either naturally aspirated using the suction from the piston to entrain the air charge or turbocharged using an exhaust-driven turbine to pressurize the charge. Turbocharged units produce a higher power output for a given engine displacement, whereas naturally aspirated units have lower initial costs and require less maintenance.

The fuel typically used in these engines is sweet, dry natural gas (sales gas, greater than 85 percent methane). In some cases, in the gas production fields, raw sour natural gas containing hydrogen sulfide (H_2S) (up to 8,000 ppm), carbon dioxide (CO_2), and nitrogen (N) is used. There is also an increase in the use of digester gas collected from sewage systems and landfill gas that is being used as fuel.

Analyzing Reciprocating Machinery [5]

Analysis of reciprocating machinery is more complex than analysis of other types of rotating equipment. To analyze the condition of reciprocating machinery, the following signals have been proven to be useful:

- Conditioned vibration versus crankshaft position: The vibrational energy associated with faults in cylinders arises from a) mechanical impacts such as valve closure or chattering rings and b) gas leakage such as cylinder blowby and leaking valves
- Cylinder pressures as a function of crankshaft position
  Knowledge of the pressure in the cylinder as a function of crank angle allows:
  - Detection of performance problems such as poor air/fuel ratio and excessive firing pressures
  - Detection of mechanical problems such as low compression pressure and carbon build-up in ports
  - Interpretation of anomalies through correlation with vibration and ultrasonic signals
- Ultrasonic energy versus crankshaft position
  These patterns will be more sensitive to gas leakage and less sensitive to mechanical impact events. If we take both ultrasonic and vibration signals on a cylinder, problem detection becomes more sensitive and reliable and diagnosis is also more reliable.

- Ignition voltages versus crankshaft position
  In spark ignited engines, the ignition system is a source of many of the problems which arise. It is common practice, therefore, to measure the ignition primary waveform and/or the ignition secondary waveform.
- Process type data: pressures, temperatures, flow rates

Gathering and Analyzing Data From Natural Gas Engines Driving Reciprocating Compressors [6]

The collection of reciprocating equipment data started when the first engine and compressor were invented. The temperatures were sampled and the noises were noted. Over the years, engines and compressors have improved and so has the data. Reciprocating equipment analysis has long been described as more of an artform than a science. The task is neither an art nor a science.

The equipment analyst has been described as an equipment doctor. The reality is the analyst can be better described as a detective than a doctor. Persistence, objectivity, creativity, and attention to detail are just some of the traits needed to be a good detective. The analyst will use these same traits to logically and methodically compile information, to identify problems, and determine the health of that equipment. An inspector, mechanic, or even an operator could be considered an equipment analyst if they can effectively identify and report potential mechanical issues. The detail of that analysis will always depend on the tools available in conjunction with experience and knowledge of the equipment systems. The analysis can be generally separated in three steps.

- Data Collection (Listening)
- Data Analysis (Understanding)
- Reporting (Resolving)

A huge advance in reciprocating equipment analysis has been the development of the portable reciprocating equipment analyzer, which added a whole new level of listening, understanding, and resolving. The creation of this new tool then created today's reciprocating equipment analyst, who is typically a mechanic with above average detective skills and computer literacy. One inherent challenge of using this new tool is the potential variability of findings based on the experience and knowledge of the analyst using it. Companies and organizations have long worked at standardizing the different aspects of the equipment analysis with most of the attention on the data analysis.

Training of the analyst focuses heavily on the data analysis. The data collection is typically only generally outlined, allowing for personal interpretation. However, new strides are being made on data collection with online systems, data loggers, and interfacing with AI. Some important variables for collection remain: Where should an end device be placed? How many sensor points should be considered? Which end device should be used? These questions have been sticking points with standardizing data collection. Each of these variables can differ because of engine design and personal preference. Changes in the variables, as might be expected from different analysts, will affect results of the analysis. A 1950's vintage two cycle natural gas engine differs drastically as it relates to individual components and sheer size in comparison with a 2019 four cycle diesel engine in terms of data collection. The data analysis and reporting methods will have more in common than the data collection.

The primary focus of field data collection is to compile enough information to definitively assess and then report the health of the equipment. A primary challenge with reciprocating equipment is the multitude of simultaneous forcing functions. Some of these events occur deep inside of the equipment and the equipment can be quite large. The analyst must compartmentalize his analysis of the unit to reach a conclusion of the overall health of that unit. Every possible forcing function must be evaluated and assessed before the analysis starts. Each valve, port, bearing, etc. can generate measurable values and give insight to health of the component. An adequate number of sensor points must be used near each component to determine the health of those components. The analyst would then compare the health of each component to the general health of the unit using these basic indicators:

- Temperature
- Pressure
- Vibration
- Acoustic Response
- Voltage

Statistically speaking, the number three is an important value. If three indicators can be found to support your analysis for the health of the unit or a specific problem, then it carries a higher degree of certainty in the analysis. The three indicators can come from just one of the basic indicators or can be combinations of them. One phased vibration plot may contain enough information to satisfy the reciprocating analysis rule of three or one data sample could be bad or misleading. The analyst must balance his time at each component to prevent having either too little or too much data to permit established condition of the unit.

The reciprocating equipment analyzer collects phased vibration, acoustic response, and pressure in small sample sets. The phased term refers to the timing of the data relative to the angular position of the crankshaft, which is displayed over 360 degrees (one revolution). The sample set size is controlled by the number of averages selected. Increasing the number of revolutions recorded helps to identify transient vibration and prevent unreliable data and collect a more accurate average of the amplitude.

The vibration is collected with an accelerometer, velocity probe, or proximity probe. The vibration frequency window can be controlled and adjusted with most analyzer's software. The acoustic response is collected with an ultrasound microphone, which most likely has a voltage gain control. It is critical that the analyst compare like cylinders to each other in the analysis. The voltage gain control and the vibration frequency windows should not be adjusted after the 1st data sample has been collected. Comparing like components would be very difficult if the ultrasound microphone gain has changed during the analysis.

There will always be components in reciprocating equipment that cannot be recorded easily or effectively from one equipment model to the next. Smaller units are harder to analyze since fewer components are accessible. Identifying forcing functions that can be recorded, the limitations of the tools available, and the components that can be combined are a never-ending challenge for the analyst. Understanding the subjects of the data and what the data should look like can help to ensure the data is being recorded properly.

Listed below are the basic components with the sensor points recommended, along with some basic tips for data collection.

- Power Cylinder Valves & Ports
    - Low and middle frequency phased vibration
    - Phased acoustic response
    - Phased pressure
        - 1st derivative
    - Temperature
        - Exhaust, fuel, PCC, and air starter
- Power Cylinder Liner (if it does not have ports and it is accessible)
    - Low and middle frequency phased vibration
    - Phased acoustic response
    - Phased pressure

- Cam Follower
    - Low and middle frequency phased vibration
- Cam Bearings
    - Low and middle frequency phased vibration
    - Frequency based vibration
- Main Bearings
    - Low and middle frequency phased vibration
    - Frequency based vibration
- Scavenging Cylinders
    - Low and middle frequency phased vibration
    - Phased acoustic response
- Crossheads
    - Low and middle frequency phased vibration
- Compressor Cylinders
    - Low and middle frequency phased vibration
        - Cylinder ends
    - Phased acoustic response
        - Cylinder ends
    - Phased pressures
        - Cylinder ends
    - Temperature
        - Suction and discharge nozzle
- Compressor Cylinder Valves
    - Middle frequency phased vibration
    - Phased acoustic response
    - Temperature
        - Valve cap
- Ignition
    - Secondary ignition
        - Phased voltage
        - Voltage characteristics
    - Primary ignition
        - Phased voltage
- Accessories
    - Frequency based vibration
    - Temperature
    - Panel readings

Vibration and acoustic response both need to be relatively close to the forcing function that is driving it. Vibration is very directional, so the end device should be in the direction of the forcing function if possible.

Acoustic response is not as directional and the higher frequencies do not travel as far as the lower frequency vibrations. Here are a few tips for phased vibration and acoustic response collection.

- The exhaust blowdown event should be the primary focus for power cylinders.
- Compressor cylinder valve openings and closings should be the primary focus.
- If possible, use head bolts by the exhaust valve or exhaust manifold.
- Try not to collect vibration data on the skirt or covers, the liners jugs and head bolts are preferred.
- Bolts that run deep into the cylinder or head are preferred.
- Use the "Autoscan" feature on the 1st like component to find the best place to record the primary subject and then use that same spot on all like components.
- Most engine and compressor main bearing internal designs have webs that are a direct path to an anchor bolt. The engine frame directly above the anchor bolt, where the web attaches to the frame, is the best place to record excessive clearance with main bearings.
- Cam bearing vibration should be recorded in between cam bearing inspection doors.
- Crosshead vibration should be recorded in the vertical direction and in the center of travel of the crosshead. An extra axial sensor point is recommended if the piston rod is 4" or over.
- The compressor valve cap or the side of the valve cap can give an excellent acoustic response.
- The base line should not be just a flat line, you should see definition in the base line.
- High-speed four-cycle engines will need higher frequency vibration filtering.
- Use the flat magnet on flat surfaces only.
- Use the rail magnet on everything else.

The internal pressures should be recorded with direct coupled pressure transducers. D.C. pressure transducers must be zeroed. Here are a few tips for the pressure transducers:

- Heat soak the pressure transducer for about a minute before zeroing it.

- Re-zeroing the pressure transducer is needed if the data exceeds +-5 PSI at atmospheric air pressure.
- Overheating the pressure transducer can damage it, give some time for the pressure transducer to cool down in between high compression ratio cylinders.
- Select a pressure transducer that will record every pressure inside of the cylinder and the internal pressure will use the largest amount of the transducer's range.

The temperatures of an engine or compressor are easy indicators that anyone can record for analysis. Here are a few tips for the temperatures:

- The compressor suction and discharge nozzle skin temperatures will be significantly different than the actual internal gas temperatures.
- The compressor suction nozzle skin temperature, on average, is about 5 degrees hotter than the internal gas temperature.
- The compressor discharge nozzle skin temperature, on average, is about 20 degrees cooler than the internal gas temperature.
- The analyzer's software needs the actual internal gas temperature for the suction and discharge nozzles to perform accurate performance calculations.
- Most of the unit panels record the actual internal discharge nozzle temperatures, but only a few record the suction nozzle temperatures
- Record the temperatures directly before or after the internal pressure is recorded

Frequency based vibration is also called spectrum analysis or FFT analysis. Typically, vibration collection and analysis data points are highly configurable. Custom data collection configurations may be required when rotating components have a different RPM than the crankshaft. Some other examples of when different vibration data configurations may be required include collecting possible rolling element frequencies, gear mesh frequencies, or blade/vane pass frequencies. Also, the run speed amplitude could be monitored in displacement for "movement" measurements. The time waveform of the frequency base vibration also offers more information on impacting than the low frequency phased vibration. Here are a few tips for assessing frequency-based vibration:

- Displacement sensor points can be used to measure how much larger objects are moving. The fundamental frequency of this displacement vibration is normally running speed (1x).
- Velocity sensor points are used to display different RPM frequencies, looseness, imbalance, and misalignment that may be present. Similarly, the fundamental frequency of this vibration is normally running speed (1x).
- Acceleration sensor points are used to measure and display vibration data over 6000 CPM, such as those generated at gear mesh frequencies and blade/vane pass frequencies. These higher order vibration frequencies are many orders (nX) above the 1st order or running speed frequency.
- Multiple measurements are collected at each point because of how reciprocating equipment analyzers record the time waveform. Comparison of the various vibration components like run speeds and vane passes are necessary with multiple sensor point measurements to allow their comparison.

How Accurate are Rotating Equipment and Reciprocating Equipment Analyst Findings? [7]

"Just how accurate are equipment analysts at finding problems?" is the question that company managers and industrial equipment owners have asked me the most in one form or another during my twenty-five-year career as an equipment analyst. This question is an important one to companies who are trying to decide how useful an equipment analyst is to their maintenance program and whether they can justify this expense as part of their maintenance budget. "What type of maintenance program are they using? What equipment will be monitored and how often? Who will be interpreting and filtering the analyst's information?" These are all questions that then define how certain the analyst must be on their "calls". The more definite the analyst must be, the more precise the information to and from the analyst must be.

There are two different kinds of equipment analysts: rotating equipment analysts and reciprocating equipment analysts. Company managers and industrial equipment owners may expect an equipment analyst to be trained to perform both kinds of analysis. However, rarely is an equipment analyst capable of performing both rotating equipment analysis and reciprocating equipment analysis. The accuracy of both equipment analysts is typically very different for a variety of reasons.

The rotating equipment analyst is often a mechanical engineer, millwright, or technical specialist. A rotating equipment analyst usually has experience aligning a driver to centrifugal equipment, changing the centrifugal equipment's bearings, or measuring its internal parts. The rotating equipment analyst uses the centrifugal equipment's technical data, field measurements, and vibration to determine the health of rotating equipment.

The rotating equipment's vibration data is measured with a vibration analyzer that can record and display accurate phase, amplitude, and frequency vibration measurements. Today's vibration analyzers also have built-in features designed to notify the analyst if there are specific faults. Most of these faults are specific to a phase, amplitude, or frequency of that vibration. Interpreting this data can be done with the software with a high probability of accuracy.

The average rotating equipment analyst that can utilize load curves, amperage and power readings, and the vibration analyzer outputs will typically diagnose equipment with a high degree of accuracy. Of course, the analyzer program itself will always hinge on an analyst's experience, training, and resources. An analyst can spend years collecting vibration data, but if they have not properly set up the analyzer software, they will end up only using vibration trend plots to detect anomalies. If an analyst simply relies on vibration trending to capture faults, the accuracy of his interpretation will be less accurate than a more skilled analyst who uses all the tools available to him. Proper training and oversight are important for any analyzer program.

The rotating equipment vibration analyst training has been ANSI accredited and ISO certified for many years. It's easy to find courses offered by highly regarded training companies each year. Vibration analyzer companies will also perform analyzer specific training or setup the analyzer for customers at a premium cost.

In summary, a novice vibration analyst specializing in rotating equipment, who has good supervision and technical help, can maintain a vibration program covering rotating equipment with above average accuracy. Generally, the more training, resources, and experience a rotating equipment vibration analyst has, the greater his accuracy and the less reliant he is on technical assistance.

The reciprocating equipment analyst is normally a natural gas equipment mechanic or technical specialist. The analyst typically only has some general knowledge of millwright, loading curves, or vibration analysis. They know the internal components of the engines and compressor and are familiar with the operating conditions of the units. The reciprocating equipment analyst uses technical data, measurements, and the reciprocating equipment analyzer data to determine the health of reciprocating equipment.

The reciprocating equipment analyzer data basically has three data categories: a) performance data, b) pattern interpretation data, and c) basic FFT data. The performance data is based on the reciprocating motion of pressure inside of the cylinders. This data is called pressure vs. time and pressure vs. volume and is where most engine and compressor performance data is derived. The pattern interpretation data is time waveform plots of vibration and ultrasound data that is phased to that component's TDC. Patterns in the timewave indicate normal conditions or certain faults. The basic FFT data is primarily looking at amplitude and frequency for trending purposes.

The reciprocating equipment analyzer software normally has some built in features to detect faults with the performance data. However, the pattern interpretation and FFT data have little to no software features to autodetect faults. Experience is critical with the reciprocating equipment analyzer.

The fact is that the average reciprocating equipment analyst with a reciprocating equipment analyzer will be less accurate than the average rotating equipment analyst with a vibration analyzer. The main reason is that the reciprocating equipment analyzer software just cannot provide the analyst similar types of the same notifications of faults that the rotating equipment analyzer software does. There are also more components per unit and more possibilities of faults with reciprocating equipment.

The average reciprocating equipment analyst typically does pick up some of the basic pattern interpretation in their career early. Leaking valves and engine piston rings in the acoustic ultrasound timewave plots are some of the first faults the analyst can detect with high accuracy in the beginning of their career.

If a novice analyst starts with no formal training in analysis, he or she must rely heavily on relationships with available experienced analysts for assistance in making only obvious calls. The result will be average accuracy on normal faults and poor accuracy on abnormal or unusual faults. Novices begin by only using a small portion of the analyzer's capabilities due to a lack of understanding of basic reciprocating analysis techniques. They must rely more heavily on finding faults that occur more regularly, such as leaking valves.

The greatest improvement to analysis accuracy will result from attending formal reciprocating compressor and engine analysis training by the analyzer manufacturer and collecting data on the same units for years and years. The reciprocating equipment analyzer training is not ANSI accredited or ISO certified. The analyzer companies do offer analyzer specific training through generic compressor/engine analysis classes and will setup the software at a premium cost.

In summary, a novice reciprocating equipment analyst with good oversight and technical help can maintain a reciprocating equipment program with only average accuracy. This analyst will typically sample data on components that have high failure rates or known issues and bypass other critical components that fail less. An example of this would be catching compressor valves but missing a broken tooth on the lubricator gear because they do not collect enough data or cannot detect the subtle changes to existing data.

A two-stroke engine can have hundreds of design variations and thousands of types of faults. An analyst would typically collect hundreds of points of data for a basic health assessment. In contrast, an 18-pole electric motor will have significantly less variations and less types of faults. An analyst would typically collect only ten points of data for a basic health assessment. Combine that with the help of the rotating equipment analyzer and it is easy to determine which analyst would be more accurate.

The analyst, mechanics, operators, and monitoring systems are the best options to detect faults before critical failure occurs. The analyst gathers a unique set of information that the others only partially utilize. Climbing the unit and the flexibility to tailor the analysis to specific issues can far outperform the other detection methods, even if the equipment analyst is at best average. The more the analyst knows your equipment, the better the accuracy and results. The experienced equipment analyst that collects data on the same units and knows the history really does make all the difference in providing recommendations that the maintenance department can rely on.

References

1. Noria Corporation Staff, Reciprocating Compressor Basics, Website: https://www.machinerylubrication.com/Read/775/reciprocating-compressor
2. Perez, R. X. Is My Machine OK? pp173-174, Industrial Press, 2012.
3. Scott, Robert, Noria, *Stationary Natural Gas Engine Lubrication*, Website: Machinerylubrication.com, accessed on 2/3/2022
4. Website: epa.gov, *Natural Gas-fired Reciprocating Engines*, page 3.2-2
5. Bryan R. Long, Ph.D., P. Eng. And David N. Schuh, P. Eng., *Fault Detection and Diagnosis in Reciprocating Machinery*, www.betamachinery.com, accessed on 2/3/2022

6. Wimberly, Tracy, Technical Services Specialist, U.P.S. MIDSTREAM, *How Accurate are Rotating Equipment and Reciprocating Equipment Analyst Findings?*, Southern Gas Association, Quarterly Newsletter, 2019.
7. Wimberly, Tracy, Technical Services Specialist, U.P.S. MIDSTREAM, How Accurate are Rotating Equipment and Reciprocating Equipment Analyst Findings?, Universal Plant Services Midstream Services Website.

7
Managing Critical Machinery Vibration Data

By Robert X. Perez

Beware of False Positives and False Negatives

Field vibration analysis errors and sensor issues can result in erroneous conclusions, which in turn can lead to unnecessary repairs or missed mechanical faults. Analysis accuracy is of paramount importance when monitoring critical machines in order to control maintenance costs and reduce risks to acceptable levels. In addition, a disciplined sensor selection, installation, and testing program are required to ensure the fidelity of critical sensors throughout their lifetime.

Figure 7.1 Regularly monitoring machinery condition is a vital part of any machinery reliability program.

Robert X. Perez (ed.) Condition Monitoring, Troubleshooting and Reliability in Rotating Machinery, (195–200) © 2023 Scrivener Publishing LLC

Organization can reduce potential vibration analysis mistakes and errors by adopting various data collection and maintenance strategies aimed at improving the validity and reliability of the vibration data (Figure 7.1). Before we discuss some recommended data collection strategies, let's explore the four possible outcomes from a machinery vibration analysis:

- **True Positive (TP)** The problem detected is real: This is the expected outcome from a vibration analysis when an actual machine fault is encountered. With the proper training and equipment, the analyst is expected to detect true machinery faults in their early stages in most cases.
- **True Negative (TN)** There is a correct indication that a machine is healthy: This is also an expected outcome from a vibration analysis. With the proper training and equipment, the analyst should be able to give a machine a "clean bill of health" when there are no internal mechanical faults.
- **False Positive (FP)** The reported problem is not real: The outcome when the analyst discovers a fault that does not really exist or the measurement system wrongly indicates a problem. This type of analysis error may be caused by 1) an inexperience analyst, 2) by using the wrong data collection methods, or 3) a malfunctioning sensing system. False positives lead to unneeded repairs.
- **False Negative (FN)** An incipient machine deficiency is not detected: This possibility is the most insidious of them all. This outcome arises when the analyst or analysis equipment is not capable of finding a machine fault. In this case, the fault remains undetected and will continue to get worse until additional machine damage or collateral property damage occurs. A false negative has the potential of putting your plant at risk by failing to detect a machine malady in its insipient stage, which could result in an undesirable secondary consequence such as a product release, fire, etc.

The first decision a reliability professional needs to make is determining the level of analysis required. There are two basic levels of vibration analysis:

Maintenance-quality level analyses are those usually performed by a mechanic or operator using a vibration data collector or vibration meter. This level of analysis is limited to trending and simple spectral analysis. Sites composed mostly of low consequence machines tend to employ maintenance-quality analysis methods. Low consequence machines are small to medium horsepower process machines that represent low levels of secondary consequences if they fail unexpectedly.

Advanced vibration level analyses are those usually performed by a certified vibration analyst using advanced vibration data collection and analysis hardware and software. This level of analysis involves spectral analysis, phase comparisons, Bode plots, correlation analysis, etc. Sites with high consequence machines typically employ laboratory quality vibration analysis methods. High consequence machines are process machines associated with high consequence levels. If these machines fail unexpectedly, bad things happen, i.e. fires, product releases, production outages, costly secondary mechanical damage, etc. This machine group normally represents about 10% or less of the site's populations.

It may make sense to employ off-site analyst capabilities for special machines that require special analysis knowledge to assess their condition. High horsepower gas turbines and integrally geared compressors are two examples of special machines where remote monitoring might make economic sense due to their complexity and sensitivity to load changes. With today's technology it is becoming more cost effective to remotely monitor critical plant machinery.

Vibration Analysis Strategies

The potential for false positives related to misinterpretations of the vibration data can be greatly reduced by getting a second opinion from an experienced analyst. Therefore, rotating machinery that is costly to repair can be initially monitored by the maintenance or operations group, but then should be analyzed independently by a trained vibration analyst whenever alert vibration levels are detected. The potential for false positives related to questionable sensor outputs requires knowledge about sensor performance and installation best practices. For example, if not installed properly, proximity probe sensors are susceptible to ground loop signal issues that result in the presences of 60 Hz electrical components to be present. Signal errors can also be caused by sensor mounting resonance issues which result in misleading frequency spectrums. An experienced vibration analyst should be able to identify most sensor installation problems before a repair recommendation is made. These strategies should greatly reduce the likelihood of unneeded repairs due to false positives.

The potential for false negatives can be reduced by using a skilled machinery analyst that employs multiple technologies and methodologies. An example of this approach would be monitoring case velocity readings as well as proximity probe readings on a machine. Another example of a multifaced monitoring approach is using case mounted accelerometers along with spectral analysis and a bearing fault monitoring system. Multiprong approaches can greatly increase your chances of detecting a machine fault on a critical machine before secondary damage can occur. False negatives can also be caused by miscalibrated, uncalibrated, or non-working sensors. The installation of multiple sensors and regular sensor calibration should reduce the likelihood of these types of measurement failures. X-Y proximity sensors, X-Y accelerometers, and dual voting probes on thrust bearings are some examples of multiple sensor arrangements.

Each site will have to determine the machine installations where significant repair costs and the potential for costly secondary damage warrant higher skill levels and multifaceted approaches to machinery monitoring.

I will leave you with a few tips to improve your vibration analysis program:

1. It is acceptable and even beneficial to use operations and maintenance personnel to monitor low and medium criticality machines. Having personnel walking around machinery increases the chances of catching other machine problems such as oil and product leaks, unusual sounds, piping vibration, and so forth.
2. Only use a highly trained analyst for monitoring critical machines.
3. Consider multiple monitoring methods on critical machines.
4. Before recommending a major repair, consider retaking data with another sensor or at a different location on the machine to confirm the initial reading.
5. When monitoring critical machines with accelerometers, check the effect of the sensor mounting method on your results.
6. Regular test and calibration vibration sensors to ensure they are working properly
7. Watch out for sensors mounting issues and electrical ground loops.
8. Encourage your data collectors and analysts to ask for help if they are unsure of their results.

It makes sense to regularly audit your entire condition monitoring program to determine its overall effectiveness. The audit should first review the condition monitoring performance history to see if any major early failures "slipped through the cracks" and resulted in costly secondary damage. You should try to understand why these early failures were not detected in order to improve your detection performance. It also makes sense to track the percentage of time repairs that were recommended and no apparent machine damage was found. It is important to investigate and understand why unneeded repair events could occur. Additionally, the audit of the condition monitoring program should include a review of sensor calibration records to ensure critical sensors are getting inspected and calibrated as required. Finally, the audit should review the training level of site analysts to ensure they are properly trained.

Part 2
TROUBLESHOOTING

8

Addressing Reciprocating Compressor Piping Vibration Problems: Design Ideas, Field Audit Tips, and Assessment Methods

By Robert X. Perez

Anyone who has worked around reciprocating compressors has probably noticed piping around reciprocating compressors vibrating (Figure 8.1). Due to the presence of gas pulsations generated by reciprocating compressors and the mechanical nature of piping, piping vibration will always be present. However, with regular field inspections around reciprocating compressors, most piping vibration issues can be detected and addressed before they can lead to piping failures.

Process piping performs a vital role in processing and fluid transport facilities by tirelessly transporting gases and liquids between different stages of processes or from point "A" to point "B." However, piping is often forgotten due to its apparent passive nature; it only garners attention if failure occurs due to corrosion or fatigue. Corrosion is an electro-chemical mechanism related to metal and fluid properties, whereas fatigue is a mechanical process related to excessive cyclic motion that results in damaging stress levels. This section addresses the risk of piping failure due to vibration-related fatigue.

All piping associated with machinery vibrates due to the shaking forces that machines generate. Some forces are transferred to the piping via the machine flanges and mounts, whereas others are fluid borne. Fluid borne pressure pulsations generated by machines can result in destructive excitation forces under certain conditions. Here are some field situations that can lead to piping failures:

1. The frequency of any machine excitation, e.g., 1 x rpm and 2 x rpm, matches a piping natural frequency.

Robert X. Perez (ed.) Condition Monitoring, Troubleshooting and Reliability in Rotating Machinery, (203–226) © 2023 Scrivener Publishing LLC

Figure 8.1 Operators must remain vigilant around reciprocating compressors and their associated piping. Poor designs and changing conditions can lead to excessive piping vibration. Piping around reciprocating compressors can vibrate as a result of any number of dynamic excitations such as pressure pulsations traveling within the piping, flow induced turbulence, excessive compressor cylinder or frame vibration, lack of support, etc.

2. The unmagnified pressure pulsations, e.g., 1 x N x rpm and 2 x N x rpm, where N is the number of impeller vanes, number of plungers, etc., generated by a machine, are sufficient to create excessive piping shaking forces.
3. The pressure pulsations generated by a machine excite an acoustic resonance within the piping, which in turn generates large shaking forces at a piping elbow.
4. The combination of an acoustic resonance and piping resonance coincide.
5. The piping system is inadequately supported or restrained to resist normal machine generated excitation forces.

Once large piping shaking forces are present, for whatever reason, piping may be prone to fail if the developed deflections, which can be correlated to stresses, are such that the endurance limit, a limiting stress, of the piping material of construction is exceeded.

Think of pressure pulsations as little hammers that constantly impact the piping at every elbow, tee, reduce, and valve. If fundamental (1x) or harmonics (2x, 3x, etc.) pulsation frequencies acting on piping coincide with any of the piping mechanical natural frequencies (MNF), a resonant condition will occur. Similar to a tuning fork, the piping will resonate when excited at any of the mechanical natural frequencies. High amplifications

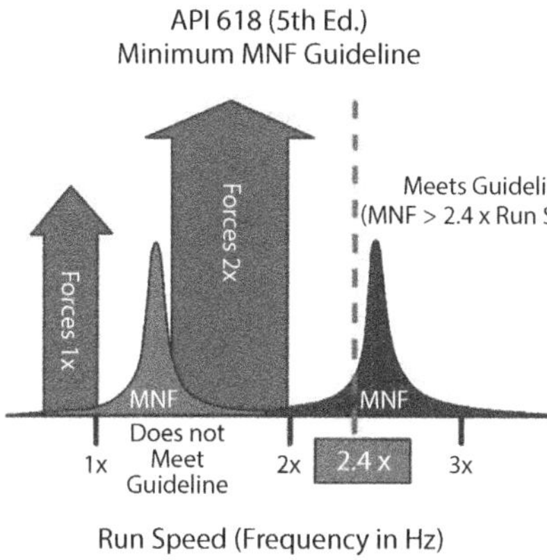

Figure 8.2 If fundamental (1x) or harmonics (2x, 3x, etc.) pulsation frequencies acting on piping coincide with any of the piping mechanical natural frequencies (MNF), a resonant condition will occur. Mechanical resonance conditions usually result in abnormal vibration levels that can result in high piping stress. In extreme cases, piping vibration can lead to metal fatigue and catastrophic failures.

levels of 20 to 40 times are often seen when a resonance condition is present. As a means of avoiding dangerous resonance conditions, API 618 (5th edition) recommends that all MNF in the piping system and structure should be designed so that they are all less than 2.4 times the compressor running speed (see Figure 8.2).

In most cases, piping systems are properly designed, however, there may be times when vibration levels rise to levels where piping fatigue could occur. There are multiple reasons why a carefully designed piping system could become problematic: a) process conditions are different from the analyzed values, b) required orifices were not installed or mis-installed, or c) the field supports are either missing or mis-installed. It is not uncommon that piping modification such as orifices and pipe restraints need to be installed after the original compressor installation to correct piping design issues or installation errors.

Piping Restraints

Pipe restraints (commonly called pipe clamps) are used in piping applications where significant pressure pulsations or shaking forces are likely to be present. These devices are used to control vibratory motion in all directions by tightly gripping the pipe. (Note: I have seen instances where pipe *guides*, which are loosely fitting restraints designed to prevent side to side piping movement while allowing for linear thermal expansion, have been erroneously installed in place of pipe clamps around reciprocating machines.) Figure 8.3a is an illustration of a pipe clamp with liner material, also referred to as gasketing, and Figure 8.3b illustrates how a pipe clamp should be installed. Notice in Figure 8.3b that there is an air gap between the clamp and the soleplate to ensure a tight fit around the pipe.

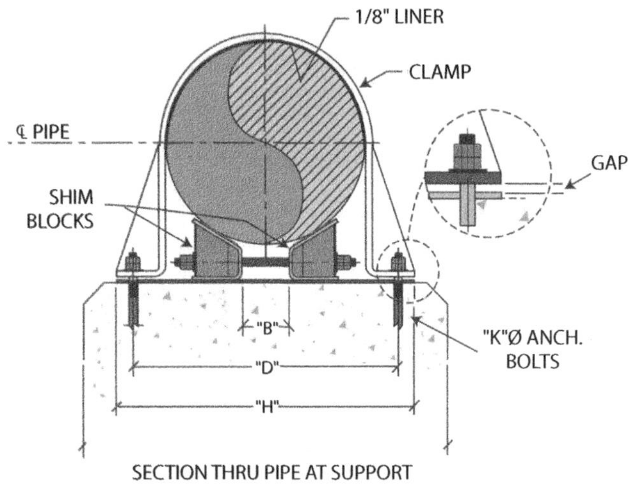

(a)

(b)

Figure 8.3 (a) Pipe clamp with liner material on clamp ID and pipe wedges. (b) Properly installed pipe clamp.

Pipe Clamping Systems

A pipe clamping system, which is made up of the pipe clamp affixed to a concrete pier or some type of rigid structural member, must be rigid enough to resist the forces they are trying to control. Keep in mind that to be effective, all pipe clamps must also be properly installed and maintained in the field. The following guidelines are aimed at insuring pipe clamps are effective for years after their installation.

Guidelines

1. Place piping supports at every elbow and concentrated mass (see Figure 8.4). Locations are usually determined by the firm conducting the dynamic analysis.
2. Always use a pipe clamp approved by the engineering department.
3. When specifying piping clamps, always check with engineering before selecting the belting or liner material. The belting or liner material, which is placed between the pipe and the clamp, ensures a tight fit between the two surfaces.
4. Anchor bolts should have a total L/D ratio (ratio of the bolt length to the bolt diameter) of 24 to 36. 1/3 of the anchor bolt should be free length, or 8 to 10 times the bolt diameter. Use the Table 8.1 to determine the recommended anchor bolt lengths for your application.
5. Anchor bolt material shall be made of either ASTM A193 Grade B7 (preferred) or ASTM A490. (Bolts and nuts should not be galvanized. However, IMF bolt coatings (Dry film lubrication) are permitted. Be sure to correct recommended target torque when using IMF coating.)
6. Anchor bolts should have a sleeve around its free length to protect it during grouting and allow it to stretch freely.
7. Minimum anchor bolt torque values are given in Tables 8.2 and 8.3 at the end of the chapter.
8. Anchor bolts should have a minimum of 6" cover in the pier.
9. Anchor bolts should have a hardened washer under the top nut. At the bottom, anchor bolts should be terminated with a washer and nut (with enough clearance around the washer and bolt to allow grout to flow freely). J-bolts should not be used.

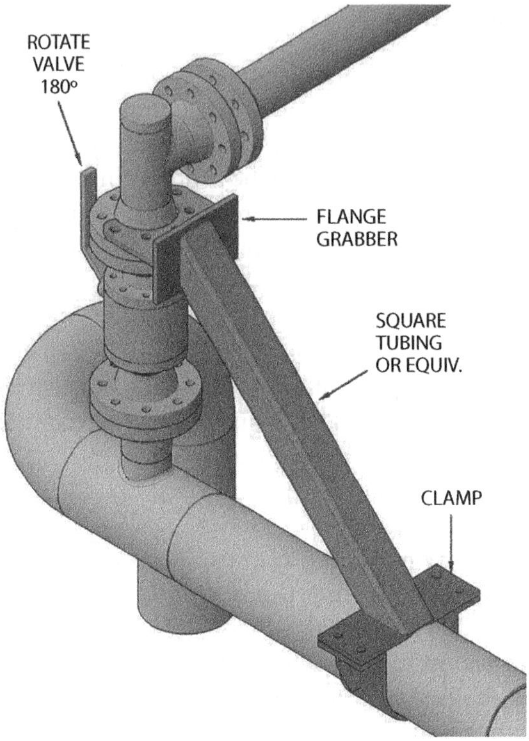

Figure 8.4 This valve represents an unsupported concentrated mass, so is should be supported off a main piping.

Table 8.1 Recommended anchor bolt lengths.

Anchor bolt size (inches)	Anchor bolt size (inches)	Total bolt length (minimum in inches)	Total bolt length (maximum in inches)	Free bolt length (minimum in inches)	Free bolt length (maximum in inches)
1/2"	0.5	12	18	4	6
3/4"	0.75	18	27	6	9
7/8"	0.875	21	31.5	7	10.5
1"	1	24	36	8	12
1-1/4"	1.25	30	45	10	15

Preloading Clamp Bolts

The purpose of pipe clamp bolting is to provide enough restraint rigidity in the presence of dynamic piping forces to minimize piping motion. Clamp bolts are designed to support large tensile forces by intentionally elongating their entire length during the torquing process. As a nut is torqued, its bolt stretches and the components between the bolt and the nut compress. The torquing process continues to increase the so-called clamp load until the proper clamping force is obtained by the bolted system.

The most common reason bolts fail (see Figure 8.5) is due to the combination of a loss preload and high dynamic forces applied to the bolt. The loss of preload eventually allows external applied forces to overcome the clamping force holding the bolted system from moving. Loss of bolt preload can be caused by either settlement, creep, relaxation, or loosening of the bolt due to vibration. In practice, designers try to maximize the level of preload by specifying a torque value that will result in a bolt stress equal to 50% to 70% of the bolting material's yield stress.

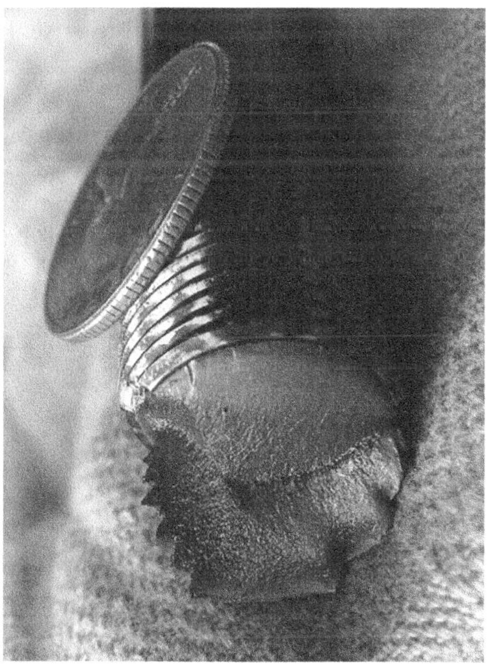

Figure 8.5 The failure surface on this bolt is indicative of fatigue due to inadequate bolt tensioning.

In general, there are no standard criteria for preloading anchor bolts. The required preload level generally depends on the specific application of the anchor bolt. In general, preloaded bolts will require a sleeve or bond breaker to permit elongation to occur along the entire length of the bolt. Without a properly designed sleeve, concrete will bond to the shank of the bolt during construction and preclude its stretch.

Anchor bolts used for equipment supports should be preloaded to the equipment manufacturer's recommendations when specified. This is especially true for bolts anchoring rotating equipment and its associated piping. When bolts are anchoring equipment or are subject to possible loosening due to high dynamic loads, a locking device should be provided. Acceptable locking devices include double nuts or jam nuts and tack welds (for weldable materials only).

Piping Assessment Steps

Most plant personnel are not involved in the piping design process. They tend to get involved after the installation and start-up of the compressor and piping. This section is meant to assist those drawn into piping vibration problems that are encountered in the field after the compressor is under full load conditions and problems begin to arise. Here is a general procedure for evaluating vibrating field piping.

First, perform the following pre-field analysis steps:

1. Ensure an acoustic and mechanical analysis has been conducted. If an acoustic and mechanical analysis has not been performed, you may want to initiate one.
2. Ensure recommended orifices and supports have been installed. Review all the recommendation drawings to understand what was recommended.
3. Ensure the operating conditions are correct:
    a. Speeds
    b. Gas analysis

Next:

1. Talk to anyone who works close to the problem areas. Ask if 1) this is a new installation, 2) if there have been any recent modifications to the piping system, and 3) if the compressor operating conditions are normal or not.

2. Identify the problem points. If there are only a few problem locations, the field analysis should be quick and easy. However, if there are numerous points of concern, you will need to record numerous points simultaneously in order to understand the phase relationships between points.
3. Measure or have a qualified professional measure the vibration levels to determine if they are acceptable or not. (Refer to the "Collecting and Assessing Piping Vibration" section in the back of this chapter for more information about assessing piping vibration levels.)
4. If the vibration levels are excessive, identify any abnormal conditions, i.e., compressor speed, pressures, gas density, etc.
5. There are typically several possible outcomes:
    a. Change operating conditions if possible
    b. Add bracing or supports
    c. Bring in experts to evaluate the situation and redesign if necessary

Small-Bore Piping

Be on the lookout for unsupported small-bore piping branches, such as those seen in Figure 8.6. (Small-bore piping sizes are defined in Figure 8.7.) Small, cantilevered piping branches and PSV branches tend to be less supported than major process piping and are often times found to be completely unsupported. These piping arrangements must be identified and braced to preclude the possibility of catastrophic piping failures and their associated product releases. Proper support can be provided by either

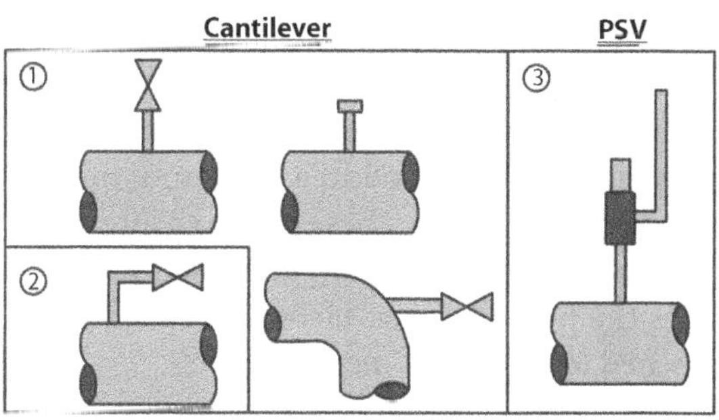

Figure 8.6 Some typical small-bore piping designs that can result in excessive vibration if not properly braced.

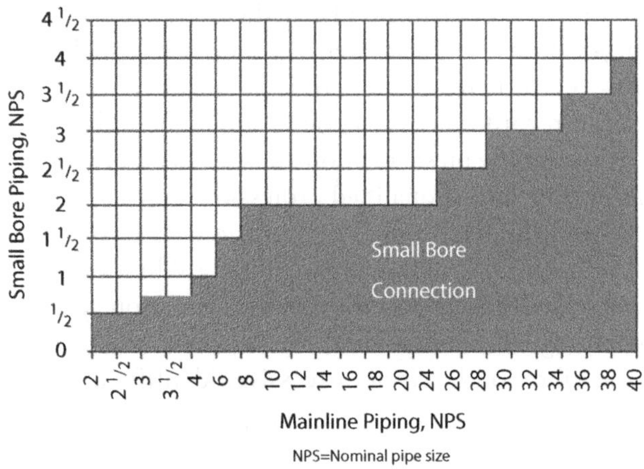

Figure 8.7 This chart defines when a piping branch can be considered small-bore piping. For example, a ½" coming off a 6" line is considered small-bore piping, while a 1" line coming off a 3" mainline is not considered small-bore piping.

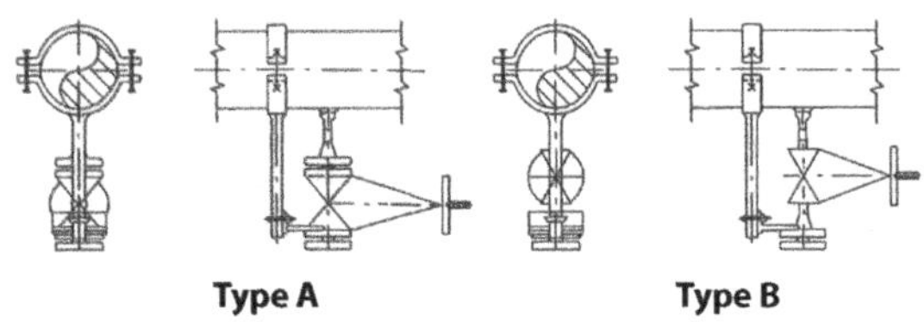

Figure 8.8 Ways that flanged and welded cantilevered valves can be supported off the main line.

bracing them off the main piping (see Figure 8.8), gusseting them, or eliminating them all together.

Attaching Pipe Clamps to Structural Members

There will be times that due to the elevation of a pipe or space limitations, a pipe clamp may need to be affixed to a structural member instead of to a concrete pier (see Figure 8.9). In these cases, ensure that the design provides adequate access to the bottom nuts and that there is a free bolt length of at least 8 times the bolt diameter. Here are few more notes concerning pipe clamps attached to structural members:

- The structural member that the clamp is attached to must be more than twice as rigid as the pipe it is restraining. If it

Figure 8.9 Pipe clamp with integral spacers using cap screws.

is not, the clamp will be totally ineffective at controlling piping vibration.

- Use bolt spacers to ensure the stretch length of a bolt is at least 8 times the bolt diameter. A spacer is an extra strong tube or pipe that fits around a clamp bolt to increase its effective stretch length.
- Cap screws are preferable.
- Use approved torque values to properly tighten the clamp bolts.

The Ideal Pipe Clamp Installation

There are several common deficiencies found when inspecting pipe clamp installations in the field. After their installation, a qualified piping design inspector should be employed to ensure the integrity of all critical piping restraints. To ensure a pipe clamp is effective, a field inspector must confirm the following after the installation of the restraint (See A-B-C-D annotations in Figure 8.10):

A. There must be full contact between the pipe and the clamp. Use 1/8" belting material between the clamp and pipe to ensure good contact is achieved.
B. There must be full contact between the bottom of the pipe and the clamp wedges or sole plate. Note: The clamp shown

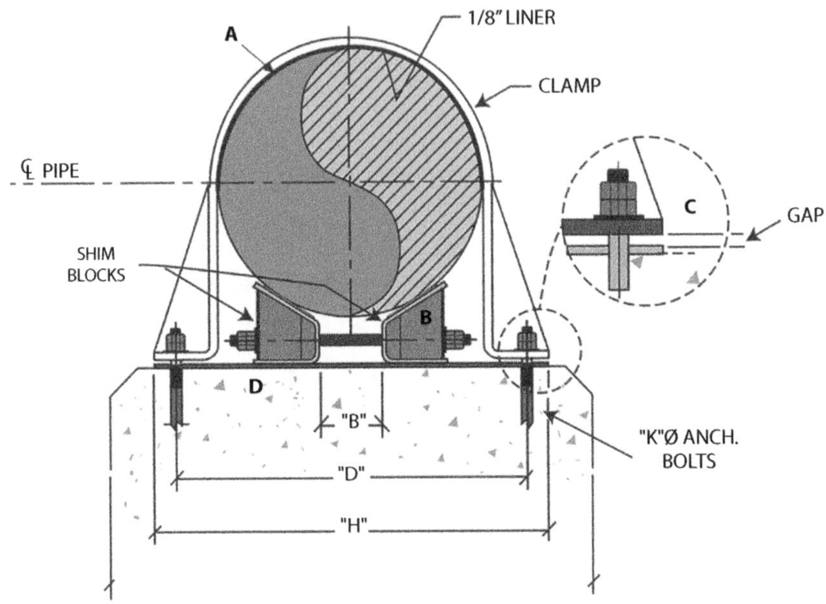

Figure 8.10 Keys to ideal pipe clamp installation.

in the figure uses wedges ("B") to prevent the pipe from vibrating. Some pipe clamps simply use a flat sole plate to support the pipe.

C. There must be a ½" (or more) gap between the clamp feet and sole plate to allow for bolt stretch.
D. The sole plate is fully grouted. Ensure that grout shims do not have sharp edges.
E. All clamp bolts are tightened as specified.

Installation Examples

Figures 8.11 through 8.14 show some good and bad examples of pipe clamp installations that have been found in the field.

Here are a few more pipe clamp tips:

1. The surface of the clamp should be parallel with the surface of the concrete pier.
2. There should be a washer under every pipe clamp nut.

Figure 8.11 A properly installed pipe clamp bolted to a concrete pier. Notice that there is a gap between the pipe clamp and the pedestal support to ensure there is adequate clamping force present.

Figure 8.12 Correctly installed pipe clamps on concrete footing.

Figure 8.13 An incorrectly installed pipe clamp. Notice that with the clamp bottomed out and there is an air gap between the pipe and the top of clamp, i.e., there is no clamping force on the pipe.

Figure 8.14 There is insufficient clamping force on this clamp due to the loss of wedge travel. (Notice that the wedges are touching.)

3. Foundations should have a minimum depth below grade and below the frost depth.
4. Concrete should have a minimum specified compressive strength.
5. Foundations should be constructed with footings.

6. If adjustable wedges are used, ensure they fit properly around the pipe. Below is an example of a pipe clamp with poorly fitted wedges.
7. The pipe must be in full contact with the sole plate liner.
8. All pipe clamps should be inspected periodically to ensure they are still effective at controlling vibration. Specifically, check that there are no broken hold-down bolts and that the clamp gasket materials are still intact and have not degraded.
9. Seating Pipe Clamps: New pipe clamps should always be checked for proper fit around the pipe. The goal is to get 50% or more contact between the gasket material in the ID of the clamp and the pipe. If you are not sure of the fit, you can use a feeler gauge to ensure you have the proper contact. You shouldn't be able to get a feeler gauge between the clamp and the pipe if the fit is correct. If you are able to get a feeler gauge between the clamp liner and the pipe, you probably have poor contact. You can use a rubber mallet on top of the clamp to seat it against the pipe. Continue to impact the pipe clamp with the rubber mallet until you are sure it is fully seated. Once the clamp has been fully seated-in, re-check all clamp bolt torques, and re-torque as required to achieve the recommended values.

Collecting and Assessing Piping Vibration

The purpose of this brief introduction to piping vibration analysis is to provide readers with a flavor for what is involved in the data collection process and how piping vibration data is evaluated.

If large piping shaking forces are present, for whatever reason, piping may be prone to fail if the experienced deflections are such that the endurance limit of the piping material of construction is exceeded. The evaluation guidelines we recommend are based on a proven methodology that has been used by vibration professionals for decades. The assessment method defines marginal, correction, and danger piping vibration levels based on the amplitudes of the spectral frequency components detected (see Figure 8.15). If vibration levels are kept below the marginal vibration levels, critical stress levels should never be exceeded.

Piping vibration is normally acquired with a magnetically mounted accelerometer (see Figures 8.16 and 8.17 below), which generates an electrical signal proportional to the piping motion. The raw signal from the accelerometer is sent to a spectral analyzer where it is broken into fundamental

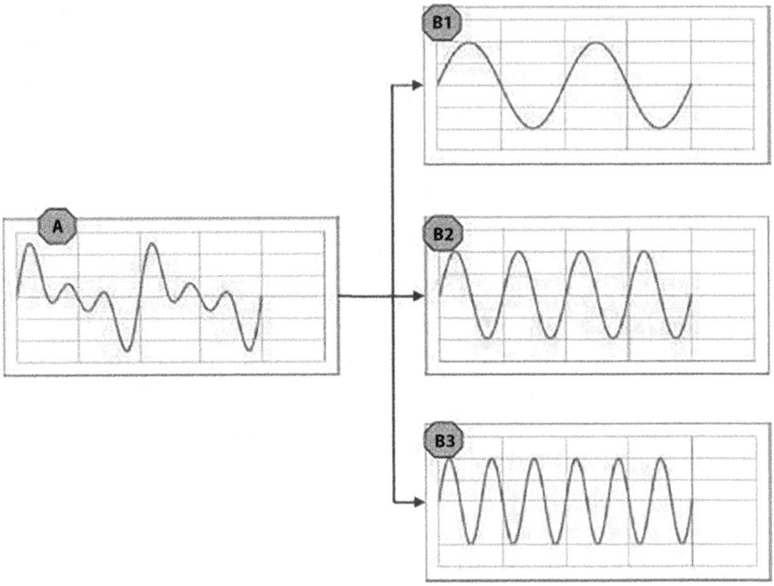

Figure 8.15 A vibration spectrum is a mathematical process where a time waveform is converted to a frequency domain plot. A spectral plot lists all the frequencies contained in the time waveform being analyzed.

frequency components (see Figure 8.18), similar to the way a prism breaks down a light beam into its fundamental components. The largest frequency component in the vibration spectrum should be compared with allowable

Figure 8.16 Mag-mounted tri-axial accelerometer on a pipe.

Figure 8.17 Mag-mounted tri-axial accelerometer on a valve.

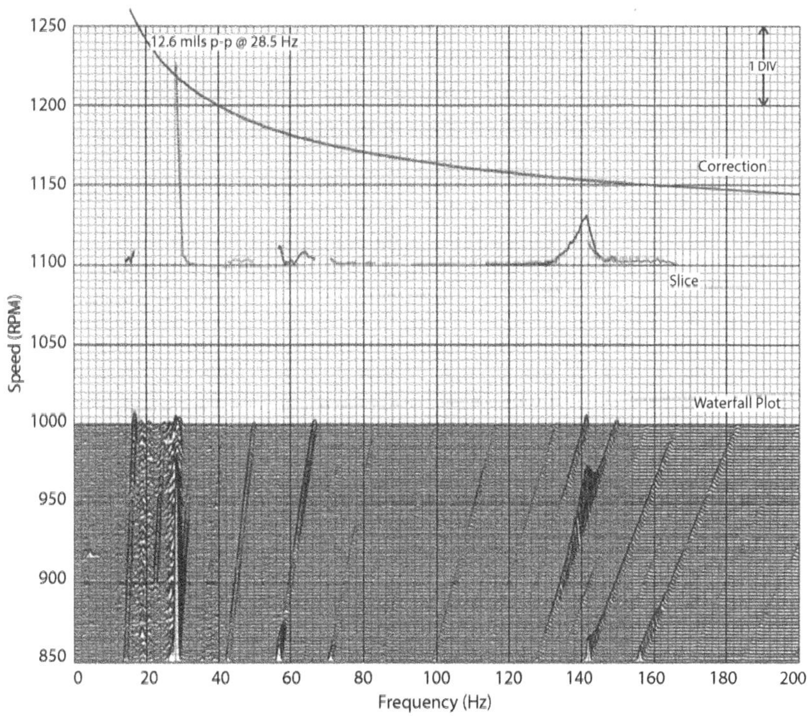

Figure 8.18 Waterfall display of piping vibration.

piping vibration guidelines (see Figure 8.19). For example, you can see by inspection in Figure 8.18 that the plot of the 2x vibration component (labeled as "12.6 mil p-p @ 28.5 Hz" in Figure 8.18) is by far the largest vibration component in these spectra. The 2x component peaks at 28.5 Hz (1710 cpm) at a level of 12.6 mils, probably due to some type of mechanical resonance. The peak amplitude of the 2x component was found to be above the "correction action required" vibration level.

Piping Analysis Steps

Before beginning any evaluation effort, confirm that the field analysts have experience performing field piping assessments. Once approval is obtained for them to enter the processing area, the field analyst team typically follows these steps to complete their evaluation:

1. They perform a field piping survey to identify vibration problem locations.
2. They gather data at the points of interest using magnetically mounted accelerometers.
3. They generate and inspect spectra for all the vibration points to identify any significant spectral peaks.
4. They evaluate all the largest spectral peaks using the "Piping Assessment Chart" (see Figure 8.19).

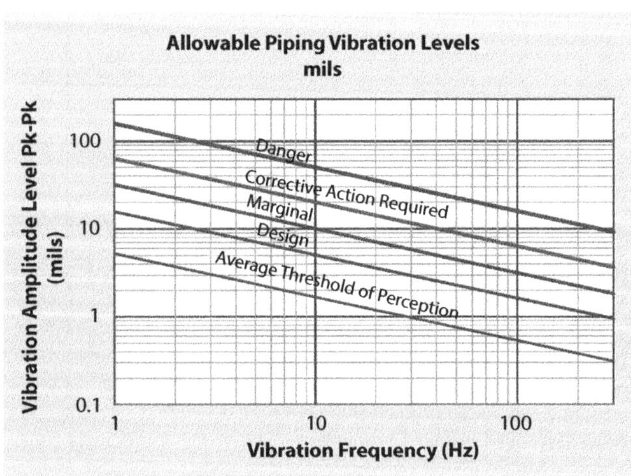

Figure 8.19 Piping assessment chart.

5. They report any of the piping vibration frequency components that fall above the "Corrective Action Required" line to operations.
6. After carefully reviewing all their results, they will provide a formal report with a summary of what was found along with recommendations based on good engineering practices.

Piping Vibration Examples

To better understand the piping assessment process, we will go through a few examples. In each case, we will refer to Figure 8.19 to determine if the piping vibration levels are acceptable or not.

Example 1:
A compressor discharge 8" pipe is vibrating 20 mils pk-pk at 2000 cpm. Since 2000 cpm is 33.33 Hz, by inspection we see this vibration amplitude falls between the "correction" and "danger" levels. Based on these findings, there should be a complete inspection of the piping and supports to see if they were properly designed and installed.

Example 2:
A 6" process line is moving 10 mils pk-pk at 600 cpm. Since 600 cpm is 10 Hz, by inspection we see this vibration amplitude falls at the lower boundary of the "marginal level. Since the vibration amplitude is at the marginal level, the line should be inspected periodically to see if levels are steady or deteriorating. If vibration amplitude levels are found to be increasing, then the root cause of the escalation should be investigated.

Example 3:
An accelerometer has been placed on a 12" compressor discharge line in the horizontal and vertical direction. The compressor runs at 900 rpm. The following spectral vibration data was acquired:

Horizontal Data

Frequency (Hz)	Amplitude (in mils pk-pk)	Assessment based on EDI criteria above
15	9	Marginal
30	40	Greater than danger
45	5	Marginal
60	2	Design

Vertical Data

Frequency (Hz)	Amplitude (in mils pk-pk)	Assessment based on EDI criteria above
15	3	Less than design
30	8	Greater than design
45	2.3	Design
60	1	Less than design

After studying the assessment results in the far-right column, we conclude that the 30 Hz component in the horizontal direction is the only spectral component of immediate concern. Additional testing and data acquisition will be required to determine if this high vibration is due to excessive compressor pulsations, lack of piping constraints, piping resonance, or some other phenomenon.

Bolt Torque Tables

Table 8.2 Recommended target torque values for low-alloy steel bolting.

The target torque values table is taken from Table 1 in ASME PCC-1-2000.		
Nominal bolt size, in.	Target torque (ft-lb)	
	Noncoated bolts [Note (1)]	Coated bolts [Notes (1), (2), (3)]
1/2	60	45
5/8	120	90
3/4	210	160
7/8	350	250
1	500	400
1 1/8	750	550
1 1/4	1,050	800
1 3/8	1,400	1,050
1 1/2	1,800	1,400
1 5/8	2,350	1,800
1 3/4	2,950	2,300
1 7/8	3,650	2,800
2	4,500	3,400

GENERAL NOTE: The values shown are based on a Target Prestress of 50 ksi (root area). This prestress level is generally considered suitable for joint systems designed using SA-193-B7 low-alloy steel bolts, except for joint systems using ring-joint gaskets. Target Torques for different Target Prestress levels may be obtained by reducing (or increasing) the values in Table 8.1 bv the ratio:

$$\frac{Target\ Prestress(ksi)}{50(ksi)}$$

Table 8.3 Torque values for lubricated and IMF coated threads for: ASTM A-193 grade B7 studs.

Diameter	Thread/in	Sae 10 oil (ft.lbs)	Anti-seize (ft. lbs)	Imf (ft.lbs)
¼"	COARSE 20	8	6	5
5/16"	COARSE 18	16	13	9
3/8"	COARSE 16	28	23	17
7/16"	COARSE 14	46	37	27
1/2"	COARSE 13	70	57	41
9/16"	COARSE 12	100	82	59
5/8"	COARSE 11	138	113	82
¾"	COARSE 10	246	201	145
7/8"	COARSE 9	396	324	234
1"	COARSE 8	594	486	351
1 1/8"	UN8 8	872	713	515
1 ¼"	UN8 8	1,225	1,002	724
1 1/3"	UN8 8	1,663	1,360	983
1 ½"	UN8 8	2,194	1,795	1,296
1 5/8"	UN8 8	2,827	2,313	1,671
1 ¾"	UN8 8	3,572	2,923	2,111
1 7/8"	UN8 8	4,438	3,631	2,622
2"	UN8 8	4,915	4,022	2,904

Note: IMF stands for Industrial Metal Finishing, with a dry lubricant coating.

Chapter Glossary

Spectrum Analyzer: An instrument that analyzes a dynamic input signal by breaking down its content in a tabular or graphic display of amplitude versus frequency within a selected frequency range.
Wedge (Pipe Clamp): One of a set of two triangularly shaped pieces used to support a pipe and prevent it from moving side to side.

Sole Plate: A flat, steel piece used to support a pipe or machine foot that is grouted into the top of a concrete support.
Clamping Load: The force provided by a torqued nut and bolt to hold a bolted connection together.
Torquing: The process of twisting a nut onto a threaded bolt until the required bolt stretch and clamping load is achieved.
Stretch (Bolt): The elongation in a bolt after it is fully torqued, typically measured in mils or thousands of an inch. The stretch in a bolt is given by the formula: $\delta = FL/EA$, where d is the elongation (or stretch) in the bolt, F is the load on the bolt, A is the cross-sectional area of the bolt, L is the length of the bolt, and E is Young's modulus of the bolt's material of construction. For example, a 6 inch long, 0.5" diameter steel bolt with a 3000 pound longitudinal preload will have a stretch of 3000 lbs x 6"/(30,000,000 lbs/in^2 x (3.1415 x (0.5/2)^2 in^2), which equates to 0.003 inches of bolt stretch.

The initial bolt stretch must be sufficient to provide the required clamping load with an additional safety margin to account for any relaxation that might occur over time within the bolting system. Note: Washers and spacers can be used to increase the effective length of a bolt in order to increase the stretch developed under load.

9

Remember to Check the Rotational Speed When Encountering Process Machinery Flow Problems

By Robert X. Perez

Rotating machinery field problems are often associated with operating speed issues. Operating at the wrong speed can result in vibration problems, pressure and flow issues, and overloading. However, the most common speed related problems that I have encountered in the field are low flow issues due to operation below the design speed.

We know that the flows generated by most types of rotating machinery are a strong function of speed. For example, the performance of centrifugal, i.e., dynamic, pumps and compressors (Figure 9.1), tend to follow the affinity laws, which are a set of mathematical relationships that predict how capacity, head, and horsepower are affected by changes in impeller diameter (D) and shaft speed (N) (see a summary in Table 9.1). We can see from studying the table that, in centrifugal machines, flow (Q) is directly proportional to speed and that head (H) is proportional to speed squared. Therefore, if operating speed falls below the design speed, then the flow will also drop below the design flow accordingly, assuming there isn't some other system issue.

In positive displacement compressors, such as reciprocating and screw compressors, gas flow is directly proportional to the rotational speed. This is because positive displacement compressor's displacement rate, i.e., volume per unit time, is proportional to its rotational speed. Therefore, with all things being equal, the higher the input shaft speed, the higher the flowrate and conversely the lower the input shaft speed, the lower the net flowrate.

When attempting to troubleshoot process machinery field flow problems, it's crucial to understand the significance of operating speed on compressor and pump performance. Along with an accurate operating speed, we need to know the design speed and the historical range of operating

Robert X. Perez (ed.) Condition Monitoring, Troubleshooting and Reliability in Rotating Machinery, (227–232) © 2023 Scrivener Publishing LLC

Figure 9.1 Multistage centrifugal gas compressor installed in processing facility.

Table 9.1 Affinity laws for centrifugal compressors, centrifugal pumps, and fans.

Diameter change only	Speed change only	Diameter and speed change
$Q_2 = Q_1\left(\frac{D_2}{D_1}\right)$	$Q_2 = Q_1\left(\frac{N_2}{N_1}\right)$	$Q_2 = Q_1\left(\frac{D_2}{D_1}\times\frac{N_2}{N_1}\right)$
$H_2 = H_1\left(\frac{D_2}{D_1}\right)^2$	$H_2 = H_1\left(\frac{N_2}{N_1}\right)^2$	$H_2 = H_1\left(\frac{D_2}{D_1}\times\frac{N_2}{N_1}\right)^2$
$BHP_2 = BHP_1\left(\frac{D_2}{D_1}\right)^3$	$BHP_2 = BHP_1\left(\frac{N_2}{N_1}\right)^3$	$BHP_2 = BHP_1\left(\frac{D_2}{D_1}\times\frac{N_2}{N_1}\right)^3$

Note: Q = flow, D = diameter, N = speed and BHP = brake horsepower. Subscript "1" denotes initial conditions and subscript "2" denotes new conditions.

speeds and flows. A comparison of this data will help determine if anything has changed and if it has, by how much.

However, there are other common field issues other than a low operating speed that can also result in flow problems. These problems include:

- An upstream line restriction, such as a plugged suction strainer, can lead to a loss of pressure and flow. Use suction pressure gauges or pressure transmitters to check the suction pressure and compare the current value to historical readings. If a recent drop in the suction pressure is found, investigate the possibility of a plugged strainer in the suction line.

- A downstream line restriction, such as a plugged line or partially closed valve, will result in a loss of flow. Use discharge pressure gauges or pressure transmitters to check the discharge pressure and compare the current value to historical readings. If a sudden rise in the discharge pressure is discovered, isolate the location of the pluggage and resolve the problem.
- An open bypass line around a pump or compressors will reduce the net forward flow. Check all automatic and manual bypasses and make sure they are all closed if a loss of flow is detected.
- Centrifugal compressors:
    - If the gas being compressed is too light, i.e., has lower than normal molecular weight, a lower than normal compressor pressure differential and gas flow will result. I have experienced a situation where a centrifugal compressor was only able to move a fraction of its rated flowrate due to a change in gas properties. It turned out that the start-up gas was pure hydrogen instead of a mixture of hydrogen with other heavier gases. Once the start-up gas was spiked with natural gas, the compressor flow returned to a more normal flow range.
    - Internal wear in centrifugal compressors can result in internal leakage and a corresponding drop in the net forward flow. For example, if the inter-stage and balance piston seals in centrifugal compressors are gradually wearing, we would see a gradual drop in the overall efficiency and drop in the net forward flow. The extent of wear can by determined by calculating the adiabatic efficiency and comparing it to what is expected.
- Reciprocating compressors:
    - If a cylinder is unintentionally unloaded due to an unloader or unloader control issue, the subsequent reduction in the compressor's volumetric efficiency will result in less gas flow. Ensure that valve and pocket unloaders are functioning properly if flow losses are experienced.

- o In reciprocating compressors, piston rings gradually wear, and cylinder valves can fail or leak due to fouling. Conducting a PV analysis under field conditions is a good way to determine what internal issues are present.
- Screw Compressors:
  a) Rotors and casing gradually wear. Review flow, pressure, and driver power data trends to see if changes have occurred over time.
  b) Flow losses can sometimes be experienced if the unloading slide is leaking or partially open.

We examined how non-speed issues can lead to low flow problems. However, a speed issue can result in a low flow condition. Because we know speed has a major effect on process machinery performance, it makes sense to first check the machine speed and compare it to the design speed. Speed is an easy machine parameter to check by either taking a reading from an installed tachometer or by using a portable tachometer.

If you find your machine is running below the design speed, I recommend that you first consider the following possible causes:

- The speed controller was manually set at the wrong speed
- The speed controller is out of calibration
- For steam turbine drivers:
  a) There is a plugged steam supply line
  b) The governor or throttle valve is sticking
- For gas turbine drivers:
  a) The fuel gas line is partially plugged and restricting the gas flow
  b) The fuel gas valve is sticking
- The driver is worn out and can no longer attain the rated speed
- In VFD applications, the input signal value to output frequency value may be improperly configured. In most cases, you want a 100% input signal to output a 60 Hz (or 50 Hz outside the US) frequency supply. I experienced a problem where a VFD controller was set up so that 100% speed requirement for a pump resulted in a 58 Hz output to the electric motor driver. The result was the motor that could

only run at 96.7% of rated speed when it received a signal to run at 100% speed. The problem was identified by determining the 1X component of the pump vibration spectrum was found to be running below the expected frequency.

- The driver is overloaded. An overloaded driver can sometimes prevent the fluid mover from reaching its rated speed. Check the process conditions to see if anything has changed. Normally, a higher flow or a higher differential pressure across a process machine will result in a higher horsepower requirement.

Note: This is only a partial list of situations that results in a loss of operating speed. There are too many possibilities to list all of them here.

If a machine is found to be operating below the manufacturer's design speed, you need to find out why. In most cases, speed issues tend to be related to a control problem. Once you return your machine to the correct speed, your flow should return to normal.

In summary, we can conclude that there are a number of possibilities why your process machinery's net forward flow might be low:

a. There is a blockage in the piping system, which is throttling the flow
b. A compressor or pump has internal wear which is affecting its performance
c. There is a control problem limiting the driver's speed
d. A driver deficiency is limiting the available horsepower
e. Process conditions have changed, causing the operating point to move. For example:
    - The gas being compressed is lighter than the compressor was designed to compress (centrifugal compressors).
    - The liquid properties have changed resulting in a change in pump performance (centrifugal pumps).
f. Process conditions have overloaded the driver and limit the steady state speed of the train from attaining the design speed.

As we have seen, there are many possible causes of low flow conditions in process machinery. So, take your time gathering and then examining the field data before determining the nature of the problem and then deciding what to do next.

Regardless of the root cause of the flow problem, checking your operating speed is the easiest place to start your field investigation. If you discover the operating speed is correct, then you can look elsewhere for the problem. If the speed is incorrect, then you can continue your investigation by investigating the non-speed related issues.

10

Troubleshooters Need to be Well Versed in the Equipment They are Evaluating

By Robert X. Perez

Let me begin by stating that there is no universally accepted approach to rotating machinery troubleshooting. Every machinery professional develops their own approach of solving field problems through years of study and practice. An experienced troubleshooter knows they must take a broad view of the machine and its subsystems if they wish to uncover the true nature of the problem. Typically, a process machine (see Figure 10.1) is merely part of a larger system this is carefully designed to meet specific process requirements. Therefore, to understand how a machine is performing, we must understand the entire process system. All system components (Figure 10.2) need to work together and function reliably to meet the production goals.

It's possible that there may be factors that give the illusion that the problem is mechanical in nature, when in fact it's not. Here is a list of common causes of machinery problems that I have encountered:

1. Machine defects, which are internal machine defects, must be corrected to restore the machine back to original working order. Examples of machine defects are failing bearings, loss of internal clearances, internal plugging, gear failure, cracked shaft, loose components, etc.
2. System issues, which are caused by external defects or temporary conditions that must be corrected to return to normal operation. Examples of system issues are line plugging, a closed valve that is normally open, open valve that is normally closed, cooling issues, etc.
3. Improper machine operation, which means either a) the intentional or unintentional operation of a machine at flows or pressures away from their design values or b) running

Robert X. Perez (ed.) Condition Monitoring, Troubleshooting and Reliability in Rotating Machinery, (233–246) © 2023 Scrivener Publishing LLC

Figure 10.1 Troubleshooters must understand the operating characters of all the elements making up the machinery train of interest. Here, a reciprocating engine, on the left, drives a centrifugal pump, on the far right, through a gearbox. Each element poses different analysis challenges.

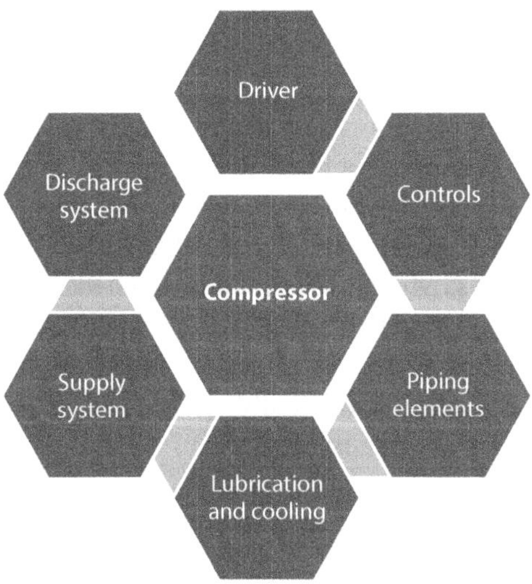

Figure 10.2 A process machine's performance cannot be understood without understanding how it fits into the process it is serving. For example, if any of the key elements of a compressor fails or is malfunctioning the entire system suffers.

with a fluid that is significantly different than the fluid expected by the designer. For example, operating a centrifugal pump below its minimum stable flow is a common cause of pump vibration that can be reversed by increasing the pump flow closer to the design flow. It is important to correct operating issues before they result in permanent damage to the equipment.
4. Design issues which are related to how the machine or system was originally designed. Examples of design issues are undersized or oversized piping, inadequate supports, wrong bearing for the application, wrong seal design for the application, undersized coolers, or misapplication of the machine for the actual process conditions.
5. Instrumentation issues: Many field problems characterized as machine issues are really instrumentation problems. Examples of instrumentation issues include miscalibrations that lead to misleading indications and alarms, instrument failures that lead to false readings, alarms, or trips, PLC programming issues, etc.
6. Assembly issues: Assembly issues are common with machine that have been in service and then repaired. Some examples of assembly issues are internal looseness due to poor fits, wrong bearing clearances, using out of special components, poor alignments, etc. A lack of training or documentation often leads to assembly errors.

Notice that a number of these possible factors listed here are unrelated to the actual mechanical condition of a machine. Since most facilities don't want to waste time and money by performing unnecessary repairs, troubleshooters try to rule out non-machine issues first. For example, electric motor trips could be caused by failing bearings or an incorrectly calibrated pump flow control system. If the real issue is a flow transmitter that's out of calibration, then repairing the motor or the pump driver will not prevent future trips.

To determine the root cause of a problem, troubleshooters must break down the entire system and analyze it methodically. They use field data, observations, and historical records to determine if it's a machine problem or system problem. A drill down approach of analysis is a common method of attacking complex systems, where the troubleshooter gathers the available data and begins ruling out what is not a problem and by the process of elimination, eventually arrives at the most likely source of the problem.

What is the Difference Between Troubleshooting and Conducting a Failure Analysis?

Troubleshooting is the exercise of evaluating an operating system that is exhibiting symptoms of distress or is not functioning as expected. Since the system is in service, field observations can be made and data can be evaluated to better understand what is happening. It is possible that minor adjustments or corrections can be made without having to perform major repairs or replace any components. Another outcome of a troubleshooting exercise may be that it is determined that no real problem exists and nothing needs to be done.

Root cause failure analyses are performed only after one or more machine components have failed, requiring the machine to be removed from service and restored to working order. The hope is that by evaluating the failed components and reviewing the machine's history, the root cause of the failure can be determined and corrected to avoid future failures. In some cases, a root cause failure analysis may uncover a latent root cause, which is an underlying organization factor, management issue, procedural issue, etc. that cannot be seen by inspecting damaged components. An example of a latent root cause is an operator using an incorrect procedure to start up a compressor due to a lack of training, causing a premature failure.

Therefore, troubleshooting involves evaluating a machine that's still in service and is thought to be malfunctioning, while a root cause failure analysis is performed after a failure has occurred and the machine has been removed from service and disassembled for inspection. In some cases, a root cause failure analysis can uncover hidden, i.e., latent, issues. Effective troubleshooting helps control maintenance costs by avoiding unnecessary work and, if done in a timely manner, reduces the number of major repairs required by allowing time to make field and machine adjustments before secondary damage can occur.

Typically, the troubleshooting process begins with a phone call (or email) where someone attempts to explain some type of field problem. The person reporting the problem will do their best to define the problem. A request for help may go something like this: *The south makeup gas compressor was working fine, but last night the flow fell off about 20%.* After reading or listening to the request, you are usually left with many questions. However, it is up to you to take it from there and start digging into the problem.

After years of troubleshooting rotating machinery, I have found that to be successful, a user needs to fully understand the machinery they are dealing with before attempting to analyze a field problem. Understanding a specific machine means knowing 1) machine design details, 2) performance characteristics, 3) the design basis, 4) system design details, 5) OEM recommendations, and 6) it's operating history. This body of knowledge will improve your chances of determining the source of the problem and gives you a real chance at correcting the true root cause. Here, I will briefly review the required body of knowledge that I have employed during my career to solve complex field problems.

Equipment Details

To effectively troubleshoot, it is imperative that the analyst know the design details of the process machine they are analyzing. Two quick examples: You need to know if you are dealing with a centrifugal or positive displacement machine or if it's a constant speed or variable speed machine. You cannot analyze what you don't understand.

Here is a list of design details that should be known:

1. What is the number of stages? (See Figure 10.3)
2. How are the reciprocating compressor cylinders staged?

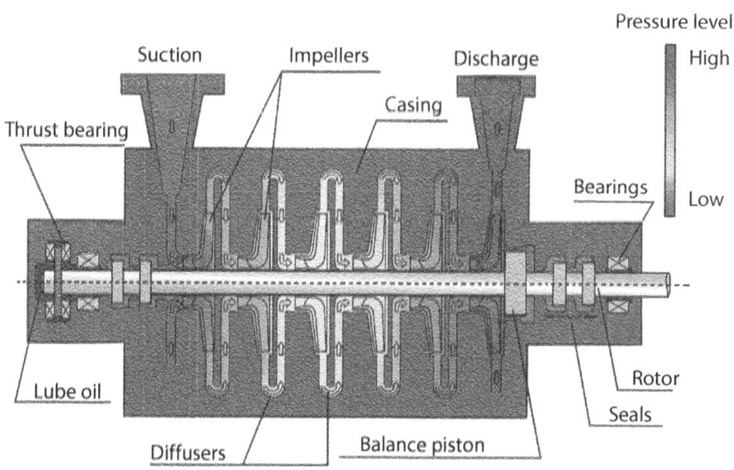

Figure 10.3 Knowing process machine construction details provides insight into performance and possible failure modes. This image of a centrifugal compressor shows the number of stages, the internal flow path of the gas, the location of the balance piston, etc.

3. Is the machine variable speed or constant speed?
4. How is the rotor constructed (centrifugal machines)?
5. How is the gear box constructed?
6. What are the horsepower and speed rating of the driver?
7. What types of bearings are used, i.e., fluid film, REB, bushings?
8. Seal details of the end seals and inter-stage seals (See Figure 10.3)
9. How many vanes do the impellers have?
10. Are the diffusers vaned or vaneless?
11. How is the axial balancing achieved? Balance holes, balance piston, balance drum? (See Figure 10.3)
12. How is the lube system configured?
13. How is the cooling system configured?
14. What field data is collected and stored in the process computer?

Performance Characteristics

Understanding a machine's performance characteristics helps you understand the cause and effect relationships that provide clues to what is going on. For example, we know that in a centrifugal pump's discharge, pressure drops as the flow increases and increases as flow decreases. We also know a centrifugal pumps' differential pressure are a function of the speed squared ($Q=RPM^2$). These simple relationships are helpful when interpreting centrifugal pump field data. If you find that a centrifugal pump's flow is low, we should probably first check its speed and differential pressure to see if the data agrees with the manufacturer's test date.

It is not possible to provide a comprehensive list of performance rules for all classes of machines, so I will provide some additional rules-of-thumb for common process machines:

Centrifugal Compressors

- As the flow decreases, the differential pressure across the compressor increases. A drop in flow with a corresponding rise in differential pressure could indicate an unwanted restriction somewhere in the piping. Conversely, as the flow increases the differential pressure across the compressor decreases.

- As the speed increases, the differential pressure across the compressor increases, which in turn raises the flow. Conversely, as the speed decreases the differential pressure across the compressor decreases, which in turn raises the flow. If there is a flow issue on a variable-speed machine, start by checking the equipment's rotational speed to ensure it is correct. A driver problem, such as a sticking speed governor on a steam turbine or a calibration error on a variable frequency motor drive could cause a machine train to run at the wrong rotational speed.
- As the gas density increases a) the differential pressure across the compressor increases and b) the horsepower required to drive the compressor increases. Conversely, as the gas density decreases a) the differential pressure across the compressor decreases and b) the horsepower required to drive the compressor decreases.

Reciprocating Compressors

- As the compressor speed increases, the net flow increases and conversely, as the compressor speed decreases the net flow decreases.
- As the cylinder suction pressure decreases the mass flow through the cylinder decreases and conversely, as the cylinder suction pressure increases the mass flow through the cylinder increases.
- As the compression ratio across a stage increases, the temperature rise across the stage increases (see Figure 10.4). Conversely, as the compression ratio across a stage decreases, the temperature rise across the stage decreases. When experiencing discharge temperature issues, check the actual compression ratio across each stage to determine if they are higher than the design values.
- A loss of cooling between stages can affect the overall flow and result in temperature problems
- A sudden loss of reciprocating compressor flow can be caused by either a suction piping restriction, an interstage cooling problem, failing valves, or a faulty spillback control system. Field data can help you understand the most likely cause of the problem.

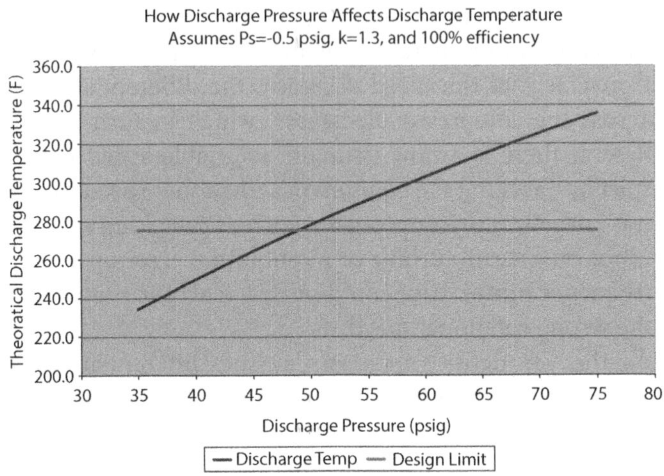

Figure 10.4 A reciprocating compressor cylinder's compression ratio (r) controls its discharge temperature. In this plot, you can see that as the discharge pressure increases, the cylinder discharge temperature increases accordingly. The operating compression ratio needs to be evaluated whenever a cylinder's discharge temperature if found to be high.

Electric Motor (Induction)

- An induction motor's speed tends to approach its synchronous speed when the load is decreased. Conversely, an induction motor's slip speed (synchronous speed - actual speed) increases as load is increased.
- Stator winding temperatures increase as the motor load increases.
- Stator winding temperature runs hotter during hot days or if cooling passages are blocked.
- The motor's amp load is an indicator of the motor being produced. However, the best means for determining the actual motor load is by using a power meter.
- Motor amps, slip speed, and stator temperatures are all indicators of motor load.

Basic Fluid Film Bearing Troubleshooting Tips

Here is a list of relationships between various bearing factors and operating temperatures:

- The operating temperature of a bearing tends to increase as the applied load increases; Conversely, the operating temperature of a bearing tends to decrease as the applied load decreases
- The operating temperature of a bearing tends to increase as the supply oil temperature increases and decrease as the supply oil temperature decreases.
- The operating temperature of a bearing tends to increase as the oil supply flow drops below the design oil flow
- The operating temperature of a bearing tends to increase slightly with an increase in oil flow, but there is a limit to how much you can drop the bearing temperature with additional oil flow
- The operating temperature of a bearing tends to increase as the speed is increased and tends to decrease as the speed is increased.
- The operating temperature of a bearing tends increase as the installed clearance is decreased and decrease as the installed clearance is increased.

Always investigate any significant step change in a bearing's operating temperature. Look for correlations between operating conditions and the bearing operating temperature. You may identify relationships between bearing temperature and the process loading, ambient temperatures, cooling water flows, etc.

Design Basis: Speed, Pressures, Flows

A good starting point for analyzing machinery performance is reviewing the design basis. The design basis is what the owner and designer agreed to and aspired to during the design, construction, and testing process. The design basis lets you know what the expectations were for your given process machine intended to provide at a stated set of conditions. However, keep in mind that things change. Process condition changes and design modifications can dramatically change how a machine functions in the field.

It is vital that the design conditions be compared with the current conditions to determine if there is a problem. Let me explain why this comparison is important. Compressor performance is especially sensitive to gas properties. For example, as the gas density in the compressor's suction increases a) the differential pressure across the compressor increases and

b) the horsepower required to drive the compressor increases. Conversely, as the gas density in the compressor's suction decreases a) the differential pressure across the compressor decreases and b) the horsepower required to drive the compressor decreases. Gas density will also move the location of the surge line.

Since the head developed in the centrifugal pump and compressors changes with flow, we need a head-flow curve for these machines to understand what is normal and what is not normal. If you are dealing with a variable speed process machine, then you will need a performance map that defines its expected performance over the design speed range.

The analyst should fully know:

1. Head versus flow characteristics and horsepower requirements. Centrifugal pump and compressor manufacturers typically provide head/flow performance curves and horsepower requirements versus flow. (Figure 10.5 is a hypothetical performance map for a centrifugal compressor.)
2. Effect of fluid properties on performance
3. Effect of speed on performance. Centrifugal pump and compressor manufacturers typically provide head/flow performance and horsepower load maps.

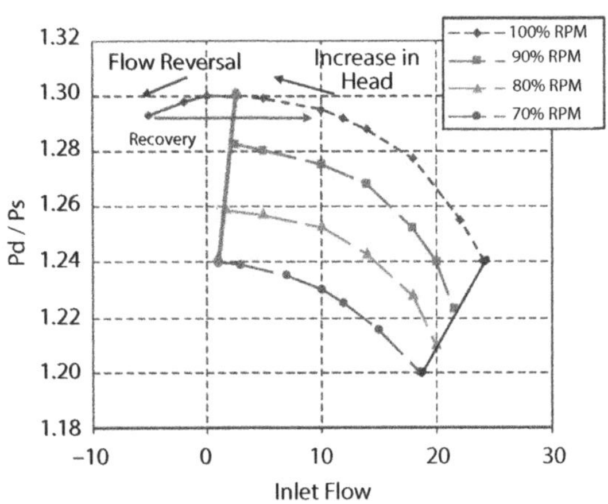

Figure 10.5 A centrifugal compressor performance map, like the one seen here, graphically illustrates the relationship between the differential pressure and flow at various speeds. Compressor performance maps also include a surge line (the red line on the left of the map) that can be used to design the surge control system.

System Design Details

To understand a process machine's performance, the analyst must know how it is intended to be operated. Is the machine in question intended to be operated in series or is it operated in parallel? In series configurations, the differential pressures of each machine are additive. In parallel installation (see Figure 10.6), flows of each machine are additive. Furthermore, process machines operating in parallel are prone to load shedding, where process loads are not equally distributed. Control schemes may be required to ensure process loads are equalized in parallel services.

Other operating possibilities are that the machine you are assessing is a partial spare that is used 50% of the time or a spare unit that is used sparingly or rarely. Idled process machines are prone to problems such as oil contamination of smaller bearing housings, rotors taking sets when not running, the gradual collection of liquids in casings and start-up lines, and fouling or corrosion in dead-leg lines. Furthermore, remote starting of spared machines without adequate safeguards can result in unwanted starting conditions that may be detrimental to mechanical reliability.

Multistage compressor applications, where cylinders or compressor bodies are in series, typically require cooling between stages. Cooling issues can lead to a loss of flow and overheating problems. Watch interstage temperatures to ensure interstage coolers are working properly.

Rotating machinery controls are vital for reliable and safe operations. There are flow controls, pressure controls, speed controls, surge controls, spillbacks, etc. Malfunctioning controls can affect the performance of the overall system and, in some cases, even lead to a premature failure. For example, a malfunctioning spillback flow system can lead to a loss of net (forward) flow, high horsepower loads, vibration, etc. A malfunctioning

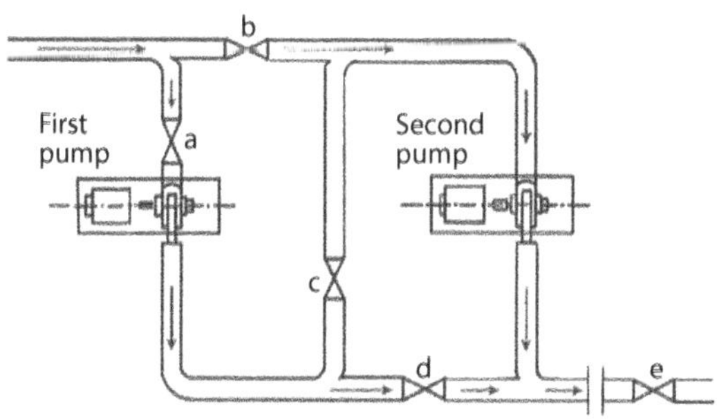

Figure 10.6 A hypothetical piping diagram two centrifugal pumps installed in parallel. It is important to understand how the pumps are intended to function within the process.

speed controller can lead to a loss of speed, which in turn will result in a loss of flow and pressure. Check that controls are functioning normally before for a mechanical problem. A friend of mine would always say, "It's not the pump!" By this, he meant always rule out other possible problems before removing the pump for inspection because it's more likely that what appears to be a pump problem is really a control, piping, or driver issue.

OEM Recommendations

It is highly recommended that the original equipment manufacturer (OEM) be consulted when addressing a new field problem. They are the most knowledgeable about their products and have seen most common field problems. In addition, they should have the most current design details and have records of any design modifications. Here is a list of OEM recommendations you should always have handy.

1. Oil recommendations
2. Preventative maintenance schedules
3. Piping recommendations
4. Spare part recommendations
5. Critical speeds and speed limits
6. Effect of wear and fouling on performance

History

One of the most valuable sources of information to a troubleshooter is historical information, which includes field data and failure reports. The best way to know if there is an issue is to compare the current operation with the past. For example, if a bearing temperature is running 150 °F and you are asked it that is normal, you should determine what it has normally been running. If you determined that it's been running between 145°F and 155°F for the last two years, then it's probably not a problem.

Equipment history also helps you understand the present by knowing what happened in the past. For example, if a compressor's inboard radial bearing fails every two years, then you can expect it to continue failure on the same schedule. Also, if a centrifugal pump has always exhibited high vane pass vibrations, then you shouldn't be surprised if you find high vane pass vibration present on the pump.

Here is a list of historical information that will help you make better reliability decisions:

1. Date of installation
2. Date of last repair
3. Common failure modes
4. Which vibration spectral components commonly appear when internal defects are present, such as a a) failing bearing, b) failing gear, c) cracked electric motor rotor bar, etc.?
5. Common signs of degradation
   a. Vibration trending upward
   b. Temperature trending upward
   c. Loss of flow
   d. Loss of pressure
   e. Pressure pulsations
   f. Loss of speed
   g. Flow or pressure fluctuations
   h. Speed fluctuations
   i. Loss of speed
6. Detailed failure history: When the repair was done, what was found, and the most likely cause of the failure.
7. Historical operating data such as flows, pressure, and vibration levels. Compare the current machine operating conditions with the historical values to see if there have been any changes.
8. Historical observations (operators). Talk to operators to get their insights into a machine's history. They may provide operating context that could help explain what's going on in the field.

Putting it All Together

When troubleshooting, knowledge is power. A minimum body of knowledge is required to understand the current operating state to determine if there is a problem with the machine, controls, or system. The troubleshooter must understand their machine and the system that it is operating in. After collecting and studying the data, the analyst has several choices: 1) Do nothing, 2) Change operating conditions, 3) Balance, lubricate, align, etc., 4) Calibrate or repair the controls, or 5) Repair the machine.

The required knowledge required to effectively troubleshoot is not obtained overnight. The accumulation of troubleshooting knowledge occurs over years of effort and determination. I obtained my troubleshooting knowledge by 1) reading numerous books to understand machine construction and their theory of operation, 2) visiting shops to learn about equipment design, and 3) attending technical conferences to hear the experiences of other users, and 4) troubleshooting whenever there's an opportunity. Practice makes perfect.

Here are some final thoughts:

- Always keep an open mind about what's going on and ask a lot of questions. The machine owners and operators are often the best source for operating and historical information. Fixating on a single probable cause before reviewing all the data can lead to false starts.
- Never hesitate to talk to the OEMs. They may have seen the very problem you are dealing with and know how to resolve it.
- To get a good picture of what's going on, you need to collect all the available data. The machinery data may be in a paper form or electronic format. If possible, collect your own data in the field to verify the problem is real.
- Studying troubleshooting case histories of equipment like those you deal with is a good way to get exposed to the various types of field issues that are possible. Users' groups are another way to get exposed to practical knowledge and advice from those responsible for keeping their rotating equipment running reliably.
- For the most complex problems, get help. It always helps to get other viewpoints. Bringing in a consultant may make the most sense for machines you have little experience with.

Remember that you don't have to possess all the answers to be an effective troubleshooter; you only need to have the desire to dig into the problem wholeheartedly and relentlessly. At the starting of your analysis, information is your friend. Don't stop digging until you have found the root cause of the problem and then work to resolve it in the most economical means possible.

11

Precise Coupling Properties are Required to Accurately Predict Torsional Natural Frequencies

By Robert X. Perez

Introduction

Torsional vibration is the twisting oscillation of a drive train about its axis, which occurs to some degree in rotating equipment trains, composed of reciprocating compressors, engines, turbines, and electric motors. Torsional vibration can be detrimental to rotating equipment because the related torsional stresses are superimposed on the static torque already present in power transmission shafts. The combined stresses can result in extremely high levels that can lead to catastrophic failures. Strain gauges and telemetry equipment can be used to measure torsional stresses in rotating equipment during operation. This brief case study describes a real torsional problem that arose from inaccurate coupling properties that resulted in poor torsional modeling results.

Case Study

An analytical torsional analysis of a two motor-driven compressor drive train was performed to ensure there would be acceptable margins (+/-10%) between mechanical and electrical torsional excitations and the frequency of the train's torsional modes of vibration at its design speed.

The natural gas compressor drive train consisted of an 8000 Hp, 720 rpm synchronous motor directly coupled to a 6 throw reciprocating compressor (see Figures 11.1 and 11.2). During the torsional vibration analysis, various coupling and compressor-mounted flywheel assembly configurations

Robert X. Perez (ed.) Condition Monitoring, Troubleshooting and Reliability in Rotating Machinery, (247–254) © 2023 Scrivener Publishing LLC

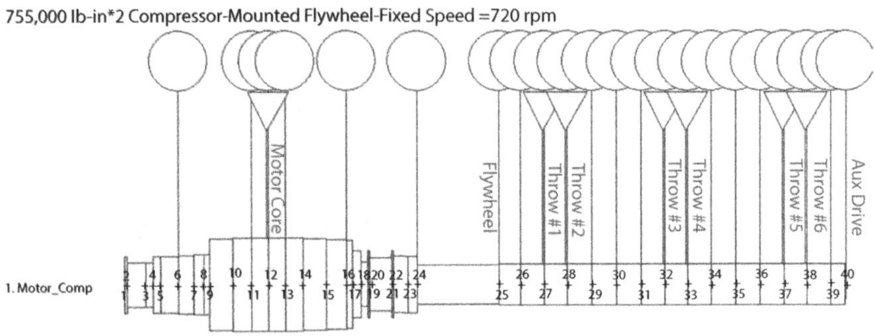

Figure 11.1 Schematic of torsional vibration model depicting the electric motor (left) and compressor (right) train.

Figure 11.2 Photo of disc pack coupling (the electric motor is on the left and the compressor is on the right.)

were evaluated to determine their effect on torsional performance. It was hoped that the recommended flexible disc-pack coupling and flywheel configuration (Figure 11.3) would provide a separation margin of greater than 10% (+/-) between the mechanical and electrical torsional excitations and the frequency of the train's fundamental torsional mode of vibration at its operating speed of 720 rpm. To ensure the accuracy of the drive train's

Figure 11.3 An internal flywheel was added to the back end of the compressor crankshaft to lower the compressor/motor train's first torsional natural frequency.

dynamic behavior, the compressor's torque-effort performance data was provided by the manufacturer and used in a forced response vibration analysis to evaluate steady-state torques and vibration amplitudes.

The lowest "fundamental" torsional twist natural frequency (Mode 1) of the compressor/motor train was predicted to be 3,203 cycles/min (cpm) with the flexible disc pack coupling selected. At the time, this prediction seemed to be an acceptable design because it placed the first torsional mode between the 4th and 5th orders of the operating speed (i.e., 4.45 times the speed of 720 rpm). A 3203 cpm first mode would have provided a 45% separation from 4x torsional excitation at rated speed, but it didn't turn out that way.

Start-Up Issues

Immediately after starting up these units, the site began experiencing coupling issues. There were repetitive disc pack failures (Figure 11.4) experienced and extensive disc pack bolt damage (Figure 11.5) was found during the coupling's disassembly. The site soon asked for help in determining the root cause of these failures.

Field Vibration Study

Since all the physical evidence pointed to a torsional vibration problem, these units were instrumented with strain gauges so that motor shaft strain

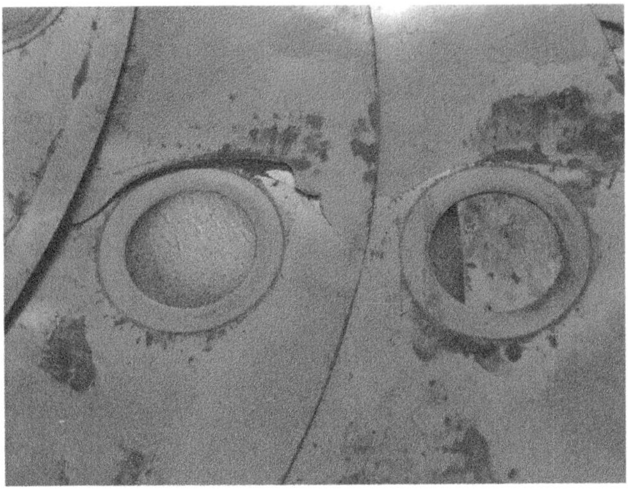

Figure 11.4 Typical disc pack damage.

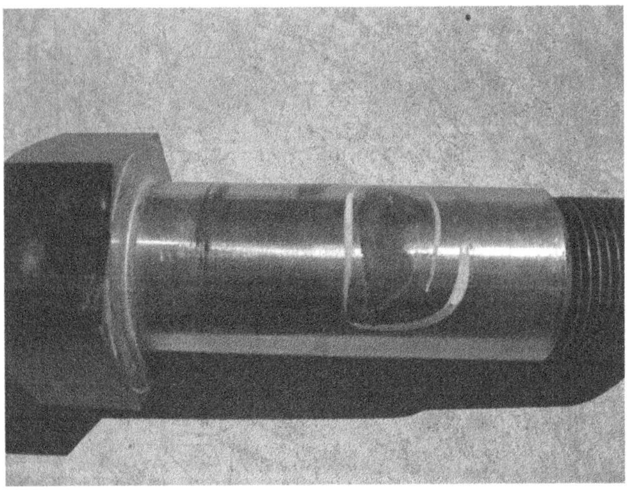

Figure 11.5 Typical coupling bolt damage.

levels could be measured and evaluated. It was discovered that the first torsional natural frequency (FTNF) of the initial train analyzed was actually at about 3600 cpm (instead of 3203 cpm as expected), which coincided with the compressor's 5x excitation frequency (see Figure 11.6). Since all compressors naturally generate 1x, 2x, 3x, 4x, and 5x torque pulses, this means that ever-present 5x compressor torque excitation would continuously excite the fundamental torsional natural frequency, which in turn would

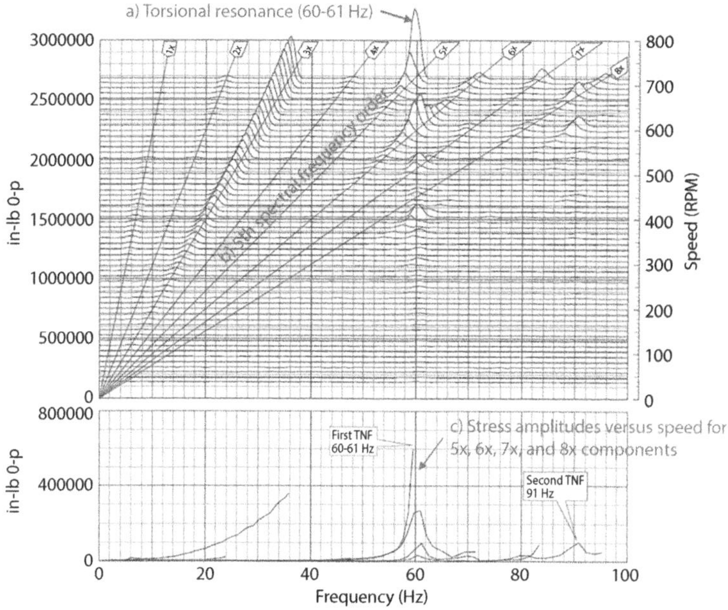

Figure 11.6 Series of torsional strain spectrums at different speeds recorded on a reciprocating compressor shaft experiencing coupling failures.

lead to high torsional defections in the coupling, motor, and compressor shafts. When the measured stress values were converted into dynamic torque values, it was found that the dynamic torque values exceeded the torque rating for the coupling based on the manufacturer's specifications. (Note: Field data taken on the second compressor/motor set showed that there was a similar coincidence of the torsional natural frequency and 5x excitation frequency.)

These findings led to the review of the original torsional models. It soon became clear that the discrepancies between actual FTNF and the predicted FTNF values were caused by inaccurate coupling stiffness values provided by the coupling manufacture. A refined torsional model was created by modifying the coupling data to fit the field data. The refined torsional model was then used to size the required internal flywheel needed at the back end of the compressor crankshaft (see Figure 11.1) to lower the compressor/motor train's first torsional natural frequency safely below five times running speed. A new flywheel was fabricated based on the torsional analysis modeling results. The subsequent field testing confirmed that the new flywheels dropped the FTNFs of these two motor/compressor trains and moved them safely away from the 5x excitation frequencies.

Lesson Learned

The main lesson learned in this case study was the importance of having precise coupling design data. Users should question how the coupling torsional stiffness values provided by manufacturers were determined. This is particularly true when evaluating larger couplings. Since it is more difficult to experimentally determine the stiffness of a large coupling, manufacturers sometimes arrive at a torsional stiffness by either extrapolating data from similar couplings (questionable approach), using finite element modeling results (somewhat better approach), or using a lab setup to measure the torsional stiffness directly (best approach). In this case study, the rotational stiffness of the coupling between the electric motor and compressors was a major parameter required for the determination of the FTNF. For this reason, the precision of the rotating stiffness value is crucial in arriving at an accurate FTNF.

Here is a procedure one manufacturer uses to measure the torsional stiffness of their couplings: 1) one end of the coupling is attached to a fixed housing, 2) the opposite end of the coupling is attached to a freely moving beam moment arm, which is capable of being loaded with various weights, and 3) a dial indicator is then used to measure the relative deflection between the fixed and freely moving ends of the coupling. This relative deflection is converted to rotational or torsional deflection (in degrees or radians). The coupling's torsional deflection under varying loads can then be used to determine the coupling's rotational stiffness. The arrangement is used to measure the coupling's overall stiffness, including the end hubs, flex elements, and spacers.

Torsional stiffness is defined as:

$$K_T = \frac{T}{\theta}$$

T = torque in in-lbf
θ = angular deflection in radians

Let's assume that the torsional stiffness of this coupling was 700 x 10^6 in-lbs/radian (the actual value has been rounded up for simplicity). If we use a 6 foot moments arm, measured from the coupling's rotational centerline, to induce the torque, then the formula to determine the torsional stiffness is given by:

$$K_T = \frac{T}{\theta} = \frac{R \times W}{\theta} = \frac{72\,in \times W\,lbf}{\theta} = 700 \times 10^6$$

where R is the radius of the moments arm.

Solving for load (W) required for a one tenth of a degree torsional deflection, we get:

$$W = \frac{700 \times 10^6\ in - lb/rad}{72\,in} \times 0.001745\,radian = 16{,}965\,lbf$$

One tenth of a degree represents an arc length, which is measured by a dial indicator of 72 in x 0.001745 or 0.1256 inches (125.6 mils) seems like a reasonable amount of deflection. However, it would take quite a bit of effort to apply 16,965 lbf in a shop to get this amount of deflection. To prove linearity, the deflection weights could be applied in 10% increments, i.e., 1696.5 lbs, 3393 lbs, 5089.5 lbs, etc. all the way up to 16,965 lbs. In practice, it probably makes more sense to load the test rig using a hydraulic jack and a load cell.

Final Thoughts

Shortcuts, miscalculations, and bad assumptions by coupling manufacturers can lead to significant discrepancies in coupling torsional stiffnesses provided to end users, which in turn could result in inaccurate torsional models and their analytical results. In this case, inaccurate coupling stiffness values provided by the manufacturer resulted in a 12.3% percent difference between the calculated (3203 cpm) and the actual (3600 cpm) FTNFs. It's always a good idea to ask coupling manufacturers how their coupling torsional stiffness was determined. If you don't get a good answer, proceed with caution.

12

Is Vibration Beating on Machinery a Problem?

By Robert X. Perez and Andrew P. Conkey

What is Vibration Beating?

When two vibration signals of nearly equal frequencies are generated by a machine or group of machines, a beat vibration frequency can be produced. The interaction between the two sinusoidal signals composed of similar frequencies results in an alternating constructive and destructive interference pattern (see Figure 12.1a), which causes the composite wave to periodically increase and then decrease in amplitude (see Figure 12.1b) regularly over time. The frequency of the variation in the composite wave is called the beat frequency, which is why this phenomenon is called beating. The count of beats per second is equivalent to the difference in frequencies of the two waves is called beat frequency. Therefore, the beat frequency is calculated as follows:

$$\text{Beat Frequency} = f_b = |f_2 - f_1|$$

In this formula, f_b is the beat frequency, f_1 is the frequency of the first vibration signal, and f_2 is the frequency of the second vibration signal (see Figure 12.2).

Beat phenomena can be found on a number of process machines found in the industry. Here are some common machinery situations where beat phenomena may be seen:

- On two poles, induction electric motors where 2x and 2x-line amplitudes can be significant. The absence of beats in the vibration signal may indicate there is only a <u>mechanical</u> problem. In contrast, the presence of beating

Robert X. Perez (ed.) Condition Monitoring, Troubleshooting and Reliability in Rotating Machinery, (255–264) © 2023 Scrivener Publishing LLC

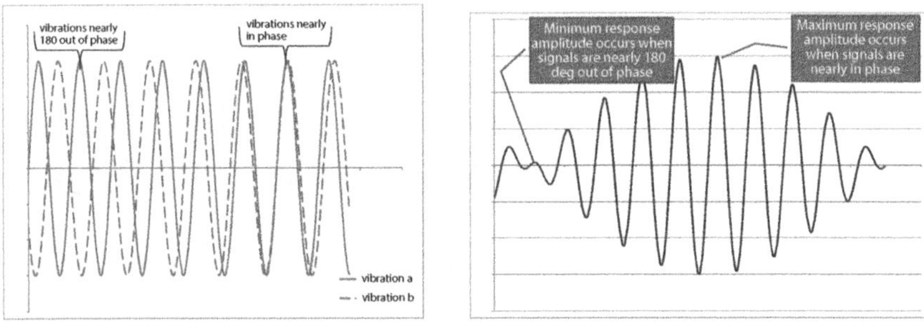

Figures 12.1a and 12.1b Image "a" shows how signals with different frequencies go in and out of phase over time. Image "b" show how constructive and destructive inference of two signals results in a beating signal.

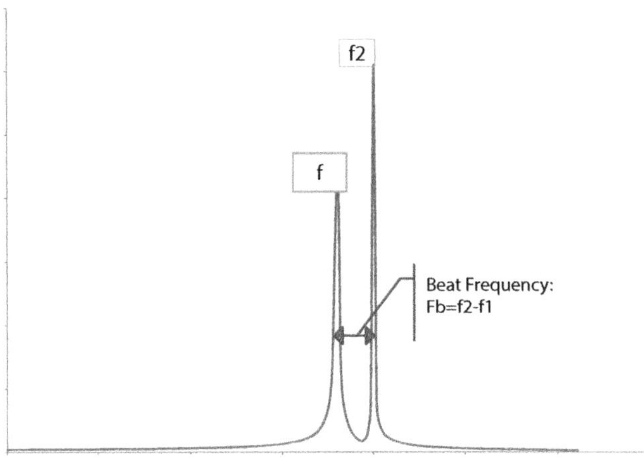

Figure 12.2 In a zoom configuration, signals with similar frequencies will appear in a spectrum as two closely spaced peaks as seen here.

can indicate a mechanical problem in addition to a magnetic issue. Beating often shows up on a 2 pole electric induction motor with closely spaced frequencies composed of a) peak at twice the rotational speed (2x), b) a peak at 2x minus twice the line frequency of the electric power (2xline), and c) a 2x peak plus twice the line frequency of the electric power. These types of problems tend to be highly load-sensitive.

- In situations where two machines with similar speeds are operating close to one another. However, for this type of problem to occur, vibrational energy needs a path to travel from one nearby machine to the other. Frequently, it's the connecting piping that enables the transmission of vibration energy between local machinery.

- In situations where a machine with two separate rotors are operating nearly identical in the same machinery casing, such as a two shaft gas turbine. We will review a case study below, there a beating problem was observed and resolved.
- In any machinery situation where there is a significant center frequency peak flanked by smaller "sidebands" or peaks spaced on either side of the center frequency. Due to their proximity, side bands can modulate the center frequency and show up as a beat effect at low frequency resolution settings. Modulation peaks flanked by side bands are commonly found in electric motor, gearbox, and rolling bearing vibration spectra.

Here are some rules of thumb that can help you understand beat vibration created by two closely spaced frequencies:

1. The beat frequency is the difference of the constituent vibration frequencies.
2. The minimum amplitude of the beat waveform is the difference of the two amplitudes
3. The maximum amplitude of the beat waveform is the sum of the two amplitudes
4. If a single spectral peak is beating, then the frequency window width may be too broad to see the two constituent beating frequencies.
5. If closely spaced frequencies are suspected, use the analyzer's frequency zoom capabilities to zoom into the frequency range of interest. It might take some trial and error to find the best frequency resolution to see all the peaks of interest. Remember that the tighter the frequency zoom, the longer it will take for each sample.

Zoom FFT (Fast Fourier Transform) Analysis [1]

When setting up a vibration analyzer, the analyst must select 1) a frequency range to measure (F_{max}) and 2) a frequency resolution, which is set by the "number of lines" to be displayed in the frequency range. The frequency resolution of the vibration spectrum is equal to the F_{max}

divided by the number of lines of resolution. (Typical vibration measurements are taken with 800, 1600, and occasionally 3200 lines of resolution depending on the Fmax.) For example, if we collect a spectrum with an F_{max} of 2000 Hz and 1000 lines of resolution, the resolution of the spectrum will be 2000/1000 or 2 Hz. This would be a problem if we have vibration peaks at 59.6 Hz and another at 60 Hz. We would not be able to distinguish between the two signals because our frequency resolution is not fine enough to resolve them. These two frequencies would simply appear as one peak in the spectrum.

In contrast to a normal spectral analysis, a Zoom FFT analysis involves selecting the narrower frequency range to achieve a finer frequency resolution than a standard FFT setting. Let's say, instead of analyzing a frequency range from zero to 2000 Hz, we select a tighter frequency, say from 100 to 200 Hz. The following equation describes the frequency resolution of the zoom spectrum:

$$\text{frequency resolution} \cong (\text{stop frequency} - \text{start frequency})/\text{number of lines}$$

If we use an analyzer with 1000 lines of resolution, then the standard frequency resolution will be 2000 Hz/1000 or 2 Hz per line. However, if we zoom into a 100 to 200 Hz range our resolution will increase to 100 Hz/1000 or 0.1 Hz.

Electric Motor Zoom Analysis

Let's consider a situation where we are trying to analyze a two pole, 60 Hz, electric induction motor that is exhibiting a vibration beating effect. We know on two pole electric motors that the 2 x vibration peak tends to have two, 2 x slip speed sidebands. If we are analyzing a motor running at 3576 rpm (59.6 Hz), we can expect two slip frequency peaks at 60 – 59.6 or 0.4 Hz, so the 2 x slip is 0.8 Hz. This means that if we want to see the center frequency and both sidebands, we need to zoom into a frequency range that includes the frequency range from 58.8 Hz to 60.4 Hz. Assuming we have an analyzer with 1600 lines of resolutions, a zoom range of 50 Hz to 70 Hz would provide analysis bins of (70-50)/1600 or 0.0125 Hz per bin. This should be adequate to locate the three peaks that we expect to see.

Field Case Study: "Beating" Effect Caused by Two Closely Spaced Mechanical Frequencies Observed on Two-Shaft, Gas Turbine Drive [2]

The following case study involved a gas turbine located at a gas processing facility in South Texas (see Figure 12.3). The gas turbine is rated at 4700 HP (63 kW) and constructed in a two-shaft design (see Figure 12.4). The gas turbine is composed of a forward section called the gas producer section, which is

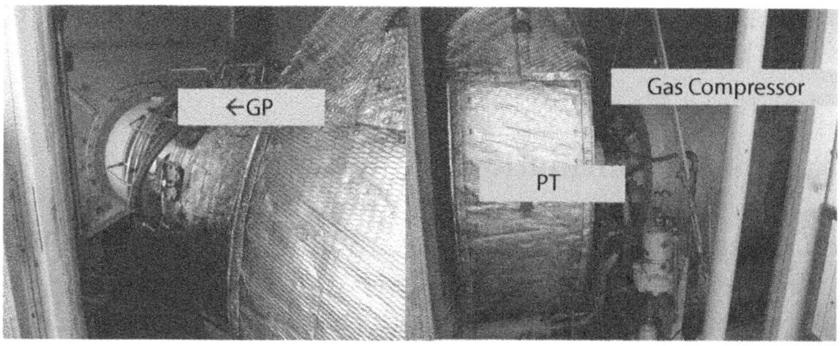

Figure 12.3 The two shaft gas turbine mentioned in this case study is seen here inside an enclosure. The gas producer (GP) is seen on the left and the power turbine (PT) is seen on the right.

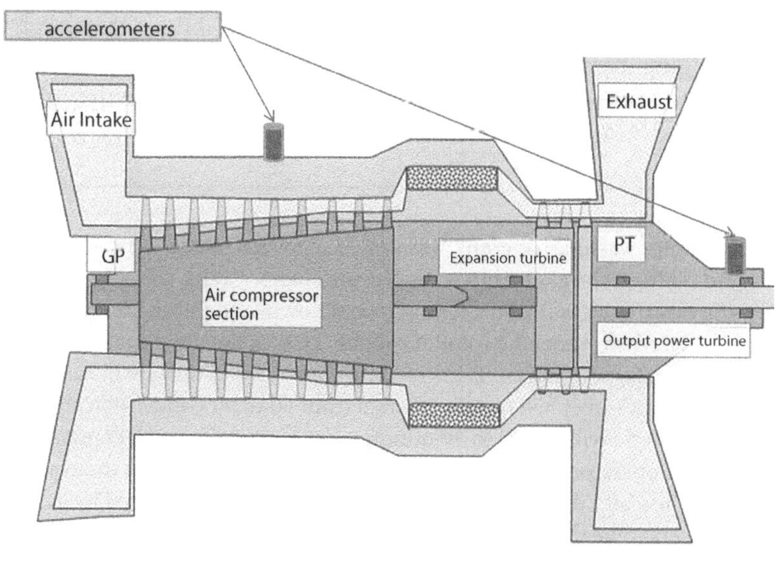

Figure 12.4 Cross section of similar two-shaft gas turbine.

labeled "GP" and a power turbine section labeled "PT." The gas turbine, which has a rated speed of 15,000 rpm (250 Hz), consists of the air compressor, combustion section, and two expansion stages. The combustion gases produced by the GP first pass through the expansion turbine and are then directed to the aft end of the gas turbine, called the power turbine (PT), which has a rated speed of 15,500 rpm (258.33 Hz). The PT, which consists of a single expansion stage and an exhaust gas diffuser, directly drives a multistage centrifugal compressor.

Notice, in Figure 12.4, that the output power turbine (PT) and gas producer (GP) turn on separate shafts and share common housing. During this case study, two accelerometers are used for monitoring. One was mounted on the gas producer casing and one was mounted on the power turbine bearing housing.

Background Information

A rebuilt gas producer (GP) and power turbine (PT) were put into service, along with a controls upgrade and a compressor overhaul. After the installation of the gas turbine, the gas turbine manufacturer's field representative trim balanced the PT. After the field balancing, vibration levels ranged from 0.14 to 0.18 ips (0.0055 to 0071 mm/s). Six months later, after a routine crank wash (see "Crank Washing" insert below) was performed on the GP to maintain gas turbine efficiency. Immediately after start-up, PT vibrations rose to 0.28 ips (0.011 mm/s) and stayed there, sometimes reaching 0.32 ips (0.013 mm/s). It was soon observed that vibration levels were swinging from 0.20 to 0.40 ips (0.0079 to 0.0157 mm/s) whenever the PT operated at a speed ranging from 91.0% to 92.7% of its maximum speed but remained steady above or below this speed range.

Crank Washing

Off-line water washing, or crank washing, is a proven means of removing dirt and oil from the gas turbine compressor blades. Crank washing involves shutting down the gas turbine, letting it cool down, and putting it on a crank cycle, which turns the rotor at roughly 20% of the rated speed. Then, a mixture of water and solvent is injected through nozzles that are installed in the compressor's inlet bell mouth at a rate from 50 to 90 GPM. Typically, the water wash cycle will last 20 to 30 minutes, but to make sure the compressor is clean, the operator should monitor the water discharge and let the system run until the water discharge is clear. Crank washing is considered by gas turbine users to be the best way to effectively clean the compressor and regain lost power due to fouling.

Vibration Response Analysis

After studying the nature of the oscillations of the overall amplitude in the 91.0% to 92.7% range of PT rated speed, it was determined by the field analyst that the vibration variations could be caused by a "beat" phenomenon. As we stated earlier, beating of the overall vibration level becomes noticeable whenever two closely spaced vibration frequencies are present.

We soon learned that the vibration spectrum for the PT spectrum was composed of two peaks that were slightly separated. The frequency of the peaks corresponded to the speed of the PT and the GP. It was noticed that whenever the two 1X vibration peaks in the PT spectrum were separated by about 1 Hz or less, the periodic variation on the control room monitor became obvious. This is because the beat frequency period became greater than the one second sampling period of the vibration monitor.

Investigation of System and Analysis

To better understand the phenomenon occurring between the GP and the PT, two curves were generated by varying the PT speed from 90% to 97.5% of the PT rated speed and recorded the spectral characteristics of the two predominant peaks in the 230 to 250 Hz range. These curves were created by collecting and then plotting the amplitudes and frequencies of the larger

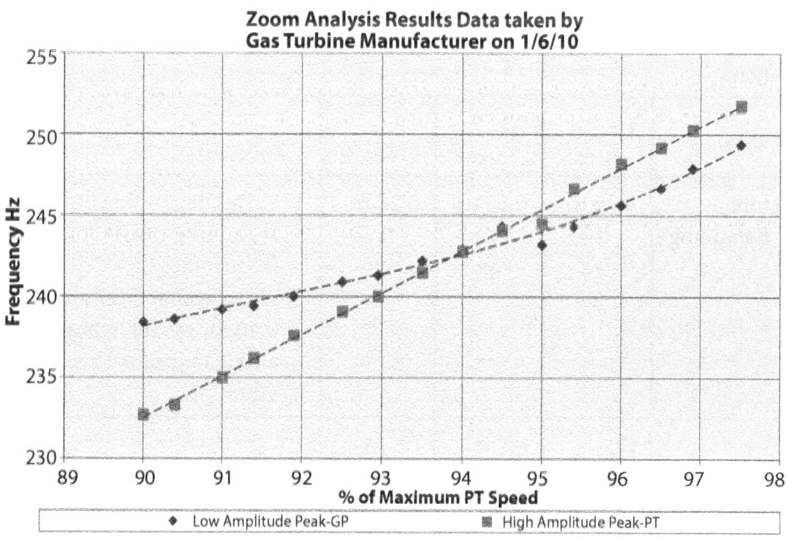

Figure 12.5 Zoom analysis results from PT speed sweep.

component and the smaller sideband with the frequency analyzer set at a 230 to 250 Hz frequency span with 1600 lines of resolution (see Figure 12.5). (Note: This provided a frequency resolution of (250-230)/1600 or 0.0125 Hz.) It was soon discovered that the larger of the two peaks related to the PT 1x frequency and the smaller sideband peak corresponded to the GP 1x frequency.

Frequency Analysis

Normally, the predominant vibration frequency on the GP casing is one times (1x) the GP rotating speed. Similarly, the predominant vibration frequency on the PT bearing housing is typically one times (1x) the PT rotating speed. We would not expect to see a significant GP vibration peak on the PT bearing housing vibration spectrum due to the fact that two shafts are independently supported. However, at that time of our initial analysis, we found the larger of the two vibration peaks corresponding to the power turbine (PT) operating speed and a smaller but significant peak corresponding to the gas producer (GP) running speed on the PT bearing housing spectrum (see Table 12.1 below). It is believed the appearance of the predominant GP peak was an indication that there had been a change in the mechanical condition of the GP's back end around the time of the recent crank wash. In addition to these factors, at the time the problem was first observed, there was unusually cool weather that enabled the gas

Table 12.1 Summary of gas turbine vibration levels before and after field balancing.

	PT peak	GP peak	Overall	Comments
Before Balancing	0.24 to 0.35 ips	0.12 to 0.15 ips	0.20 to 0.40 ips	1. Varies depending on separation of GP & PT frequencies 2. The manufacturer's limit is 0.40 for overall vibration and 0.33 ips for 1x vibration.
After Balancing	0.11 ips	0.15 ips	0.234 ips	PT @ 88%
After Balancing	0.19 ips	0.12 ips	0.280 ips	PT @ 92%

turbine to generate more power, allowing the GP and PT rotors to run at higher speeds where the beating was likely to occur.

Case Study Solution

To resolve this issue, the manufacturer's field representatives recommended the power turbine (PT) be trim balanced to reduce the overall PT vibration level to acceptable levels. After several balancing attempts, vibration levels dropped to acceptable levels (see the results in Table 12.1).

After trim balancing, vibration levels fell below the manufacturer's limit of 0.40 ips for overall vibration in the PT vertical direction and below the manufacturer's limit of 0.33 ips for 1x vibration in the PT vertical direction. These results allow the gas turbine to remain in service. Field personnel continued to closely monitor both the PT and GP vibration amplitudes to ensure they remain steady and below recommended manufacturer's guidelines.

Case Study Conclusions and Lessons Learned

The observed beating effect in this case study was caused by closely spaced GP and PT vibration peaks. This effect will likely only be observed on newly refurbished engines on colder days that are able to run at higher operating speeds. The combination of a rebuilt engine installation, wintry weather, and crank washing all led to the coincidence of GP and PT speeds at the time beating phenomenon was first observed.

The transmission of GP 1x vibration amplitudes to the PT varies significantly from one engine to the next. In this case, an elevated level of GP x1 vibration at the PT end of this engine contributed to the severity of the beating effect. Once the sources of the two frequencies were identified, a course of action became clear. In this case, we saw that field trim balancing of the PT mitigated the effects of this phenomenon by significantly reducing the PT vibration level back into the acceptable range.

Final Comments

We can summarize our brief review of beating phenomena in machinery by saying that occurrences of beating should be investigated to determine the underlying cause or causes of the beating. This is especially true if beating arises suddenly and unexpectedly. Some underlying reasons for beating may be innocuous, however, in some cases, beating may be providing clues to hidden issues such as gear defects or deterioration in an electric motor.

The factors that play significant roles in beating are the machine's construction, how its operated, and its mechanical conditions. Always perform a detailed vibration analysis of the machine to fully comprehend what is occurring. The analysis should also include data at various load conditions and speeds (if possible). Finally, to fully understand what's happening, the vibration analyst needs to have the right spectral analyzer and be competent in performing a zoom frequency analysis.

In most cases, a machine with a beating vibration signal is trying to tell you something. Listen carefully and you might determine what's actually going on.

References

1. Reliability Web, How Many Lines of Resolution?", https://reliabilityweb.com/tips/article/how_many_lines_of_resolution, accessed on 12/10/2011.
2. Perez, R. X., Conkey, A., P., "Beating" Effect Caused by Two Closely Spaced Mechanical Frequencies Observed on Two-Shaft, Gas Turbine Drive," Texas A&M Turbomachinery Symposium Case Study, 2011.

Andrew P. Conkey, PhD is currently an Associate Instructional Professor at Texas A&M University McAllen Higher Education Center. Dr. Conkey has also served as an Adjunct Professor at Texas A&M University-Kingsville, a Visiting Assistant Professor at Texas A&M University-Corpus Christi, and an Assistant Professor at Texas A&M at Qatar. He earned his MSME degree in Mechanical Engineering from Texas A&M University-Kingsville and his PhD degree in Interdisciplinary Engineering from Texas A&M University. He co-authored *'Is My Machine OK?,"* Industrial Press, 2012 and *"Troubleshooting Rotating Machinery,"* Scrivener Publishing, 2016.

Part 3

RELIABILITY

13

Using Standby Machinery to Improve Process Reliability

By Robert X. Perez

Introduction

The goal of this chapter is to introduce readers to the concept of reliability modeling in processes using machinery. Reliability modeling methods are a proven way of predicting the availability and reliability of process systems made up of machinery and flow elements. Some common questions that arise when designing machinery installation are: 1) Do I need build redundancy into my machinery installations? and 2) If so, what is the most cost effective way to build redundancy into my installation? There are always tradeoffs between process reliability, initial capitals costs, operating complexity, and maintenance costs (see Figure 13.1). The right design configuration for your particular system depends on the cost of downtime, equipment, and maintenance. I will use Raptor, a reliability modeling tool, to illustrate the benefits and tradeoffs of different machinery sparing configurations.

Basic Reliability Theory [1]

Before we begin talking about reliability modeling, we need to understand the difference between system availability and equipment reliability. Availability is a metric that measures the probability that a system (or equipment) has not failed or is undergoing a repair action when it needs to be used. This usually equates to the financial performance of the system or process. Availability is defined as: Uptime/Total time (Uptime + Downtime). For example, let's say we want to calculate the availability of a critical production asset that ran for 704 hours in a single month. During that time, the asset experienced 8 hours of unplanned downtime because of a breakdown and 8 hours of downtime for weekly PMs for a total of 16 hours downtime. For this example, the availability of the asset is:

Robert X. Perez (ed.) Condition Monitoring, Troubleshooting and Reliability in Rotating Machinery, (267–288) © 2023 Scrivener Publishing LLC

Figure 13.1 Process designers know that there are always tradeoffs between process reliability, initial capital costs, operating complexity, and maintenance costs.

Availability = 704 ÷ (704 + 16) = Availability = 704 ÷ 720 = 0.97.8 = 97.8%

A piece of equipment may be considered available but may not be highly reliable. For example, let's say a pump is down every month for two days due to a seal failure. This translates into an availability of 93.4% (365-24/365) but has a mean time between failure of 1 month or a failure rate of 12 failures a year. In most plants, this pump would be considered a bad actor.

Reliability, on the other hand, is a measure of the probability that a piece of equipment will perform its intended function for a specified interval under stated conditions. For example, let's say you have an unspared reciprocating compressor providing hydrogen for a process that requires catalyst changeout every 24 months. You might want to know what the likelihood is that the compressor can run continuously for 24 months. The chance of survival (R(t)) from time zero till time t of interest is given by:

$$R(t) = e^{-\left(\frac{t}{\eta}\right)^{\beta}}$$

The two parameter Weibull formula is defined by the following parameters:

- β is the shape parameter, also known as the Weibull slope
- η is the scale parameter

Note: Weibull distribution can be represented by 2 or 3 parameters. For simplicity, I have chosen to use two parameter Weibull distributions for an example in this chapter.

If we use this two parameter Weibull formula and find that the probability of making a 24 month run is unacceptable, we can either decrease the interval between the catalyst changeouts or improve the reliability of the compressor. We will talk more about the ways to improve reliability through redundancy later in this chapter.

The Weibull shape factor (b) is an important aspect of the Weibull distribution. Weibull distributions with $\beta < 1$ have a failure rate that decreases with time, also known as infantile or early-life failures (see Figure 13.2). Weibull distributions with β close to or equal to 1 have a fairly constant failure rate, indicative of useful life or random failures. Weibull distributions with $\beta > 1$ have a failure rate that increases with time, also known as wear-out failures. These comprise the three sections of the classic "bathtub curve," as seen in Figure 13.2.

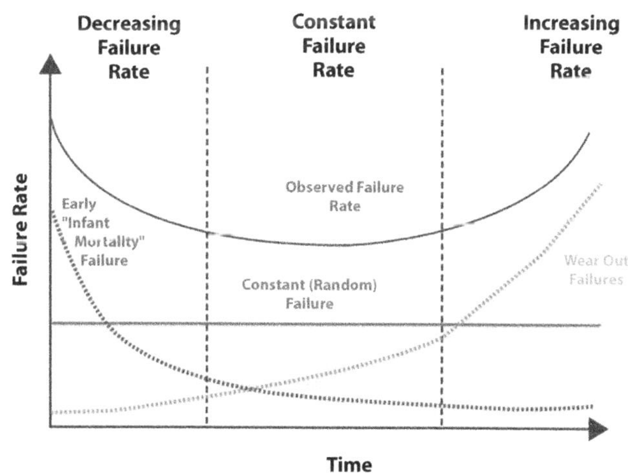

Figure 13.2 This image depicts the three types of beta (β) Weibull distributions: β<1 which represent infant mortality distributions, β=1 which represent random failure distributions, and β>1 which represents wear out failure distributions.

η (Eta), seen in the two parameter Weibull formula, is called the "scale parameter" because it scales the value of mission time, t. That is, it stretches or contracts the failure distribution along the age axis based on the units of time being used. A scale parameter can have units of time, i.e., hours, days, months, years, etc. For random failure distributions, i.e., β=1, h is approximately equal to the component's mean time between failures. In general, if two components have identical shape factors, the one with the larger scale factor will tend to survive longer in service.

A good starting point in selecting a shape factor is to assume a given machine exhibits a shape factor (β) between 1.2 and 1.9. Since machines are made up of many moving parts and wearing services, they tend to display a failure distribution that fall somewhere between random and slightly time related. For example, reciprocating compressors have multiple wearing components that tend to wear out over time and tend to experience an overall wear-out distribution of failure. Table 13.1 contains a collection of Weibull parameters for various machine types.

Here's an example of how to use the Weibull function to determine the probability of survival. Let's say you want to know what the probability of a centrifugal compressor running continuously 4, 6, and 8 year intervals is. Assume you are trying to set the outage schedule for the plant and you are not sure if your compressor can reasonably run from outage to outage. We will use the Weibull date from Table 13.1, which shows a typical beta of 1.9 and a typical ETA of 2500 days.

The probability of making a 4 year run is:

$$R(t) = e^{-\left(\frac{t}{\eta}\right)^{\beta}} = e^{-\left(\frac{1460\ days}{2500\ days}\right)^{1.9}} = e^{-0.36} = 0.697 = 69.7\%$$

The probability of making a 6 year run is:

$$R(t) = e^{-\left(\frac{t}{\eta}\right)^{\beta}} = e^{-\left(\frac{2190\ days}{2500\ days}\right)^{1.9}} = e^{-0.7776} = 0.4595 = 45.95\%$$

The probability of making an 8 year run is:

$$R(t) = e^{-\left(\frac{t}{\eta}\right)^{\beta}} = e^{-\left(\frac{2920\ days}{2500\ days}\right)^{1.9}} = e^{-1.343} = 0.261 = 26.1\%$$

Table 13.1 [2] Here are the typical Weibull parameters for common process machinery and their components. For example, in absence of any failure history, we can assume that centrifugal compressors have an average beta value of 1.9, which means failures increase with time, and an average scale factor (η) of 2500 days (6.85 years). The scale factor is the time interval where we can expect 63.2% of the population will fail. Notice that mechanical seals have an average beta value of 1.4, which represents a failure distribution that is somewhere between random and slightly time dependent and an average scale factors of 1042 days (2.95 years). *Be aware that the range of values in this table are very broad and should therefore be considered a starting point for detailed analyses. An advanced practitioner may choose to either derive Weibull parameters from actual field data or get more accurate values from their original equipment manufacturers.*

	Beta (β)	**Scale Factor (η)**		
Components	**Average**	**Low (days)**	**Typical (days)**	**High (days)**
Compressors Centrifugal	1.9	833	2500	5000
Recip. Compressor	1.9	730	1460	2190
Pumps Centrifugal	1.2	42	1458	5208
Valves Recip. Comp.	1.4	125	1667	3333
Packings Recip. Comp. Rod	1.1	208	833	1375
Seals Mechanical	1.4	125	1042	2083
Reciprocating Engines	1.4	833	3125	7083
Steam Turbines	1.7	458	2708	7083
Motors AC	1.2	42	4167	8333

These Weibull calculations show that as the run length between outages increases, the probability of this compressor running without failing through the time interval of interest decreases. If there is no penalty for a premature compressor outage, then it doesn't matter what the outage length is. However, if the cost of an outage is high, then the user must weigh the cost of improving the scale factor (h) through design improvements or decreasing the time between unit outages to obtain an acceptable level of risk.

Chemical processing facilities are usually more interested in process availability than equipment reliability. Poor process availability results in reduced processing time and operational costs, while poor equipment reliability results in more maintenance headaches and higher maintenance costs. Since a process unit's availability is highly dependent on machine reliability, designers are keenly interested in the overall reliability of individual and multiple machine installations. Unspared machines are used only if they have been designed to be highly reliable or if the process downtime has no or very little economic consequence.

Garden variety centrifugal pumps represent a category of process machines that are not considered to be highly reliable. We can see in Table 13.2 that pumps tend to have a random failures distribution (β=1.2) and a mean life of about 4 years. (In my experience, I have found that mean times between failures on pumps can range between 2 to 5 years.) The questionable reliability of centrifugal pumps has forced system designers to develop strategies aimed at obtaining acceptable system reliability.

A common strategy used to improve the availability of process systems is installing twin pumps in critical pumping services, where one pump is in service and the second pump is idle. A high level of reliability can be achieved by always having an installed spare ready to go in the event the pump in service fails unexpectantly. Some sites refer to these twin pumps as "A & B" pumps, while others refer to them as "main and spare" pumps. These pumps are normally installed next to each other (see Figure 13.3) and are tied into common suction and discharge headers. If the pumps are installed properly, similar operating performance can be expected, regardless of which pump is in service. Rapid switching between these twin pumps by capable operators can ensure that the process flow is never interrupted long enough to result in a production upset or outage.

Figure 13.3 Typical twin pump installation.
Notice that both pumps are tied into common suction and discharge headers.

Exercising Spared Machinery [3]

The key benefit of parallel machinery installations is that the probability of failure-free operation of a parallel system is always higher than the probability of failure free operation of the best element in the system. However, the key to achieving high mechanical reliability is exercising installed machinery regularly to ensure their serviceability. Selecting an optimum pump switching interval can prevent bearings and mechanical seals from seizing in idle pumps. Furthermore, regular pump switching allows the mechanical condition of both pumps to be evaluated periodically under normal operating conditions, i.e., vibration data can be collected, bearing temperature can be assessed, etc.

Alternating Twin, Non-Critical, Process Pumps

Swapping frequencies for common, API centrifugal pumps, ranging from 50 to 500 horsepower, typically range from 30 days to up to 90 days. Every processing facility is different, but most facilities tend to base their swapping intervals based on the skill levels of their operators and their past experiences. In my opinion, swapping pumps monthly is ideal, while swapping pumps quarterly should be considered the maximum switching interval. Here is an example of a generally used pump swapping procedure:

Definitions for the following procedure:

- *The "A" pump* is the pump that is in operation at the beginning of the switching procedure.
- *The "B" pump* is the installed, spare pump to be placed in service.

1. With the *"A" pump* still running, start–up the *"B"* pump, following your normal start-up procedure.
2. Once you are satisfied the *"B" pump* is healthy, shut down the *"A" pump.* Ensure that the check valve on the pump you just shut down is holding.
3. (A leaking or stuck check valve will lead to a loss of flow and possible pump damage due to reverse rotation.)

    a. If the check valve is not holding, completely close the *"A" pump's* discharge valve.
    b. If the check valve is holding, then keep the discharge valve on the *"A" pump* partially closed (25%) so that it's ready for the next start-up.

4. Keep the down pump's seal flush flowing. Adjust flow as required.
5. Check the down pump's seal pot level to ensure it's at the normal level before walking away.

Recommended Swapping Procedures for Critical Motors, Pumps, Blowers, Compressors, Generators, and Steam Turbines

There are many ideas concerning the how or when spared installed machinery should be swapped to ensure reliable performance. Here are some general switching interval guidelines proposed by Alklowaiter [4]:

1. Maintain one of the operating blowers, compressors, or pumps designated as a main machine.
2. For A, B configuration, the standby compressor or pump is only started once monthly for one hour with maximum recycle flow, then shut down without bringing it online. The standby machine switches roles with the main machine and enters full operation at six-month intervals. *Note: This procedure requires the installation of a properly sized recycle line to allow a maximum flow condition.*
3. For larger machines with API-614 Special Purpose Lube Oil Systems: The lubrication pump is started for a half-hour weekly or twice monthly, which also becomes a lubrication system test.
4. Do not shut down the main operating compressors or pumps except every 6-months for switchover to standby mode. This saves ten startup-shutdown damaging cycles per year.
5. For compressor dry gas seals, maintain the tertiary seal nitrogen continuously, without interruption.
6. Centrifugal or Screw Instrument Air Compressors: Apply the same method above of maintaining the operating air compressors online without switching off, while starting the standby compressors for 30 minutes monthly offline. This increases reliability as frequent starting and shutdowns tend to increase machinery failures. Test the standby air compressors at monthly intervals. Start the lube oil system weekly.
7. Lubrication of self-contained bearing housing: If necessary, apply synthetic oils to allow extension of oil replacement

interval to 6 months and apply bearing housing isolation seals to minimize ingress of water and dirt.
8. Main Firewater Pumps: These are a special case. Always follow Fire Protection NFPA standards that apply.

Recommended Swapping Procedures for Reciprocating Process Plant Machinery above 200 HP

1. **Emergency Diesel Generators:** These are not normally in an A, B main and standby configuration as they are only utilized upon loss of plant power. Typically, they should be start tested weekly or twice monthly. It is not recommended that they be subjected to 52 starts per year. Applying a twice per month test starting period is sufficient to guarantee their availability, meaning they are subjected to 26 starts per year. Some plants have extended this to once monthly, but that is the maximum possible period for such critical service equipment.
2. **Reciprocating Engines or Compressors:** These machines are also best not stopped and switched over every month. It is advised to utilize six-month switching and operating intervals for reciprocating engines or gas compressors. The standby machine should be started once a month and run for 1 hour under a light load recycle mode, but not actually placed into service. The unit, if equipped with electric lube oil motors, should have the lubrication system started weekly for 30 minutes. Water cooled cylinders system piping, heat exchangers, pumps, and cooling fluid should be properly selected and modified to resist six month downtime corrosion.
3. **Reciprocating Pumps:** These pumps can be operated similarly to the reciprocating gas compressors using six-month duty cycles, with standby pump once-monthly startup for 1 hour operation on recycle mode, not actually placed into service. However, reciprocating pumps handling thicker polymerizing, crystalizing, or caking fluids should be addressed on a special case by case basis.

Remember that if you have installed spare machinery in the field, you need to develop switching and PM procedures that fits the different machine types that will be idled. When it comes to equipment switching and PM procedures, one size does not fit all. It's also important that your repair shops stay vigilant for potential clues that your current switching intervals are not optimal for your machinery. Signs such as false brinelling, pluggage, accelerated corrosion damage, sticking O-rings, etc. are just a few clues that suggest that a machine has been idle too long.

Raptor Modeling Software [5]

RAPTOR (Rapid Availability Prototyping for Testing Operational Readiness) is a generic simulation product developed by ARINC to predict the reliability, maintainability, and availability performance of systems and components[3]. It employs sophisticated Monte Carlo simulation algorithms to analyze complex systems that are described by a reliability block diagram (RBD). In RAPTOR, discrete failure and repair simulation are determined by using random number generators. Simulations can be run based on total run time or total number of systems failures. After the simulation runs are complete, RAPTOR will then provide a summary report containing:

- Availability
- Mean time between downing events (MTBDE)
- Mean down time (MDT)
- Reliability

RAPTOR can also analyze spare parts use, costs, and capacity issues.

Disclaimer: I have personally used RAPTOR for a number of years and found it to be easy to use and very intuitive. I consider it to be uniquely suited to evaluate the reliability of real-world machinery installations. Although there are a number of reliability modeling software on the market that we could have used, I have chosen to use RAPTOR for all the examples in this chapter.

Figure 13.4 shows an example of a reliability block diagram that can be constructed using RAPTOR. A block diagram is a graphical representation of a system's components arranged in a manner to convey how system-level failures occur. Included within this representation are the logistical relationships between the components and the system itself. Reliability block diagrams can be constructed with components in series and in parallel. The block diagram in Figure 13.4 represents an installation with three process machines in parallel. Notice on the right there is a node labeled "2/3, Discharge," meaning that 2 out of 3 machines must be running for the discharge node for the system to be functional. Any combination of "k-out-of-n" options can be selected for parallel components.

The Raptor software is used to analyze reliability block diagrams (RBD) that represent a system for which Reliability, Availability, or Maintainability (RAM) characteristics are desired (Figure 13.4). RBDs can be comprised

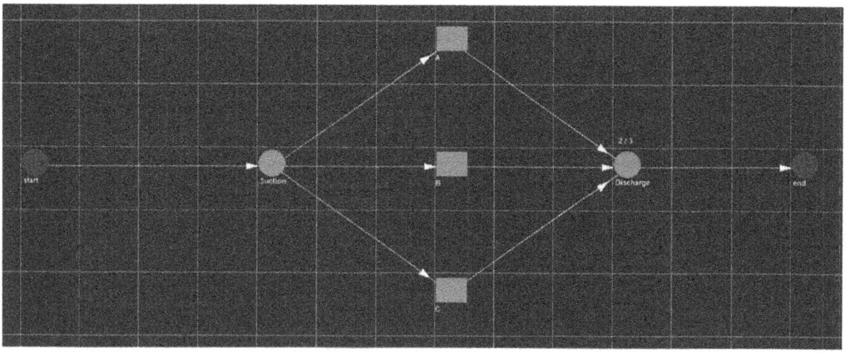

Figure 13.4 Simple raptor block diagram with 3 parallel elements.

of six different types of elements: blocks, events, nodes, links, hierarchies, and markers. Each block usually represents a physical piece of equipment. All blocks fail and repair based on probabilistic distributions as time passes through a simulation. When a block initiates its repair cycle due to a failure, it consumes resources, spares, and time.

Once the initial RBD system is constructed, it can be used to examine ways to enhance the overall reliability of the system. The beauty of constructing a RAPTOR model is having the capability of easily modifying the model to perform "what-if" analyses. The user can tweak failure and maintenance distribution parameters and add or delete components from the model to see what effect changes have on reliability performance.

Modeling Examples

To illustrate how equipment redundancy affects the reliability of machinery sets, I will evaluate the following configurations:

1. A single machine, i.e., no spares
2. A main and spare machine (1oo2), i.e., one installed spare
3. Three machines, where two 50% machines are needed to keep the system running (2oo3), i.e., one installed spare.

In these analyses, I have chosen to assume we are running a reciprocating compressor for all the elements. Let's assume the reciprocating compressor has a shape factor of 1.9 and a scale factor of 1460 days (4 years).

Note that in all of the reliability models presented here, I have assumed that we have an infinite number of spare parts available from stores. Raptor can be used to evaluate spare strategies, however this type of analysis if beyond the scope of this chapter.

Example 1: Unspared Compressor

To establish a baseline, let's run a simulation with a single unspared reciprocating compressor, as seen in Figure 13.5. Here are the input values. I have chosen the repairs to follow a log normal distribution with a mean repair time of 30 days.

What is the difference between normal and lognormal distribution?
The most commonly used distributions for maintainability analysis are the normal, lognormal, and exponential. However, the lognormal distribution is the most commonly used for equipment/system maintainability analysis. A lognormal distribution looks like a right-skewed bell curve, when compared to a normal distribution.

I ran 10 simulations with each one lasting 146000 hours. Here are the final results for the simulation:

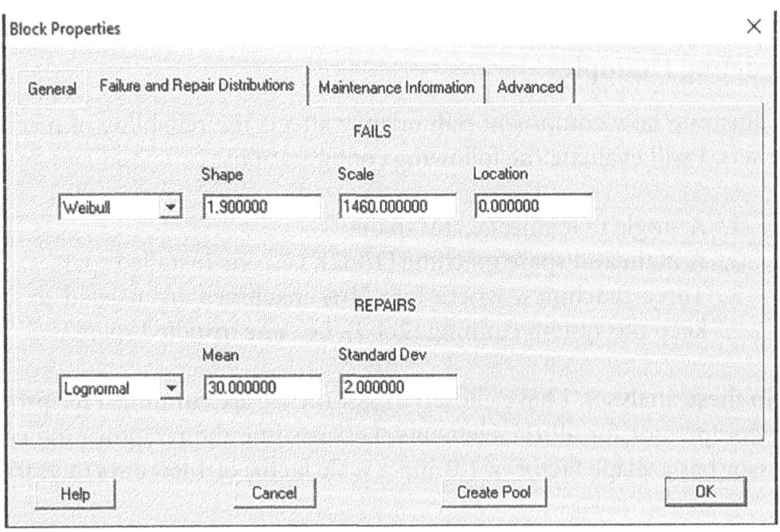

Figure 13.5 Inputs for example 1.

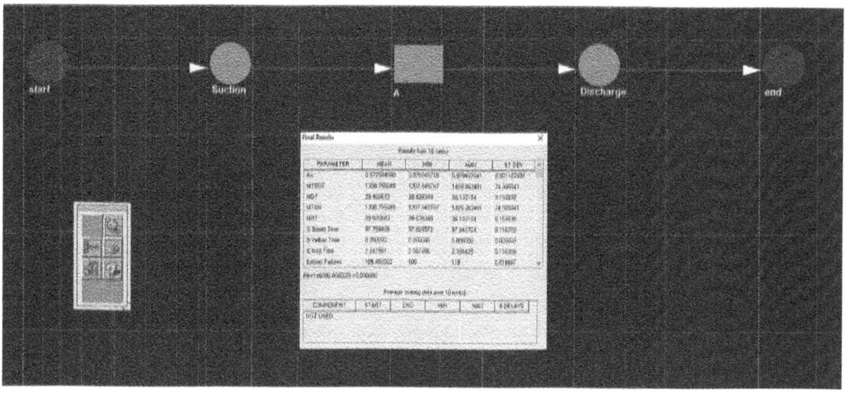

Figure 13.6 Raptor results from running Example 1.

From these results (Figure 13.6), we see that the mean availability (A_o) for the 10 runs is 97.758%. This might seem acceptable until you notice that the system is down 2.24% of the time, which is equivalent to a mean down time (MDT) of 29.92 days every 1308.755 days. Therefore, every 3.5856 years there will be an average of 29.92 days of downtime required for compressor repairs. If there is no consequence for downtime, then this will not be a major concern. However, if we assume the cost of being down is $100,000 per day, then on average, we will lose $100,000 per day x 356 x (1-0.97758) or $818,330 per year due to compressor downtime.

Example 2: Main and Spare Compressor Installation

Now, we'll look at two compressors in a main with an installed 100% spare configuration (Figure 13.7). Notice the "Discharge" node has the "1/2" notation, meaning that only one of the parallel elements is required to be functional for the entire system to be functional. If only one compressor is down, the system is said to be in a "yellow" condition and if two compressors are down, the system is said to be in a "red" condition, meaning a complete system failure has occurred.

After running 10 simulations with each one lasting 146000 hours, I arrived at these final results:

From these results (Figure 13.8), we see that the mean availability (A_o) for the 10 runs is 99.9454%, which means the system is down only 0.0546% of the time. Therefore, this arrangement will result in a mean down time (MDT) of 15.837 days every 32,973 days. So, every 90.33 years there will be a 15.837 day outage to repair the compressor. If there is not a consequence of failure, then this amount of downtime will not be a concern. However,

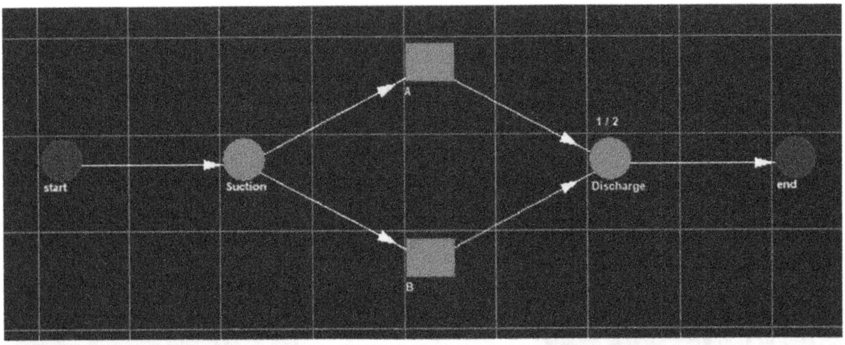

Figure 13.7 Block diagram of main and spare compressor installation.

Final Results

Results from 10 run(s):

PARAMETER	MEAN	MIN	MAX	ST DEV
Ao	0.999454533	0.999082091	0.999734375	0.000213843
MTBDE	32973.210590	18233.248163	48653.739594	11969.785724
MDT	15.837029	10.024665	21.841124	3.056597
MTBM	670.379968	623.358912	701.736629	26.015403
MRT	29.965680	29.834376	30.157455	0.113558
% Green Time	95.581856	95.308578	95.764326	0.164344
% Yellow Time	4.363597	4.187024	4.599631	0.151955
% Red Time	0.054547	0.026562	0.091791	0.021384
System Failures	5.100000	3	8	1.920937

R(t=146000.000000) =0.000000

Figure 13.8 Raptor results from running Example 2.

if we assume the cost of being down is $100,000 per day, then on average, we will lose $100,000 per day x 365 x 0.000546 or $19,929 per year due to compressor downtime. We can conclude that by installing a second compressor we can prevent $ 818,330-$19,929 or $798,401 per year in process losses with an installed spare.

Example 3: Two out of Three (2oo3) Compressor Configuration

In this analysis, we have three machines, where two 50% machines are needed to keep the system running (2oo3), i.e., one installed spare

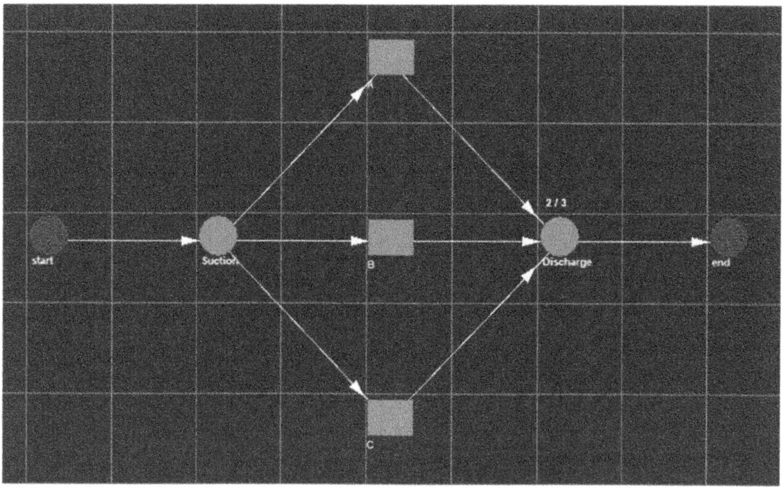

Figure 13.9 Block diagram of three compressors in parallel.
2 out of the 3 compressors must be functional to keep the system running.

(Figure 13.9). Notice the "Discharge" node has the "2/3" notation, meaning that 2 of three parallel elements must be running for the system to be functional. If any compressor is down, the system is said to be in a "yellow" condition and if two compressors are down, the system is said to be in a "red" condition, meaning a complete system failure has occurred.

After running 10 simulations with each one lasting 146000 hours, I arrived at these final results (Figure 13.10):

By examining these results, we see that the mean availability (A_0) for the 10 runs is 99.839%. This means the system is down only 0.161% of the time. There is a mean down time (MDT) of 14.644 days every 10568.25 days (MTBDE). Therefore, every 28.95 years there will be a 14.644 day outage to repair the compressor. If there is no consequence of failure, then this will not be a concern. However, if we assume the cost of being down is $100,000 per day, then on average, we will lose $100,000 per day x 14.644 days/28.95 of $50,584 per year. We can conclude that by installing a second compressor we can prevent $ 834,449 - $50,584 or $783,865 per year in process losses by implementing a 2 out of 3 machine configuration.

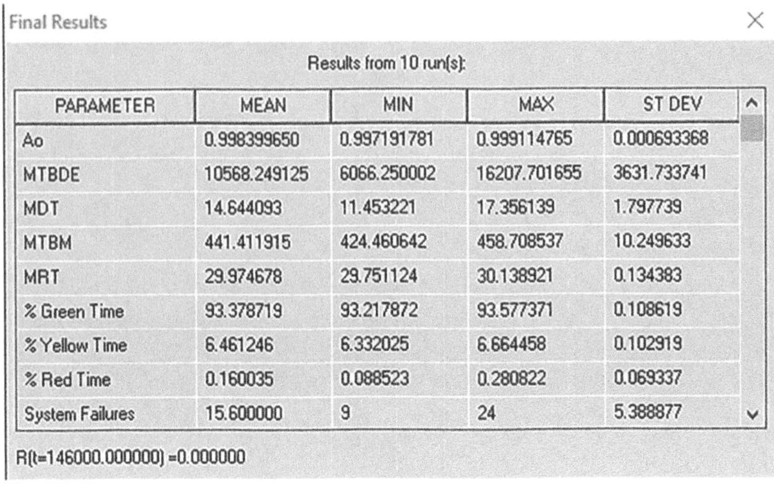
Final Results

Results from 10 run(s):

PARAMETER	MEAN	MIN	MAX	ST DEV
Ao	0.998399650	0.997191781	0.999114765	0.000693368
MTBDE	10568.249125	6066.250002	16207.701655	3631.733741
MDT	14.644093	11.453221	17.356139	1.797739
MTBM	441.411915	424.460642	458.708537	10.249633
MRT	29.974678	29.751124	30.138921	0.134383
% Green Time	93.378719	93.217872	93.577371	0.108619
% Yellow Time	6.461246	6.332025	6.664458	0.102919
% Red Time	0.160035	0.088523	0.280822	0.069337
System Failures	15.600000	9	24	5.388877

R(t=146000.000000) =0.000000

Figure 13.10 Raptor results from running Example 3.

The Cost of Redundancy

The costs associated with redundancy are summarized in Table 13.2. From these results, we can conclude:

1) Increased capital cost due to the additional machinery and auxiliaries is required.
2) The additional maintenance cost is related to the fact that more equipment is installed.

In general, more machinery in the field means more maintenance will be required. Notice that as we go from one machine to three machines, maintenance becomes more and more frequent. The MTBM goes from 1308.75 days down to 441.41 days, while the MTBDE actually drops when compared to the 1oo2 case.

Table 13.2 Summary of three simulations.

Configuration	MTBDE (days)	MTBM (days)
1oo1	1308.75	1308.75
1oo2	32973.21	670.38
2oo3	10568.24	441.41

It could be argued that the three 50% compressors could cost less to repair since they are smaller. However, if reliability is the goal then these results seem to suggest that a 100% installed spare is preferred over a single 50% spare.

Example 4: Cost of Unreliability

Let's look at the 1oo2 model again to see what effect individual machine unreliability has on system availability (Figure 13.11). We will run the model again, but this time we will assume the Weibull scale factor to be 365 days for the compressors. The point of this analysis is to see what effect poorly performing compressors have on the overall reliability of the compressor set.

Here are results (Figure 13.12) after running 10 simulations with each one lasting 146000 hours:

A comparison of the two output tables shows that the system availability (Ao) dropped from 99.9445% to 99.3%, which decreases the time between downing events (MTBDE) from 32,973 days (90.33 years) to 2162.45 days (5.92 years). Furthermore, the mean time between maintenance dropped from 670 days down to 178.7 days. This represents a 375% increase in maintenance costs for this compressor set when compared with a compressors set having 1460 day scale factors.

Example 4 illustrates that simply installing standby spares is not a cure-all. Facilities can greatly improve the availably of their critical processes by installing spare units. However, both the main and spare units must be reliable to ensure system reliability and keep maintenance costs at a reasonable level.

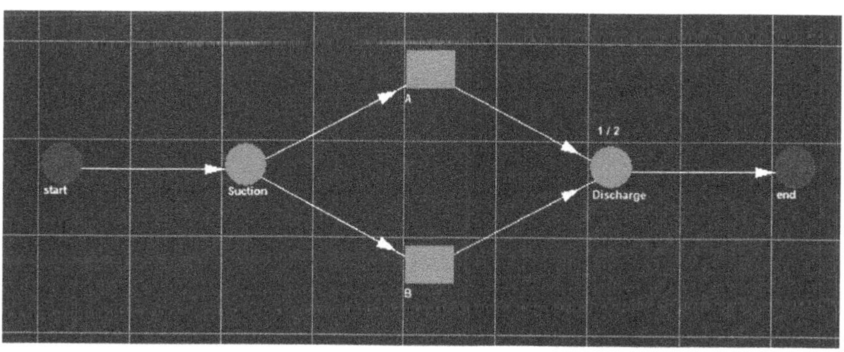

Figure 13.11 Example of less reliable compressors operating in parallel.

Final Results

Results from 10 run(s):

PARAMETER	MEAN	MIN	MAX	ST DEV
Ao	0.993090690	0.991483657	0.995014974	0.001074599
MTBDE	2162.454093	1787.375230	2548.634848	241.228093
MDT	14.821211	12.768661	17.785007	1.338484
MTBM	178.695458	173.361214	182.014890	2.894989
MRT	29.970205	29.906146	30.142947	0.063032
% Green Time	84.034578	83.599993	84.441278	0.265173
% Yellow Time	15.274491	14.794237	15.702688	0.309177
% Red Time	0.690931	0.498503	0.851634	0.107460
System Failures	67.900000	57	81	7.660940

R(t=146000.000000) =0.000000

Figure 13.12 Raptor results from running Example 4.

The moral of this example is: *Forgetting to maintain spared equipment will only lead to maintenance headaches and can potentially lead to more frequent process outages.*

Economics

Once you have run the simulations of interest, the analyst must assess the benefits and liabilities resulting from the various configurations. Here are the general formulas for determining the net benefits of installed spares:

a) Net Benefit = reduction in process losses – increase in maintenance costs
b) Base case process losses (annualized) = (1 – $A_{o(\text{base case})}$) x 365 x daily outage costs
c) Process losses with redundancy (annualized) = (1 - $A_{o(\text{new case})}$) x 365 x daily outage costs

Combining these terms, we get:

Annualized Net Benefit = ((1 – $A_{o(\text{base case})}$) - (1 - $A_{o(\text{new case})}$)) x 365 x daily outage costs – annualized maintenance costs (base case) + annualized maintenance costs (new case)

Simplifying, we arrive at the following relationship:

Annualized Net Benefit = ($A_{o(new\ case)}$ – $A_{o(base\ case)}$) x 365 x daily outage costs – annualized maintenance costs (base case) + annualized maintenance costs (new case)

It is vital that you include the terms from the base case in the analysis to determine the true benefit of the new configuration. Once the annualized revenue stream is quantified, you can perform a proper economics analysis and determine the investor rate of return (IRR) and the net present value of the proposed upgrade. If you can demonstrate that the upgrade has favorable economic returns, then there will be a higher probability that management will approve the recommendation.

Justifying of a Spare Compressor

We will now show how to use the annualized net benefits equation to justify a spare compressor. First, we will rewrite the annualized net benefits equations so that it's easier to read and understand.

Annualized Net Benefit = ($A_{o(new\ case)}$ – $A_{o(base\ case)}$) x 365 x daily outage costs – annualized maintenance costs (base case) + annualized maintenance costs (new case)

ANB = ($A_{o(new\ case)}$-$A_{o(base\ case)}$) x 365 x DOC)-R x (365/$MTBM_{(base\ case)}$-365/$MTBM_{(new\ case)}$)

ANB = Annualized net benefits

Ao = availability as a percentage

DOC = daily outage costs

R = repair cost

MTBM = mean time between maintenance

Now, let's compare two scenarios to see what savings would be realized by using the outputs from the RAPTOR 1oo2 model. We'll use the same Weibull parameters as before, (β=1.4 and η=1460) and assume each repair has a cost of $50,000. In this exercise, we will determine the annualized benefits for cases where the daily outage cost are $100,000 per day in the first example and $10,000 per day in the second example. We will use the RAPTOR outputs from Examples 1 and 2 (see Figure 13.13) since the Weibull parameters are identical for all of these cases.

When we calculate annualized net benefits for both cases, we get the results seen in Table 13.3.

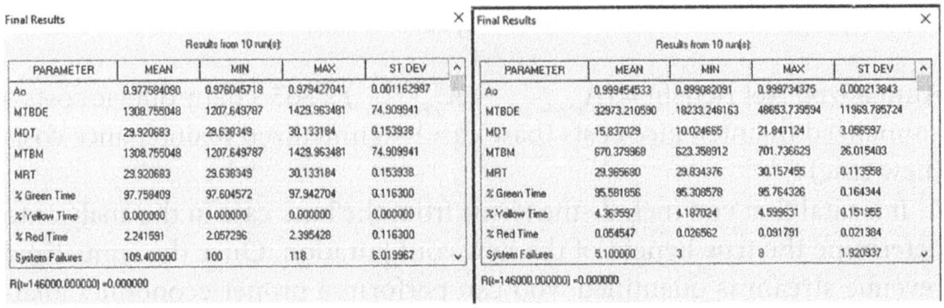
Final Results

Results from 10 run(s):

PARAMETER	MEAN	MIN	MAX	ST DEV
Ao	0.977584090	0.976045718	0.979427041	0.001162987
MTBDE	1308.755048	1207.649787	1429.963481	74.909941
MDT	29.920683	29.638349	30.133184	0.153938
MTBM	1308.755048	1207.649787	1429.963481	74.909941
MRT	29.920683	29.638349	30.133184	0.153938
% Green Time	97.758409	97.604572	97.942704	0.116300
% Yellow Time	0.000000	0.000000	0.000000	0.000000
% Red Time	2.241591	2.057296	2.395428	0.116300
System Failures	109.400000	100	118	6.019967

R(t=146000.000000) =0.000000

Final Results

Results from 10 run(s):

PARAMETER	MEAN	MIN	MAX	ST DEV
Ao	0.999454533	0.999082091	0.999734375	0.000213843
MTBDE	32973.210590	18233.248163	48653.739594	11969.785724
MDT	15.837029	10.024665	21.841124	3.056597
MTBM	670.379968	623.358912	701.736629	26.015403
MRT	29.965630	29.834376	30.157455	0.113558
% Green Time	95.581856	95.308578	95.764326	0.164344
% Yellow Time	4.363597	4.187024	4.599631	0.151955
% Red Time	0.054547	0.026562	0.091791	0.021384
System Failures	5.100000	3	8	1.920937

R(t=146000.000000) =0.000000

Figure 13.13 Raptor results from examples 1 and 2.

Table 13.3 RAPTOR results from two analysis cases.

Daily process losses	Production benefits	Maintenance benefits	Total annualized net benefits
$10,0000	(0.999454-0.97758) x 365 x $100,000 = $798,401	$50,000 x (0.544-0.279)=$13725	$798,401-$13, 725 = $784,676
$10,000	$78,840 per year	$13,725 per year	$78,8840-13725=$65,115 per year

If the cost of an installed spare compressor is estimated to be $1,000,000, then the payout period for the $100,000 per day loss scenario would be $1,000,000/$784,676 or 1.2755 years. The payout for the second case would be $1,000,000/$65,115 or 15.35 years. The $100,000 per day outage option would surely be approved, but the $10,000 per day outage option would almost certainly not be approved.

These two examples suggest that as the daily outage cost for a single compressor installation increases, the economic justification for a spare compressor gets better and better. On the flip-side of this idea, there is also a daily outage cost case where a spare compressor is no longer economically justified. However, before making a final decision, a detailed cash flow analysis to determine the project's investors rate of return (IRR) and present value (PV) should be performed to determine the actual benefits of any upgrade project.

Closing Thoughts

We have reviewed some common ways to improve the reliability of machinery installations through redundancy. However, to evaluate the economic benefits of redundancy, the machinery specialist or project engineer must incorporate all the relevant factors into their final analysis, such as:

1. Cost of process downtime
2. Cost of additional machinery installations
3. Cost of additional maintenance
4. Individual machinery reliability parameters. As we have noted earlier, poor individual machine reliability can significantly affect the maintenance costs associated to a set of machinery.
5. Cost of spare parts
6. Availability of labor
7. Expected repair times

Only by incorporating all these inputs into the reliability and economic analysis model can an analyst hope to make an informed economic decision. We have seen here that installing spare machines can decrease downtime, but will also increase the maintenance required due to the additional machines in the field. Machinery users must weigh the cost of capital and maintenance with the value of process downtime. Redundancy only makes sense if the cost of additional capital is fully justified on its own merits.

Remember that the goal in this chapter was to introduce readers to the RAPTOR modeling software and demonstrate its versatility to analyze multiple machinery installation and not to recommend any particular machinery installation configuration. The reader should not take these results presented here as absolute, but as a starting point for future analyses. I encourage readers to use the RAPTOR modeling software to analyze and optimize their own machinery installations.

References

1. Weibull.com website, Characteristics of the Weibull Distribution, Accessed in September 2021.
2. GE Digital Solutions website, Reliability Analytics: Distribution Types, Accessed in September 2021.

3. Perez, R. X, and Bloch, Heinz, P., Pump Wisdom, 2nd Edition, Chapter 10, Pump Operations.
4. Perez, R. X., Rotating Machinery Fundamentals and Advances, Volume 2, Chapter 5: Rotating Equipment Switch-Over Frequency Maximum Reliability Time Interval Study, by Abdulrahman Alkhowaiter.
5. Murphy, Kenneth, E., Manager, Simulation & Analysis RAPTOR Team Leader ARINC Engineering Services, LLC, *et al*, RAPTOR 7.0 Tutorial Workbook, Copyright © 2004 by ARINC.

Raptor simulation and analysis reliability software and training is available through ARINC.

14

Gas Turbine Drivers: What Users Need to Know

By Robert X. Perez

Overview

Mechanical-drive gas turbines are used primarily to provide rotational power to a fluid mover, such as a centrifugal pump or a centrifugal gas compressor. The majority of gas turbine-powered mechanical drive machines are used in the oil and gas industry for pumping, injection, re-injection, and compression applications (see Figure 14.1) and for the transfer of oil or gas through pipelines. Most of these gas turbines utilize natural gas as their primary fuel.

Industrial gas turbines can either be used solely as mechanical drives or can be used in association with a recovery steam generator. Mechanical drives differ from electrical power generating sets in that they are often smaller and feature a dual shaft design as opposed to a single shaft. Their rated power range varies from 1 megawatt (1341 hp) up to 50 megawatts (67051 hp) [1]. In drive applications, gas turbine engines can be connected directly or through a gearbox when driving pumps or compressors.

Single Shaft versus Two Shaft Gas Turbines

Industrial gas turbines are available in two main varieties: single shaft and double shaft designs.

- Single shaft turbines, as their name implies, have a single shaft running through the entire turbine. Therefore, the gas turbine compressor, turbine, and the output drive shaft all turn at the same speed.
- In two shaft gas turbine applications, the gas generator, i.e., gas producer, and the power turbine turn at different speeds,

Robert X. Perez (ed.) Condition Monitoring, Troubleshooting and Reliability in Rotating Machinery, (289–324) © 2023 Scrivener Publishing LLC

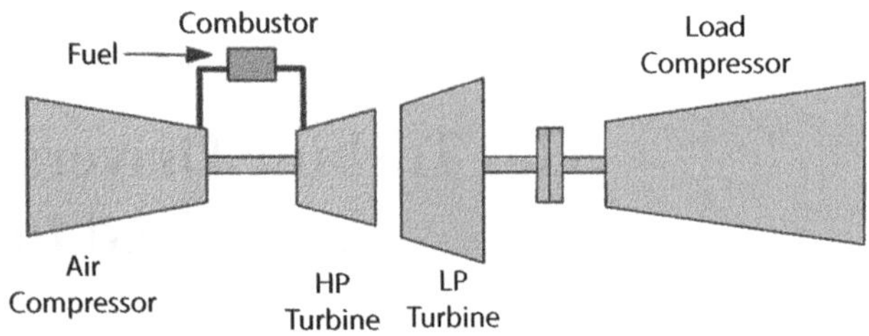

Figure 14.1 This is a schematic of a two-shaft gas turbine (on the left) driving a centrifugal compressor (on the right).

which enables the power turbine to be designed to match the power needs of the driven machine.

Ultimately, the choice of whether to use a single-shaft or two-shaft gas turbine is mostly determined by the operating characteristics of the driven load. If the load speed is constant, as in the case of an electric generator, a single-shaft unit is often specified. Therefore, a gas turbine chosen for electric power generation would typically make use of a single-shaft design configuration, whereas two-shaft engines are used when the applied load needs to be driven with varying speeds. The variable drive speed capabilities of two shaft gas turbines makes them ideal for driving centrifugal pumps and gas compressors.

Figure 14.2 shows a cross sectional view of a typical single shaft gas turbine mechanical drive. Atmospheric air enters through the left inlet volute and is then compressed by an axial compressor (shown here on the left of this image). The compressed air is mixed with fuel and then ignited. The high pressure mixture of combustion products and excess air is used to drive a series of turbine wheels. Most of the power extracted from the products of combustion is used to power the air compressor. The remainder of the power extracted by the turbine wheels can be used to either drive an electrical generator or drive a pump or compressor. After exiting the power turbine wheels, the combustion gases flow through the exhaust volute, which is shown on the far right of Figure 14.2 and then out to either the atmosphere or a heat recovery unit.

In contrast, two shaft gas turbines, as the name implies, have two distinct rotating shafts (see Figure 14.3). One rotating shaft makes up the gas producer, which is composed of the air compressor and compressor turbine rotor. The second rotating shaft, which turns independently from the gas producer, is driven by the power turbine.

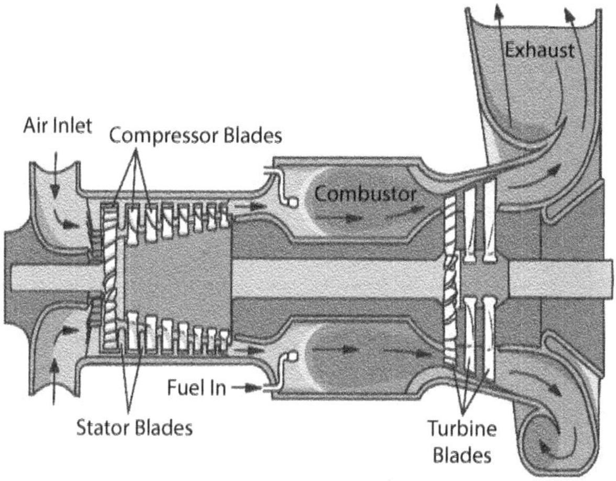

Figure 14.2 Single shaft gas turbine cross section.
Notice that the gas producer (on the left) and the turbine blades (on the right) are both on the same shaft.

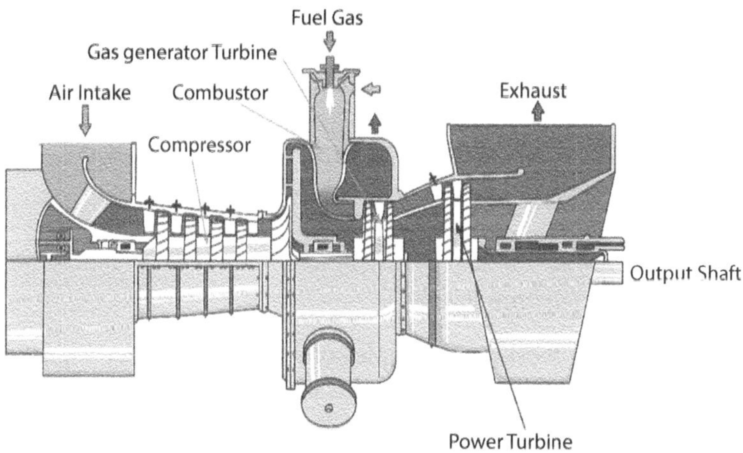

Figure 14.3 Cross section of two shaft gas turbine.
Notice that the gas producer on the left is not physically connected to the power turbine on the right.

In the two shaft gas turbine, ambient air is introduced into the compressor section of the turbine where the blades compress the inlet air. The compressed air then moves to the combustion chamber of the turbine where it gets mixed with the fuel and then burned to produce high temperature and high pressure gases. Those high pressure and high temperature gases

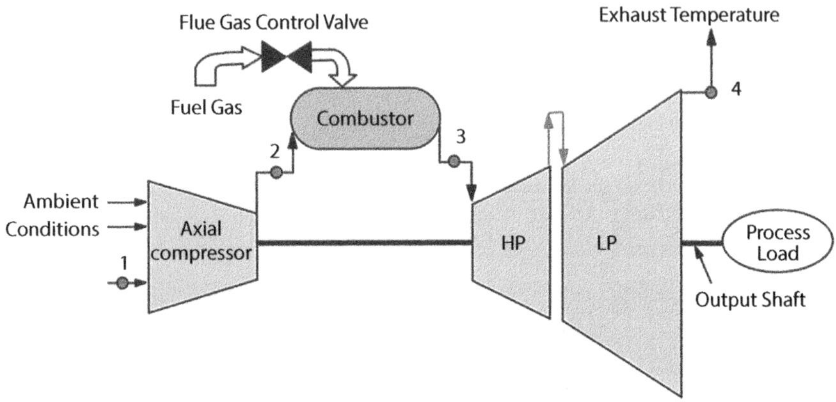

Figure 14.4 Schematic of two shaft gas turbine.
HP is the high pressure expansion section and LP is the low pressure expansion section of the gas turbine.

first flow into the compression turbine section (HP) and then flow into the turbine/power generation section (LP), which captures power from the expanding gases. The captured power then rotates the power turbine shaft, which can be used to drive a pump or compressor via a coupling or gearbox and coupling. If there is no gearbox between the gas turbine and compressor, the compressor will rotate at the speed of the power turbine, similar to the train depicted in Figure 14.4.

Theory of Operation [2]

How Does a Gas Turbine Work?

Explanations of the working principles of a gas turbine (see Figure 14.5) have to start with the thermodynamic principles of the Brayton cycle (Figure 14.6), which essentially defines the requirements for the gas turbine components.

A basic gas turbine engine consists of an air compressor, a mixing chamber, where the fuel is burned, and a turbine section. The Brayton cycle is characterized by the following two constant-pressure processes and two isentropic processes:

- First, there is an adiabatic process (point 1 to point 2 in Figure 14.6) where air is drawn in and compressed in the compressor. Within the compressor the air volume decreases and the air pressure increases.

Figure 14.5 Gas turbine view showing the inlet air volute on the left, air compressor with inlet guide vanes and variable guide vanes (left) and the hot section on the right.

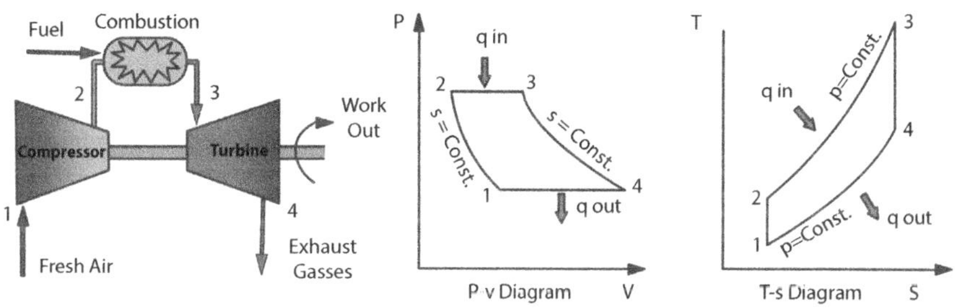

Figure 14.6 On the far left is a typical gas turbine configuration. In the middle is the pressure volume diagram for the Brayton cycle. On the far right is the temperature-entropy cycle of the Brayton cycle.

- Next, there is an isobaric process (point 2 to point 3 in Figure 14.6) where heat is added in the burner. Here, the volume of the combustion gases increases, while the pressure remains constant.
- Then, there is a second adiabatic process (point 3 to point 4 in Figure 14.6) where gas expands in the turbine section. As the combustion gases expand, there is an increase in volume and a drop in pressure.
- Finally, there is a second isobaric process (point 4 to point 1 in Figure 14.6) where heat is either captured in a heat

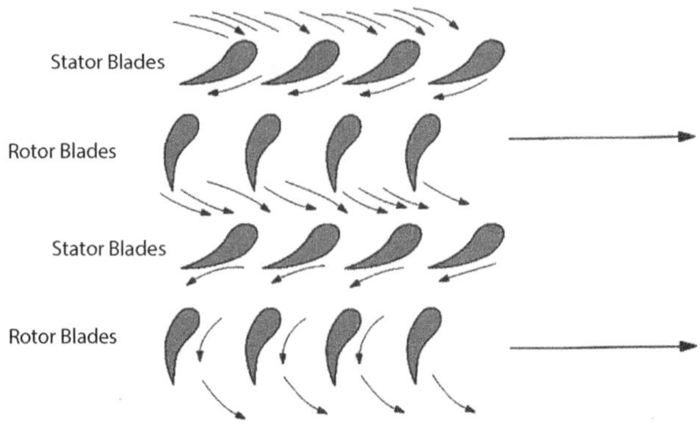

Figure 14.7 Inside the air compressor, air moves through a series of stationary and rotating blade sets. The lift developed across each stage increases the pressure incrementally until it exits the compressor. The goal of the designer is to match aerodynamic blades to the design conditions so that optimum efficiency is obtained.

recovery unit or is returned to the atmosphere. In this final process, the volume decreases and the pressure remains constant.

The energy conversion from mechanical work into the gas (in the compressor) and from energy in the gas back to mechanical energy (in the turbine) is performed by the means of aerodynamics, by appropriately controlling gas flows. Newton's principles of action and reaction and Bernoulli's law explain the energy transfer in various gas turbine stages (see Figure 14.7) [3]. In the compressor, energy is added to the air by the rotor blades, then converted to static pressure by the stator vanes. In the turbine wheels, the stator vanes increase gas velocity and then the rotor blades extract energy.

Air Compressor

In the axial air flow compressor, air enters in an axial direction (parallel with the axis of rotation) and exits in the axial direction. The compressor compresses the air by first accelerating it by a row of rotating airfoils (blades) and then diffusing (decelerating) it to obtain a pressure increase (see Figures 14.7 and 14.8). The stator vanes slow the air by means of their divergent duct shape, which converts the accelerated air velocity into a higher static pressure. The stator vanes are positioned at angles so that the exiting air is directed into the rotor blades of the next stage at the most efficient angle.

Figure 14.8 Close-up of axial compressor blades which rotate between sets of fixed stator assemblies.

As air passes from one stage to the next within the air compressor, its pressure rises slightly through each successive stage. By producing a series of low pressure increases, very high efficiencies can be obtained. Typically, the pressure ratios achieved on each stage range from 1.1:1 to 1.4:1 (Note pressure ratio absolute discharge pressure divided by the absolute suction pressure (P_D/P_S)). The use of multiple stages permits overall pressure increases of up to 30:1 in industrial applications.

Most air compressors are composed of several sections:

a) The rotor and stator assembly make up a major portion of a compressor. They perform most of the conversion of shaft horsepower to gas horsepower.
b) An additional row of stationary blades that are frequently used at the compressor inlet. These *inlet guide vanes* (IGV) are used to ensure that air enters the first-stage blading at the desired flow angle. These vanes can have a variable pitch so they can be adjusted to handle varying load conditions.
c) Multiple stationary stators with varying pitch, called *variable guide vanes* (VGV), are commonly used at the inlet compressor stages to handle changing load requirements.
d) A final set of vanes is at the exit of the compressor to diffuse the air and control its velocity as it enters the combustion section of the turbine. This final diffuser is often referred to as the exit guide vanes (EGV).

Combustors

There are three basic combustion chamber designs in use today. They are 1) the annular combustion chamber, 2) the can type, and 3) the combination of the two called the can-annular. Variations of these basic systems are used in a number of engines. Here, we will discuss combustors using combustion cans to burn the fuel and direct the combustion gases into the expansion stages of the gas turbine.

Once the air flows through the compressor diffuser, it enters the combustion section, also called the combustor (see Figure 14.9). The combustion section has the difficult task of controlling the burning of large amounts of fuel and air. It must release heat so that the air is expanded and accelerated to give a smooth and stable stream of uniformly-heated gas at all starting and operating conditions. This task must be accomplished with minimum pressure loss and maximum heat release. In addition, the combustion liners must position and control the fire to prevent flame contact with any metal parts. Combustion takes place in the forward end or primary zone of the cans.

Primary air, which amounts to about one fourth of the total engine's total airflow, is used to support the combustion process. The remaining air, referred to as secondary or dilution air, is admitted into the liners in a controlled manner. The secondary air controls the flame pattern, cools the liner walls, dilutes the temperature of the core gasses, and provides mass. This cooling air is critical, as the flame temperature is above 1930 °C (3500 °F), which is higher than the metals in the engine can endure. It is important that the fuel nozzles and combustion liners control the burning and mixing of fuel and air under all conditions so that excess temperatures are avoided within the combustor or power turbine sections. Maximum

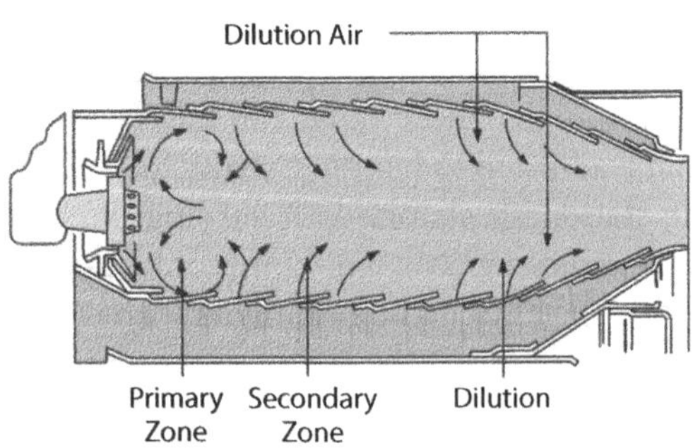

Figure 14.9 Gas turbine combustor.

combustion section outlet temperature (turbine inlet temperature) in this engine is about 1070 °C (>1950 °F).

The rear third of the combustion liners is the transition section. The transition section has a very convergent duct shape, which begins accelerating the gas stream and reducing the static pressure in preparation for entrance to the turbine section.

Another important combustion process topic is emissions control. Because the gas turbine combustion is continuous, it can be made to be very efficient in maintaining very low levels of incomplete combustion products like carbon monoxide (CO) or unburned hydrocarbons (UHC). However, the other major emissions component, oxides of nitrogen (NOx), is not related to combustion efficiency, but strictly to the temperature levels in the flame (and the amount of nitrogen in the fuel). One common solution to NOx emissions involves injecting steam or water into the combustion zone in order to suppress NOx emissions by "cooling" local flame temperatures.

Transition Pieces

Larger industrial gas turbines often have a set of flow distribution pieces between the combustors and the first stage nozzle of the power turbine. These *transition pieces* (TPs) (Figure 14.10) enables the hot combustion gases to flow smoothly from the combustors into the first stage nozzle, meaning there must be one transition piece per combustion chamber. It is common to see TPs internally coated with a thermal barrier coating (TBC) to protect them from the high temperature gas flowing through them.

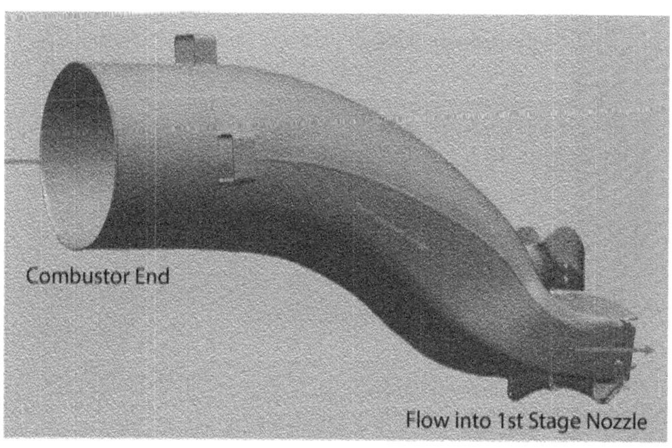

Figure 14.10 Gas turbine transition piece.

Due to the high pressure and temperature conditions TPs experience, they need to be carefully inspected during the hot gas path and major inspections. Wall thinning, thermal cracking, and loss of the thermal bearing coating are common issues detected during inspection. A cool spot in the exhaust temperature spread can sometimes be an indication of a cracked TP that is allowing cooler compressor air to leak into a TP and cool the gas entering the 1st stage nozzle. It is highly recommended that a spare set of new or refurbished TPs be kept in stores at all times in the event they are needed.

During the refurbishment of TPs, dimensional distortion around the picture frame, which is located at the. 1st stage nozzle end, due to time and temperature, can result in excessive gaps between the end seal slots and the adjoining transition pieces. Shops use inspection fixtures to ensure proper dimensional fits of all the transition pieces against the mating surfaces of the turbine.

Expansion Turbine

The expansion turbine (see Figure 14.11) is found at the back of a single-shaft gas turbine. The expansion turbine converts the gaseous energy of the burned air/fuel mixture out of the combustor into mechanical energy that can power an electrical generator or drive a compressor. Each stage of the turbine consists of a row of stationary vanes followed by a row of rotating

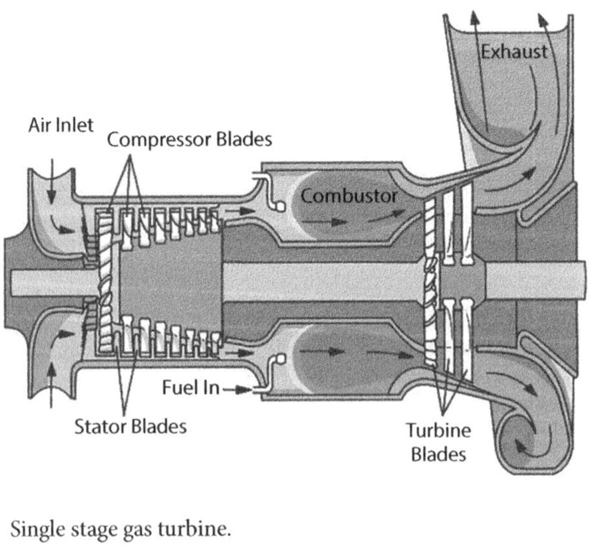

Figure 14.11 Single stage gas turbine.
Notice the three rows of expansion turbine blades inside the red dashed oval.

blades. This is the reverse of the order in the compressor. In the compressor, energy is added to the gas by the rotor blades, then converted to static pressure by the stator vanes. In the turbine, the stator vanes increase gas velocity and then the rotor blades extract energy.

The expansion vanes and blades are airfoils that provide for a smooth flow of the gases. As the burned air/fuel mixture enters the turbine section from the combustion section, it is accelerated through the first stage stator vanes. The stator vanes (also called nozzles) form convergent ducts that convert the gaseous heat and pressure energy into higher velocity gas flow (P_i). In addition to accelerating the gas, the vanes "turn" the flow to direct it into the rotor blades at the optimum angle. All the combustion gases must flow across the airfoils to achieve maximum efficiency in the turbine. In order to ensure this, seals are used at the base of the vanes to minimize gas flow around the vanes instead of through the intended gas path.

Turbine Section Challenges and Solutions

The most critical sections of a turbine are the first few stages where nozzles and rotating blades are exposed to "red-hot" gases. The highest temperature in the gas turbine, called the "firing temperature," is where the gas stream exits the combustors and enters the first stage of the expansion turbine. Over time (see Figure 14.12), typical firing temperatures have climbed from 1100 F to over 2600 F. Improved turbine section materials, coatings, and cooling methods have all been employed to increase firing temperatures. The reason for the push to increase firing temperatures is because a higher firing temperature means an improvement in power output and overall efficiency.

The design firing temperatures in this section far exceeds the creep-fatigue limits for the current alloys employed. To enable components in the hot section to survive:

- Internal cooling passages are cast and machined into discs, nozzles, and blading to allow cooler compressor bleed air to pass through them in order to maintain material temperatures at acceptable limits (see Figure 14.13).
- Creep-resistant directionally-solidified (DS) and single-crystal (SC) blade production technology are employed to improved ductility and thermal fatigue life.
- Thermal barrier coatings (TBC) are not routinely used to protect aerodynamic surfaces and materials from corrosion, oxidization, and erosion.

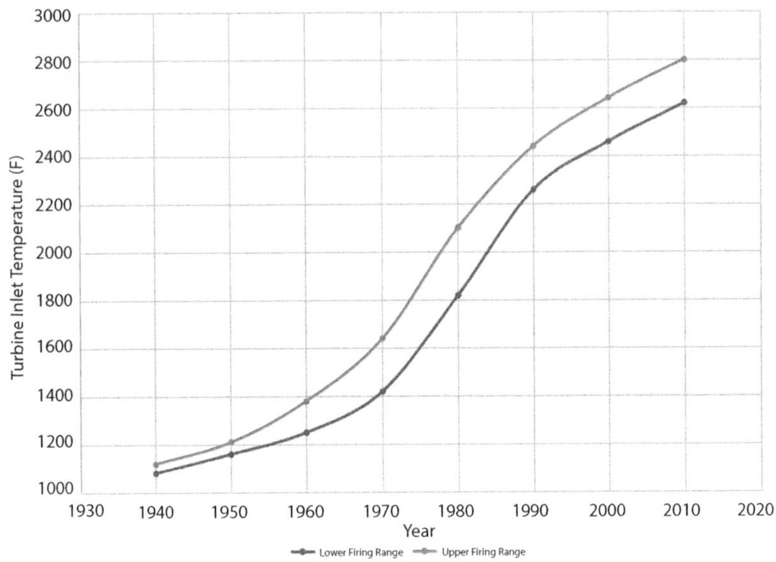

Figure 14.12 Gas turbine firing temperature trend.

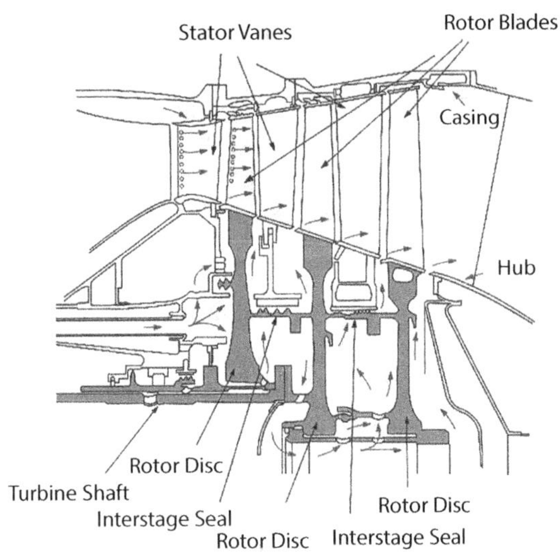

Figure 14.13 Cross section of hot section of gas turbine.
Flow arrows (blue) depict how air from the compressor is used to cool discs and blades in the hot section.

- Rotating blades and discs are carefully designed to survive at elevated temperatures, under high centrifugal and mechanical stresses.

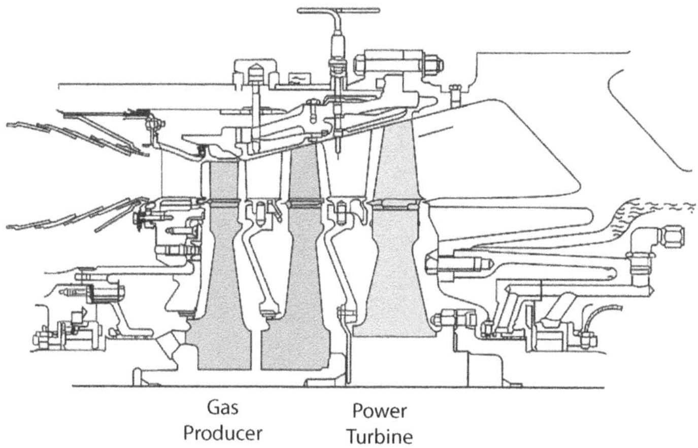

Figure 14.14 Gas producer blades and power turbine blades in two shaft gas turbine.

Two Shaft Gas Turbine Construction Details

As we have seen, two shaft gas turbines have a different internal construction than single shaft gas turbines. To understand their operation, we need to know more about the design and function of the gas producer and power turbine sections inside of two shaft gas turbines (Figure 14.14).

Gas Producer

In two-shaft gas turbines, the assembly of the air compressor section, diffuser section, and combustion section and the first few stages of the expansion process is commonly referred to as the Gas Producer.

The purpose of the gas producer turbine is to convert a portion of the energy of the expanding hot gases into mechanical rotary energy. As the hot gases pass through the gas producer turbine, it causes the gas producer turbine rotor to rotate. The rotational force drives the gas producer air compressor section and accessory drive.

Lower Pressure Power Turbine (LP)

The hot gases subsequently flow through the power turbine (Figure 14.15) after expanding through the gas producer turbine. Further expansion of the hot gases in the power turbine converts energy from the expanding gases into rotational force, causing the power turbine to rotate the driven piece of equipment.

Figure 14.15 Power turbine wheel.

Flow nozzles are used to redirect hot gases efficiently from one expansion stage to the next expansion stage. The gas pressure drops but velocity increases as the hot gas passes through each nozzle set. As the hot gas passes through the second stationary nozzles located between the wheels, it directs the flow at an optimal angle to maximize the expansion process. The flow pushes against the second stage rotating vanes of the second stage wheel.

In summary, two-shaft gas turbines consist of two rotors, turning at different speeds, each with one or more hot gas expansion wheels. The first turbine (the high-pressure turbine) is designed to extract sufficient horsepower from the hot gases to drive the air compressor. The second turbine, called the power turbine (Figure 14.15), which is not mechanically coupled to the high-pressure turbine, generates power though the expansion of the combustion gases expelled by the gas producer. The power turbine is designed to provide sufficient output power to drive a fluid machine, such as a pump or compressor, over a range of operating speeds.

Typical Conditions Inside an Industrial Gas Turbine

Now that we have reviewed the main sections of a gas turbine, let's briefly review some typical operating conditions. Gas turbine design pressure ratios vary as follows:

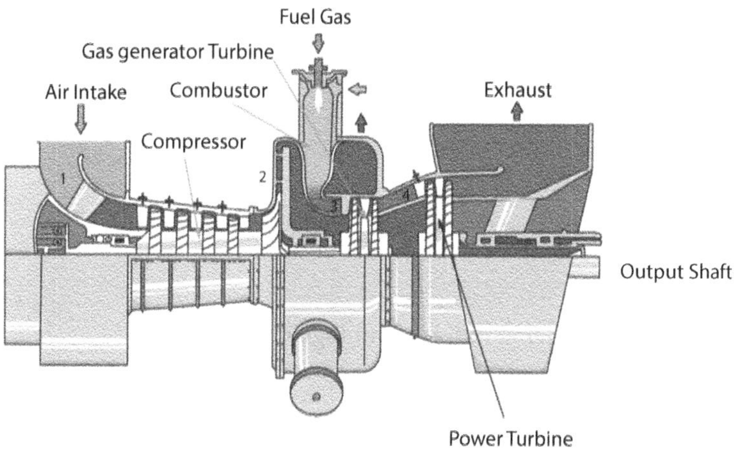

Figure 14.16 Locations listed in Table 14.1 are defined in this image.

Table 14.1 Typical conditions inside a gas turbine (see Figure 14.16).

GT location	Symbol	F/psia	C/psia	K/kPa
1. Air Compressor Inlet	T1	60.8 F	16 C	289 K
	P1	14.65 psia	14.65 psia	101 kPa
2. Air Compressor Outlet	T2	581 F	305 C	578 K
	P2	219.73 psia	219.73 psia	1515 kPa
3. Gas Generator Turbine Inlet	T3	1832 F	1000 C	1273 K
	P3	219.73 psia	219.73 psia	1515 kPa
4. Power Turbine Inlet	T4	1295.6 F	702 C	975 K
	P4	79.2 psia	79.2 psia	546 kPa
5. Power Turbine Outlet	T5	669.2 F	354 C	627 K
	P5	14.65 psia	14.65 psia	101 kpa

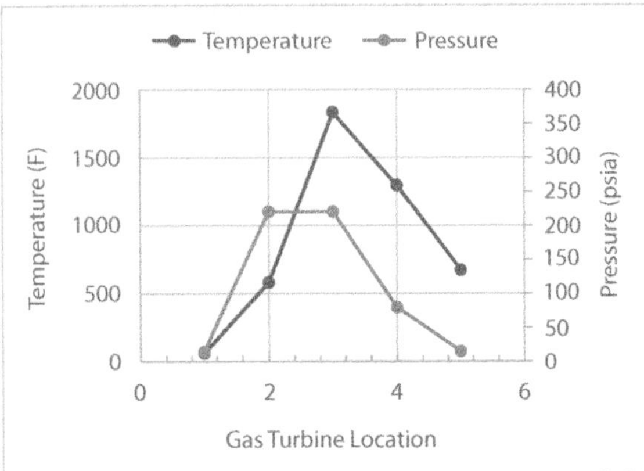

Figure 14.17 How temperature and pressure change inside a gas turbine. The location number corresponds to the turbine in Figure 14.15.

- 7.5:1 – smaller & older technology GT's,
- 35:1 ~ 40:1 – recent, most advanced GT's

Table 14.1 summarizes typical conditions you could expect to see along the gas turbine flow stream. Figure 14.17 graphically depicts these conditions.

Effect of Atmospheric Conditions

The performance of a gas turbine engine is highly dependent on the mass flow of air entering the engine. At a constant speed, the compressor pumps a constant volume of air into the engine with no regard for air mass or density. If the density of the air decreases, the same volume of air will contain less mass, so consequently less power is produced. Conversely, if air density increases, power output also increases as the air mass flow increases for the same volume of air. Atmospheric conditions affect the performance of the engine since the density of the air will be different under different conditions. On a cold day, the air density is higher, so the mass of the air entering the compressor is increased. As a result, higher horsepower is produced. Conversely, on a hot day or at high altitude, air density is decreased, resulting in a decrease of output shaft power.

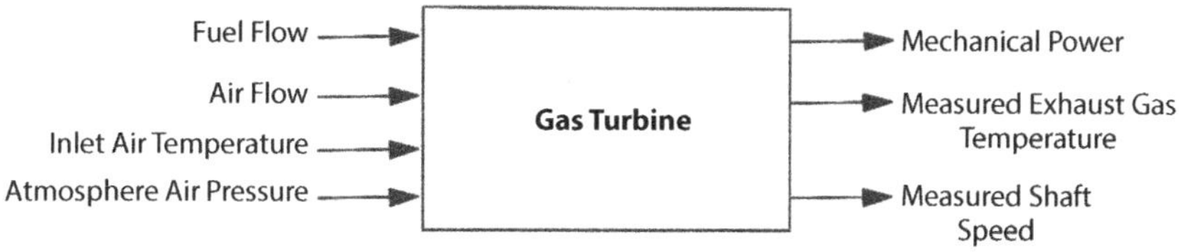

Figure 14.18 Gas turbine control variables.

Gas Turbine Controls

The principal objective of a gas turbine fuel controller is to vary the fuel flow as needed to maintain the desired power turbine speed regardless of load or fuel quality variations (see Figure 14.18). In some applications, that speed set point is held constant, but it is more commonly varied to achieve a cascade control objective. It is common for most gas turbine control systems to integrate performance and anti-surge control applications to provide integrated control and protection of the entire machine train. The set point for the fuel control application would then be controlled by an external performance controller.

Protection

The most important objective of the gas turbine controller is to ensure the safety of your machine and operating personnel. The controller is programmed with specific operating limits to protect gas turbine internals from damage and reduce maintenance. Here are the common protection control features for gas turbines:

- Fuel Regulation for controlling a selected shaft speed or the power output, while protecting against speed, temperature, and pressure extremes
- Overspeed Protection for the power turbine and gas generator by using a combination of closed- and open-loop responses. There must be protection to both high-and low-pressure rotors, depending on the configuration of the gas turbine.
- Overtemperature protection and pressure limiting
- Flameout and over-fueling protection, including scheduled acceleration and deceleration limits
- Surge protection for the gas producer compressor

- Alarms and/or shutdown if any control element deviates too far from its intended position

Fuel and Fuel Treatment

Industrial gas turbines allow operation with a wide variety of gaseous and liquid fuels, while maintaining very low emissions. Today's designs allow the use of LNG as fuel gas and the operation of gas turbines with fuel gas from a number of different sources. Lean premix combustion systems, widely used for emissions control, require particular considerations regarding fuel quality and composition. To determine the suitability for operation with a gas fuel system, various physical parameters of the proposed fuel need to be considered such as the heating value, dew point, Wobbe index, and others. Attention is given to the impact of fuel constituents on combustion characteristics and the problem of determining the dew point of the potential fuel gas at various pressure levels.

Gas Fuels

The quality and composition of fuel burned in a gas turbine impacts the life of the turbine, particularly its combustion system and turbine section. The fuel specified for a given application is usually based on availability and price. Natural gas is a typical fuel of choice for gas turbines due to its relatively low cost, widespread availability, and low resulting emissions. However, the composition of fuel gas can widely vary from gas with significant amounts of heavier hydrocarbons (butane and heavier), to pipeline quality gas consisting mostly of methane, to fuel gas with significant amounts of noncombustible gases (such as nitrogen or carbon dioxide).

Degradation and Water Washing

The fouling of axial flow compressors is a serious operating problem, therefore its control is of supreme importance to gas turbine operators, especially in the deregulated and highly competitive power market. It is also significant in the mechanical drive market where a loss in gas turbine output directly affects plant throughput. Foulants in the ppm range can cause deposits on blading, resulting in severe performance deterioration.

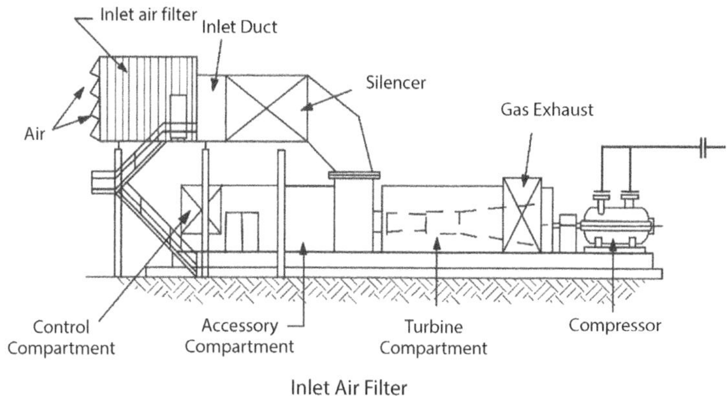

Figure 14.19 High quality air filtration system for a gas turbine.

The effect of compressor fouling is a drop in airflow and compressor isentropic efficiency, which results in a "rematching" of the gas turbine and compressor causing a drop in power output and thermal efficiency. In extreme cases, fouling can also result in surge problems, as it tends to move the compressor surge line to the right, i.e., towards the operating line.

Estimates have placed fouling as being responsible for 70 to 85 percent of all gas turbine performance losses accumulated during operation. Output losses between 2 percent (under favorable conditions) and 15 to 20 percent (under adverse conditions) have been experienced.

Fouling is best controlled by a combination of air filtration and washing: The first line of defense is to employ a high quality air filtration system (see Figure 14.19). However, as fouling will inevitably occur, compressor online and offline washing should be used to control its impact.

Advanced Materials for Land Based Gas Turbines [4]

The gas turbine (Brayton) cycle is a steady flow cycle, where the fuel is burned in air and the peak temperature directly depends upon the material capabilities of the parts in contact with the combustion gases. In the gas turbine, the combustion and turbine parts are continuously in contact with high temperature gases. The higher the firing temperature, the higher the turbine efficiency and output. Therefore, increasing turbine inlet temperature (firing temperature) has been the most significant thrust for gas turbines over the past few decades and is continuing in pursuing higher power rating without much increase in the weight or size of the turbine (see Figure 14.17). Firing temperature capability has increased from 800 °C (1472 °F) in first generation gas turbines to 2912 °F (1,600 °C) in the latest models of gas turbines. Higher firing temperatures can only be achieved by employing the improved materials for components such as combustor, nozzles, buckets (rotating blades), turbine wheels, and spacers. These critical components encounter different operating conditions with respect to temperature, transient loads, and environment. The temperature of the hot gas path components (combustor, nozzles, and buckets) of a gas turbine is beyond the capabilities of the materials used in steam turbines, thus requiring the use of much superior materials like superalloys, which can withstand severe corrosive/oxidizing environments, high temperatures, and stresses. However, for thick section components such as turbine wheels, which require good fracture toughness, low crack growth rate, and low coefficient of thermal expansion, alloy steels are extensively used. But, the wheels of the latest models of gas turbines operating at very high firing temperatures (around 1,300–1,600 °C (2912-2372 F) are made of superalloy, which offers a significant improvement in stress rupture, tensile and yield strength, and fracture toughness required for the application.

Blade Degradation

The high-pressure turbine blades experience one of the most severe environments faced by man-made materials. In addition to temperatures approaching the substrated melting point, one must add the considerable stresses caused by shaft rotation. Thermal barrier coats are a necessity in today's gas turbines. The function of all coatings is to provide a surface reservoir of elements that will form very protective and adherent oxide layers, thus protecting the underlying base material from oxidation and corrosion attack and degradation. Rapid blade failures would result if thermal barrier coatings were not applied. In addition, thermal barrier coatings can prevent accelerated oxidation and, depending on the environment, hot corrosion degradation.

> *Oxidation* is the chemical reaction at high temperatures between a component and the oxygen in its surrounding gaseous environment.
>
> *Hot corrosion* (see Figure 14.20) is a form of accelerated oxidation that is produced by the chemical reaction between a component and molten salts deposited on its surface.

Figure 14.20 Hot corrosion on first stage power turbine nozzle.

Coatings can considerably enhance the oxidation/hot corrosion resistance of hot gas path components. Metal loss due to oxidation can be reduced by the formation of protective oxide scales. Chromium, aluminum, and silicon are the only chemical elements known to form protective oxide scales at the temperatures encountered in gas turbine engine hot sections.

In addition to oxidation and hot corrosion, coatings will degrade and diffuse into the substrate alloy as they are not in thermodynamic equilibrium with the latter. This is of concern not only because it may modify the carefully designed mechanical properties of the substrate, but also because loss of aluminum to the substrate reduces the oxidation life of the coating.[3]

Condition Monitoring Approaches

The concept of a slow degradation of component strength or performance is the basis for condition monitoring systems. The condition monitoring system relies on a variety of sensors and analysis methods that help in the observation of component degradation.

Aerothermal Performance Analysis

The aerothermal performance of turbomachinery provides invaluable insight into a gas turbine's operating health. Performance analysis can vary from relatively simple calculations to advanced gas path analysis techniques to pinpoint internal faults. These analyses involve computations and correlations of all performance variables in the gas path. An aerothermal performance analysis system should be capable of modeling and detecting degraded performance.

It is often valuable to integrate performance monitoring with vibration analysis, as several vibration problems are manifestations of underlying aerothermal problems. This monitoring strategy can also provide insight into how efficiently fuel is being utilized and thus facilitates significant fuel savings if degradation is controlled. Included within "performance analysis" are items such as power turbine inlet temperature or exhaust gas temperature spread monitoring, which is a critical and valuable indicator of hot section health. These monitoring approaches can significantly extend hot section life. Excessive spreads can occur due to a variety of reasons including excessive air leakages, blockage of nozzles, and cracks in the combustor liner/transitions pieces.

Vibration Analysis

With the correct choice of sensors and analysis techniques, vibration analysis is an excellent condition monitoring tool when used in conjunction with other condition monitoring techniques. Table 14.2 summarizes the sensor configurations recommended for gas turbines.

Table 14.2 Recommended sensor configurations for gas turbines.

Measurement	Installation details
Radial Bearings	Two proximity probes should be mounted orthogonally, i.e., in an XY configuration at each bearing.
Thrust Bearings	Two proximity probes that sense a smooth flat surface that is perpendicular to the shaft's rotation, such as a thrust collar.
Shaft Speed/Phase Reference	Use a proximity probe to provide speed and phase measurement on each shaft on the machinery train to allow phase analysis of the entire train.
Casing Vibration	For machines with flexible bearings, housing seismic vibration transducers can be mounted near radial bearings.

Transient Analysis

Condition monitoring based on transient performance and vibration data, i.e., data taken during startups and shutdowns, is an important facet of any analysis system. In a gas turbine, valuable condition monitoring information is available by examining the profile of startup acceleration, coast down times, exhaust gas temperature (EGT) response during light off, and other transient behavior.

Mechanical Transient Analysis

Compared to aerothermal transient analysis, mechanical transient analysis has been an area that has been used extensively in condition monitoring systems over the years. Transient techniques include:

1. Vibration Cascade Analysis: This type of analysis permits a visual representation of the frequency content of vibration signals during startup or shutdown. Vibration cascades can be formed either on a Delta time or Delta RPM basis.
2. The Bode and Nyquist plots are classic rotor dynamic displays of particular use for displaying the location of critical speeds and their amplification factor. The separation margin from the critical speed to the operating speed range can be reviewed for compliance with the manufacturer's recommendations or industry guidelines.
3. Graphical plots of the vectorial summation of proximity probe gap voltages are also useful. These plots can be used to display the movement of the shaft within bearings during startup/shutdown.
4. For startup analysis of other mechanical parameters such as lube oil temperatures and pressure, it is possible to prepare maps showing these parameters as a function of time, rotor RPM, and load as appropriate.

Some problems that may be detected by using a mechanical transient analysis include:

1. Bearing wear and damage, which can be detected by evaluating the transient data for changes in amplification factors (Q factor).

2. Thermal bows occurring in the rotor. Very often, thermal bows are the result of a hot restart when the rotor has taken on a bow and the casing has cooled non-uniformly.
3. Rotor rubs can be detected by analysis of vibration signatures

Dynamic Pressure Analysis

The use of dynamic pressure transducers has worked well to detect certain blading and compressor flow instabilities. This is an important facet of condition monitoring that has not received much attention. When applied to the combustors, the dynamic pressure signatures can provide information regarding combustion instabilities and can explain observed combustor liner wear. With the use of dry low emission combustors, dynamic pressure measurement is a very valuable condition monitoring tool to ensure operability and reliability.

Lube Oil Debris Analysis

Debris is continuously generated by rolling and sliding surfaces that are associated with normal, accelerated, and incipient failure modes. This debris can be generated at differing production rates and be made up of different material sizes and distributions. Depending on the failure mode, debris production can increase in certain size ranges. In rolling element bearings (such as used in aero-derivative gas turbines) under full elasto-hydrodynamic lubrication (EHL) conditions, where film thickness is large compared to average bearing surface roughness, the predominant failure mode is spalling or macro pitting induced by surface fatigue. Particles produced are typically in the 100 - 1000μ range. In the boundary lubricated and mixed mode (partial EHL) regimes, particles are typically smaller (< 100μ). Under these lubrication conditions, abrasive and adhesive type accelerated wear models occur. Bearing skidding can occur when bearing loads are light and can produce small debris (< 25μ).

Borescope Inspection

One off-line condition monitoring technique that has proven to be useful is the borescope inspection. (A borescope is an optical instrument consisting of a rigid or flexible tube with an eyepiece or display on one end and an objective lens or camera on the other that allows the inspection of narrow, difficult-to-reach cavities.) These inspections are usually carried out at fixed intervals dependent on the machine and a video camera is used to record results. Borescope inspections are usually very quick and result in a minimum loss of turbine availability. For borescope inspections, it is important to have well trained personnel and clear cut procedures defined in concurrence with the machinery manufacturer to ensure full coverage of the critical components. By using a video camera to record the inspection, one can enlist expert outside help to interpret the data.

Condition Monitoring as a System

As previously discussed, there are specific methods, technologies, and processes that can be done manually and ad hoc. However, there is a trend towards automating condition monitoring and remote condition monitoring, especially in the context of fleet management. Remote monitoring and diagnostics, in conjunction with centralized data storage, creates the ability to perform comparative analysis to similar engines (Fleet Statistical Analysis). Including ancillary system performance in the periodic analysis can allow the identification of risk factors to engine durability (filter fouling rates, lube oil temperature and pressure control, fuel temperature, machine event diagnostics, etc.).

Rigorous tracking of availability and reliability as key performance indicators and downtime analysis enables understanding and analysis of what systems and situations are having the most negative impact on these metrics. Finally, it must be emphasized that a key requirement for successful gas turbine condition monitoring is that all preventative measures have to be performed regularly and that there is sufficient expertise available to analyze all the field data collected. Once the data has been collected and analyzed, it must be transformed into actionable information so that site decision makers can develop a plan of action as required.

Gas Turbine Maintenance Inspections [5]

Gas turbines require a well-defined program of planned periodic inspections along with a complete set of replacement parts and regular repairs to achieve optimum availability and reliability. The inspection and repair requirements of the gas turbine lend themselves to establishing a pattern of inspections, starting with very minor work and increasing in magnitude to a major overhaul, and then repeating the cycle based on the experiences of the manufacturer and end users. These periodic inspections can be optimized to reduce unit outages and maintenance costs for the user's specific mode of operation, with a goal of maintaining maximum availability and reliability. Inspections can be classified as operational or shut down type inspections. The operational inspections are used as indicators of the general condition of the equipment and as guides for planning the disassembly maintenance program. The entire scope of inspections can be described as a) standby, b) running, c) combustion, d) hot-gas path, and e) major.

Standby Inspections

Standby inspections pertain particularly to gas turbines used in intermittent service, such as peaking and emergency duty units. Starting reliability is of prime concern since a delay in starting usually means that the demand for the unit might have passed. This type of inspection includes routine servicing of the battery system, lubrication, changing of filters, checking oil and water levels, cleaning relays, checking device calibrations, and other general preventive maintenance. This servicing can be performed in off-peak hours without interrupting the availability of the turbine. A periodic test run is an essential part of the standby inspection.

Running Inspections

Running inspections consist of the observations made while a unit is in service. The turbine should be inspected on a defined schedule, which should be established as part of the site's maintenance program consistent with the operator's requirements. Operating data should be recorded to permit an evaluation of equipment performance and maintenance requirements. Typical running inspections (shown below) include load versus exhaust temperature, vibration, fuel flow and pressure, exhaust temperature control and variation, and startup time.

Typical Running Inspections Recommended for Gas Turbines

- Load versus exhaust temperature
- Vibration levels
- Fuel flow and pressure
- Exhaust temperature control
- Exhaust temperature variation or spread
- Start-up time

The general relationship between load and exhaust temperature should be observed and compared with previous data. Ambient temperature and barometric pressure will have some effect on the absolute temperature levels. High exhaust temperature can be an indicator of deterioration of internal parts, excessive leaks, axial-flow compressor fouling, or improper control settings. Initial startup data should be used as the reference point for checking. Power loss resulting from deteriorated parts or leaks may require disassembly of the turbine to restore power.

Power losses due to fouling of the axial flow compressor can usually be restored by cleaning the compressor while in service. Online wet cleaning involves the injection of a mixture of water and chemical detergent using atomizing spray nozzles positioned around the compressor air intake plenum. This is followed by a flushing period using pure water. With online cleaning, it is mandatory to use demineralized water for preparing the cleaning fluid and for flushing. A successful cleaning will reduce the exhaust temperature for a given load and will increase the compressor discharge pressure. Most gas turbine users also use crank washes (see *Crank Washing* insert below) as a means of regaining and maintaining compressor performance. If the need to clean the compressor is frequent, the causes of the fouling condition should be determined and corrected.

Crank Washing

Offline water washing, or crank washing, is a proven means of removing dirt and oil from the gas turbine compressor blades. Crank washing involves shutting down the gas turbine, letting it cool down, and putting it on the crank cycle, which turns the rotor at roughly 20% of the rated speed. Then, a mixture of water and solvent is injected through nozzles that are installed in the compressor's inlet bell mouth at a rate from 50 to 90 GPM. Typically, the water wash cycle will last 20 to 30 minutes, but to make sure the compressor is clean, the operator should monitor the water discharge and let the system run until the water discharge is clear. Crank washing is considered by gas turbine users to be the best way to effectively clean the compressor and regain lost power due to fouling.

The vibration levels on the unit should be observed and recorded. Minor changes will occur with changes in operating conditions. However, major changes, or a continuous trend to increase, indicates that corrective action is required.

The fuel system should be observed for general fuel flow versus load relationship. Fuel pressures through the system should be observed. Changes in fuel pressure can indicate that fuel nozzle passages are plugged or fuel metering elements are damaged or out of calibration.

The variation in turbine exhaust temperature should be measured. An increase in temperature spread across the transition pieces indicates combustion deterioration or fuel distribution problems. If not corrected, reduced life of downstream parts can be expected.

Figure 14.21 Combustion inspections are required to inspect combustion liners and fuel nozzles, which are recognized as the first parts requiring replacement and repair.

Combustion Inspections

Combustion Inspections involve 1) the removal and replacement of the combustion liners and fuel nozzles (see Figure 14.21) and 2) a visual inspection of the transition pieces and nozzles (first-stage). This visual inspection is accomplished by the use of a borescope, which is inserted through the combustion liner area to allow examination of the transition pieces and first-stage nozzle.

Hot Gas Path Inspections

The hot gas path inspection includes the full scope inspection of the combustion system components and a detailed inspection of the turbine nozzles, shrouds, and buckets.

Major Inspections

Major inspections include detailed inspections of the combustion system components, all the hot gas components, and a thorough inspection of the compressor blades, turbine buckets, bearings, rotor, and casing conditions. The major inspection is a flange-to-flange inspection of gas turbine components which are subject to erosion, cracking, and fouling during normal turbine operation (see Figure 14.22). A major gas turbine outage involves the following activities:

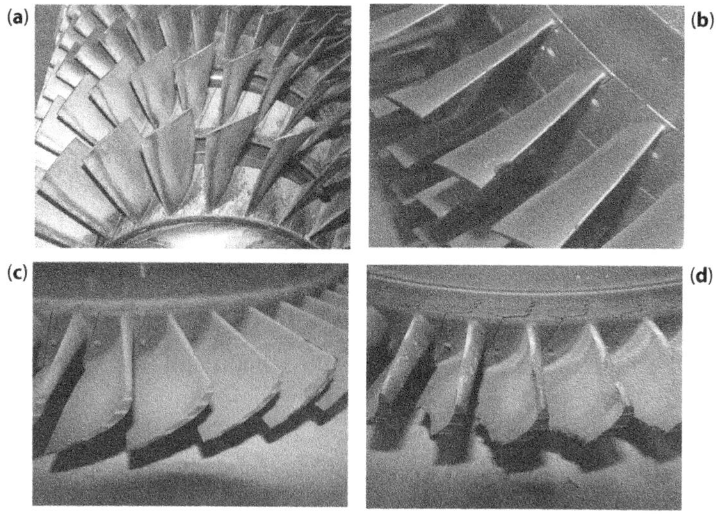

Figure 14.22 Some common inspection findings during a major inspection. (a) Air compressor fouling, (b) Air compressor foreign object damage (FOD), (c) Power turbine blade tip rubbing, (d) Hot corrosion on blades.

- Removal of the upper inlet case, mid, and discharge casings
- Inspection of the compressor and inlet cases to assess the severity of fouling, erosion, corrosion, and leakage
- Checking the rotor and stator compressor blades tip clearances and inspecting for rubs, bowing, cracking, and warping
- Checking the turbine nozzles and shroud clearances and inspecting for foreign object damage, cracking, erosion, and corrosion
- Checking the fits of seals and nozzle fits
- Checking nozzles and diaphragms for rubs, erosion, fretting, or thermal deterioration
- Removing turbine buckets and inspecting them for foreign object damage, cracking, erosion, and corrosion
- Checking bearings and seals clearances and inspecting them for damage and wear
- Checking all internal clearance and comparing them against original values

All major hot gas path component inspections involve a detailed inspection of the shrouds and turbine nozzles. The top half of the turbine will have to be removed in order to perform this inspection properly. The shrouds, diaphragms, and nozzles are removed and the turbine buckets are inspected but not removed. Users should be mindful of the environment in which the turbine is used. Some things such as the airfoils can be really affected by their environment. Other factors that should be taken into consideration when you are drawing up a maintenance plan of your hot gas path components turbine is the downtime, type of fuel, environment, utilization need, reliability need, on-site capability, manufacturer's recommended maintenance program, duty cycle, and specific design features.

Life Cycle Management [6]

Each gas turbine manufacturer provides a schedule of inspections recommended to determine the condition of the components and, depending on their condition, the repairs/replacement activities that should be carried out to enable continued safe operation for a specified period of time. In practice, we find that the recommended maintenance intervals are similar for most makes of gas turbines. These recommended interval are as follows:

- Combustor inspection at 8,000 equivalent operating hours (EOH) (see note below)
- Hot gas path component at 16,000 to 24,000 EOH
- Major inspection at around 32,000 to 48,000 EOH

Note: EOH, or equivalent operation hours, is a calculation that OEMs use to determine how much of the metallurgical life remains in a gas turbine. Factors such as number of starts, the quality of the fuel, loading, etc. determine how much of the gas turbine's useful life remains. The most important concern is how much creep has occurred in the turbine wheels and buckets due to actual firing conditions.

However, different manufacturers apply different rules with respect to arriving at the EOH, which determines the intervals between the inspections and the type of the inspection to be carried out. Table 14.3 illustrates how gas turbine users can build a long-term inspection and maintenance plan based on the various types of required inspections. In this example, we see that:

a) Combustion inspections (CIs) occur at 8000, 24,000, 40,000, 56,000, 72,000, and 88,000 hrs.,
b) Hot gas path inspections (HGPIs) occur at 16,000, 48,000, and 80,000 hrs.
c) Major inspections (MIs) occur at 32,000, 64,000, and 96,000 hrs.

Table 14.3 Hypothetical gas turbine inspection and replacement schedule.

	Operating Hours											
Inspection & Replacements	**8000**	**16000**	**24000**	**32000**	**40000**	**48000**	**56000**	**64000**	**72000**	**80000**	**88000**	**96000**
Combustion	×		×		×		×		×		×	
Hot Gas Path		×				×				×		
Major				×				×				×
Time based replacements of major hot gas path components												×

Note: This schedule assumes that certain hot gas path components need to be replaced or refurbished after 100,000 hrs. of operation.

Notice that, at the 96,000 hour mark, I have included a "time based replacement of major gas path components" event. I have added this milestone because manufacturers and users have learned from experience that certain components in the hot end of gas turbines have a maximum operating life due to erosion, corrosion, creep, loss of thermal barrier coating, etc. Of these failure modes, creep is the most insidious and challenging to assess in the field. For these reasons, manufacturers set maximum time limits on components, such as transition pieces, power turbine discs, and power turbine blades. The inspection pattern seen in Table 14.3 is repeated during the next 104,000 hr. interval until the gas turbine is retired or replaced. This type of visual display of the inspections is helpful for outlining and presenting spare part and labor cost requirements over the life of the turbine.

In practice, the various inspection intervals are ultimately based on a mixture of manufacturer's recommendations and user experience. The inspection timetable presented here has been provided to give the reader an insight into the rationale for gas turbine maintenance inspection schedules.

Non-Destructive Testing (NDT)

The following non-destructive testing (NDT) methods are highly recommended during major inspections to ensure gas turbine components are fit for service: a) A visual inspection, which involves looking for cracks in all suspected high stress locations and signs of thermal distress in high temperature locations; b) Penetrant testing (PT) and magnetic particle testing (MT), which are well suited for finding surface and slight subsurface discontinuities. Keep in mind that penetrant testing requires that the discontinuity be open to the surface for detection and that magnetic particle testing can only be used on magnetic materials; c) Longitudinal and shear wave ultrasonic testing (UT) which can be used for full volumetric inspection of a component, while eddy current testing (ET), an electromagnetic method, is sensitive to small surface or slightly subsurface indications in many materials; d) *In-situ* metallography analysis, where surface replications are taken and can be used to evaluate the metallurgical condition of aging rotors. All these inspections have been proven to be useful in ensuring that crucial gas turbine components do not run past their maximum predicted life.

Spare Parts

Due to their criticality, the gas turbine repair times need to be minimized and the total downtime needs to be limited to the maintenance time windows allocated by the process. There are many philosophies regarding what spare parts should be kept in stores to ensure they are ready when an outage occurs. Here are some spare part ideas that may help users minimize the downtimes during gas turbine inspection events:

- Stock all replaceable spare parts, assembles, and consumables that are likely to be needed during a normal repair but that are not readily available from the supplier.
- Replace unrenewable parts removed during repair with new parts from stores (or from manufacturer). Then, restock unrenewable parts and consumables that were removed from stores.
- During repairs or scheduled inspections, replace all damaged parts from stores (or from manufacturer) and scrap unrenewable parts removed during repair. (This recommendation assumes removed spare parts have no salvage value.) Then, restock all parts removed from stores.
- Renewable parts: Either a) remove and repair major renewable parts from stores during outage, or b) have renewed spares in stores that are ready for use. If the renewal time for spare parts is longer than the repair time, then it makes sense to keep renewed spares in stock.
- In most cases, it makes sense to have two sets of renewable spares: one set in the running machine and another set kept in stores. During a repair, remove and replace renewable parts with those previously repaired and held in stores. When parts are no longer renewable, buy new replacements parts for the warehouse.
- Many gas turbine users store an entire gas turbine rotor (single shaft designs) or a compressor and power turbine rotor (two shaft designs) as insurance spares. These spares are priceless in the event of a catastrophic failure or if major damage is discovered during a major inspection. Another advantage of having a spare rotor or rotors is that it allows detailed metallurgical inspections on a routine basis. If irreparable rotor damage is found during the metallurgic inspection, then replacement parts can be ordered as required so

that the rotor can be restored to serviceable condition and placed back into stores for the next major outage.

Final Words of Advice

The sensible operation and maintenance of gas turbines (see Figure 14.23) requires an understanding of the working principles of the gas turbine and their critical components. To ensure long-term reliable and efficient gas turbine performance, users need to:

- Be mindful that fuel composition and air filtration can have a substantial impact on the operating characteristics and life of gas turbines
- Continuously monitor mechanical and thermodynamic performance to detect changes that could signal internal deterioration
- Investigate all alarms and keep all instrumentation fully functional
- Keep all spare part designs, metallurgies, and coatings up to date
- Work with OEMs to develop an optimum inspection strategy for the hot gas path components

Figure 14.23 Industrial gas turbine in the field.

It makes good sense to develop a solid working relationship with the gas turbine supplier so that you can work together to resolve problems as they arise and to continuously improve the mechanical reliability of your fleet. Together, you can continuously improve gas turbine performance and reliability and avoid major unexpected outages.

References

1. Wikipedia, Gas turbines, "Industrial gas turbines for mechanical drive," Accessed 11/5/2021.
2. Kurz, R., Brun, K., *et al.*, "Gas Turbine Performance and Maintenance," Proceedings of the Forty-Second Turbomachinery Symposium, October 1-3, 2013, Houston, Texas.
3. The Commercial Aviation Safety Team Website, "Fundamental of Gas Turbine Engines," accessed on 11/18/2021.
4. Bucci, D., and Schaeffer, J., GECC-1: "Proven corrosion prevention for over 30 years," G. E. Materials & Process Engineering Flyer, GE Gas Power.
5. www. Industrial-electronics.com website, "Gas Turbine Cycles/Life Cycle Usage/Specific Maintenance Inspections," accessed on 12/5/2021.
6. www.twi-global website, "Factors Influencing Maintenance Intervals in Gas Turbines," accessed on 11/24/2021.

15

Reliability Improvement Ideas for Integrally Geared Plant Air Compressors

By Abdulrahman Alkhowaiter

Integrally Geared Plant Air Compression Packages

The heart of most packaged plant air compressors covered by the American Petroleum Institute (API) 672 standard (*Packaged, Integrally Geared Centrifugal Air Compressors for Petroleum, Chemical, and Gas Industry Services, Fifth Edition*) is the integrally geared compressor (see Figure 15.1). An integrally geared compressor is a multi-shaft design where each stage utilizes a three-dimensional (3D) impeller running at its optimum rotational speed. Typically, this compressor type consists of a gear unit with a central bull gear and different gear pinions driving a number of impeller stages (see Figure 15.2).

Refer to Figure 15.3 as I explain the basic operation principles of packaged air systems utilizing integrally geared compressors: (Note the system described here has four stages of compression, but three stages of compression are common.)

1. Atmospheric air enters the first stage impeller where it is compressed and discharged into the first intercooler system. The intercooler system must cool the first stage, discharge air, and remove any liquid water or scale in the air before it enters the second stage of compression.
2. Air exiting the first intercooler system enters the second stage impeller, where it is compressed and discharged into the second intercooler system. Again, the intercooler system must cool the second stage, discharge air, and remove any liquid water or scale in the air before it enters the third stage of compression.

Robert X. Perez (ed.) Condition Monitoring, Troubleshooting and Reliability in Rotating Machinery, (325–340) © 2023 Scrivener Publishing LLC

Figure 15.1 Typical integrally geared, instrument air compressor.

Figure 15.2 Major rotating components in an integrally geared compressor.

3. Air exiting the second intercooler system enters the third stage impeller, where it is compressed and discharged into the third intercooler system. The next intercooler system must cool the discharge air and remove any liquid water or scale in the air before it enters the third stage of compression.
4. Air exiting the third intercooler system enters the fourth stage impeller where it is compressed and discharged into the plant air system.

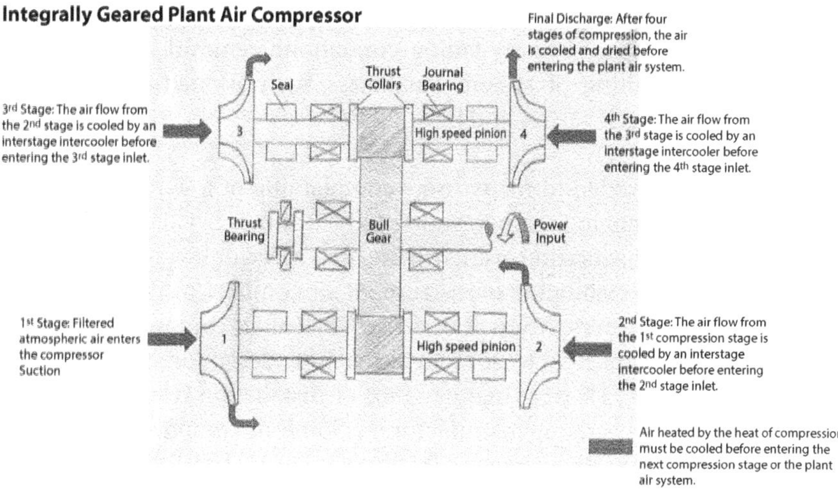

Figure 15.3 Schematic of an integrally geared air compression system.

5. It is common that the air discharged from the final compression stage is cooled and dried before entering the plant air system.

The overall reliability of an integrally geared air compressor package is based on the reliability of its numerous subsystems and components. The reliability of shaft and pinion bearings, interstage coolers, water traps, and gearing all affect the overall reliability of the system. To achieve a high overall system reliability, all of these systems and components must have exceedingly high levels of reliability. The goal of this chapter is to a) define best repair practices and b) identify key elements that can be upgraded to obtain a high level of plant air system reliability.

Reliability Concerns

The integrally geared instrument air compressors at various facilities around the world operate in challenging applications due to:

1. High suction and resulting high discharge summer air temperatures,
2. Very high operating speeds that can reach up to 55,000 rpm,
3. Dusty inlet air environment with high humidity, producing condensate that can erode impellers or liquid slugs that cause sudden seizures.

4. High temperature corrosion attack on all steel and cast iron surfaces caused by a) failing condensate removal systems and b) failing of system auxiliaries such as inlet control valves, check valves, and blow off valves.

These issues have led to many users complaining of low reliability, frequent surging events in summer, shortened run times between overhauls, and numerous high vibration events on the high speed or low speed pinions.

From long term evaluation of instrument air compressor failures covering most manufacturers, it was found that 75% of all compressor failures and trips in the years 1990 to 2015 were machine design or system design related, while only 25% were maintenance or installation related. The reliability varied greatly among manufacturers. Some air compression systems only achieved an MTBF average of 0.50 years, i.e., 6 months, while others could easily reach 5.0 years with identical PM maintenance quality and frequency. This is a highly significant difference.

This reliability evidence seems to suggest that much of the unreliability is related to the design of machines. To counter these frequent failures, this report was developed and the *Air Compressors Purchasing* standard of the company was improved from 2001 to 2015 by including the latest additions to the API-672 standard, thus insuring optimum levels of system reliability. The improvements discussed here resulted in a significant improvement in the MTBO (mean time between outages) from the original system design. The MTBO rose from 2.5 years for the original design to 8.0 years for upgraded units. (See Table 15.1: Summary of Package Design Improvements Added to API 672 to Achieve High Reliability.)

Table 15.1 Summary of recommended package design improvements to the API-672 Standard that are aimed at achieving a high level of mechanical reliability.

1. **Reliability against Surge** Compressors design summer inlet air temperature shall be the average of the highest recorded and summer design dry bulb temperature @ 1%. Design shall be rated for minimum 50% summer inlet air moisture content. 2. **Compressor Rated Discharge Pressure** shall be the actual plant air pressure control valve setting, plus 5.0 psid minimum to compensate for pressure drops between compressor and pressure control valve or dryers.

(Continued)

Table 15.1 Summary of recommended package design improvements to the API-672 Standard that are aimed at achieving a high level of mechanical reliability. (*Continued*)

3. **Blow-off valve** shall be of ball type only with stainless steel internals and Teflon soft seating.
4. **Discharge check** valve of piston type or dual plate, i.e., dual-disk, design only. Valve shall be of full stainless steel construction.
5. **The compressor head-capacity** characteristic curve shall rise continuously from maximum capacity to surge point with 107.5% minimum head rise to surge. Turndown capacity shall be at least 15% less than the capacity for all specified operating conditions.
6. **Diffuser Housings** shall be internally epoxy or phenolic coated. Vanes shall be stainless steel.
7. **Impellers shall be machined** from solid forgings or bar stock. Castings are not considered to be the best option.
8. **The required lubrication oil** is ISO-VG-46. Other oil viscosities are not acceptable.
9. **The lubrication system** including the oil reservoir, all piping, valve internals, and filter housing shall be of 304L or 316L stainless steel. The lube oil pump is not required to be stainless steel.
10. **Main drive shaft coupling** spacer length shall be at least 8.0 inches in length.
11. **Inlet Guide Vane Capacity control** shall be provided to maintain constant discharge pressure. Inlet guide vanes are required and inlet throttling valve designs are not acceptable.
12. **Intercoolers and Aftercooler** Fabricated vessels shall be ASME U-stamped or equivalent. The shells shall be of stainless steel or epoxy coated cast iron only. Tubes/tube-sheets/baffles and other internals shall be of stainless steel, Admiralty Brass, or 90-10 CU-Ni for treated water applications and of full 70-30 CU-NI for applications using sea water. The minimum tube wall thickness is 18 BWG. The drain port valve shall be fitted so that the inlet is at the lowest point on the intercooler shell base. Intercoolers shall be placed below air compressor casing.
13. **High-Performance Automatic SS ball valve** drain traps shall be provided. Additional automatic timer operated solenoid drain valves may also be provided as backup for the drainers. Solenoid valves shall have H-rated coils with full stainless steel body and Viton or PTFE soft internals.
14. **Intercooler shells shall be of stainless steel** or coated cast iron only. Aftercooler shells shall be of stainless steel only.

Developing Enhancements for Air Compressor Reliability and Performance

I have witnessed many plant engineers with unreliable units who have spent over twenty years or more trying to resolve the continuous failures experienced by air compression systems without success. The reason for this is that the *Machinery Root Cause Analysis* method was not used and a comprehensive task force was not applied to capture all defects, leading to surging and high rotor vibration. Many reliability deficiencies were found in the air compressor's aerodynamic rating, intercoolers, drainers, cooler shells, condensate recovery methods, check valves, air filtration, and other aspects. The deficiencies were primarily design related, maintenance related, and for some manufacturers, actual manufacturing defects were found on the core air compressors. The errors included incorrect machining of the gearbox and incorrect balancing of Bull gears, LS, and HS rotors. This chapter's findings and recommendations are based on the Reliability Investigations performed.

To solve these issues, a comprehensive solution program was applied from years 2005 onward on about 100 units varying from 500 HP to 4000 HP ratings at twenty locations with exceptional success. Most plants observed a life increase between overhauls of 300% to 500% and complete elimination of surge events in summer conditions. The plants, which applied the comprehensive improvements, had older units from the early 1990's or 2000's and were finally able to achieve Mean Time Between Failures above 4.0 years, Mean Time between Overhauls of 6.0 to 8.0 years, and elimination of all previous surging events. There are some operating above 6 years MTBF and reaching 10 years MTBO as a result of this total reliability program.

Reliability Improvement Program to Achieve Reliability and Eliminate Frequent Failures

The following are recommended upgrades for air compressors to achieve a major improvement in reliability and performance. The same improvements shown in Table 15.1 should also be incorporated into air compressor purchasing standards. The improvements are applicable to most manufacturers but some of them already incorporate some of the high reliability features. Neglecting one or more of these items will lead to increasing failures as an air compression machine is composed of many components operating in series. The failure of a single component or auxiliary will lead to overall failure of the machines to perform correctly.

Reliability Improvements (based on 2008 Report) Made to Five (5) 850 HP Air Compressor Failures by Engineering and Maintenance:

1. **Air Compressors Overhaul at Professional Shop:** It is not recommended that major overhauls of high speed air compressors be performed on site. This incorrect practice leads to short operating life between overhauls, while other plants that follow correct guidelines of performing full repair at shops achieve five year life minimum. The full repair-overhaul in terms of manpower training and machining skills is only available at OEM or highly specialized shops.
2. **Intercooler Shells High Corrosion:** The intercooler shells utilize internal coatings on carbon steel or Cast Iron shell wall. This coating has an average lifetime of 4 years. When the coating fails, parts of the coating or corrosion products enter impellers and cause erosion unbalance or seizures. The coating is damaged from the hot air-moisture service and must be recoated with special high temperature coatings. The solution: apply high quality coatings or purchase full stainless steel shells. The following are proven Epoxy Coating systems for the intercooler shells. Apply a single prime coat and a minimum of two top-coats. Total coat system thickness, including three coats, shall range from 275 micrometers to 400 micrometers.
3. **Cast Iron Volute casings become heavily eroded-corroded.** The air compressor usually has three compression stages. Each stage has an inlet guide and volute that corrode with time, which leads to high internal losses and therefore poor performance and surging (see Figure 15.4). Those of eight years' life and more become heavily corroded on the inner wall and that facing the impeller. This corrosion and loss of metal causes internal air recirculation and heavy loss of performance. One solution is to replace the diffusers with new ductile cast iron standard spare parts from OEM and coat them internally with HT Epoxy such as Belzona coating. Another permanent solution is to cast new volutes at a foundry from Bronze or Duplex Stainless steel. When casting in Bronze, add 25% to wall thickness of Volute. Ensure that new casting pressure parts are hydrotested at 1.30 x maximum working pressure before installation.

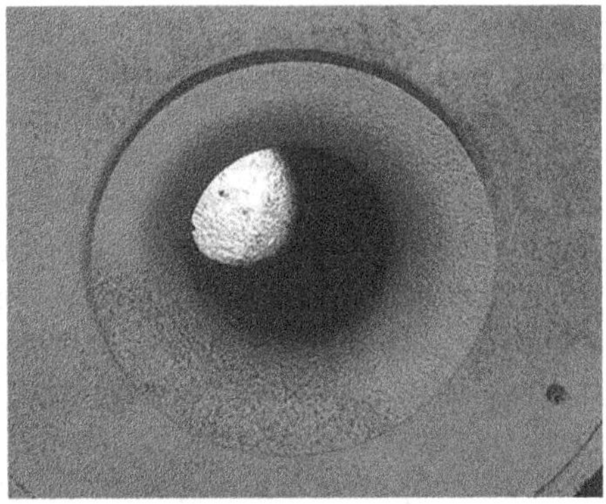

Figure 15.4 Signs of corrosion in the compressor inlet.

4. **Diffuser Vanes:** Retrofit aluminum vane diffusers with stainless steel vane diffuser assemblies on all stages of air compressors units. These are superior to the typical OEM aluminum diffusers. The aluminum diffuser material is soft and abraded by small dirt particles and moisture droplets. We have seen erosion on aluminum diffusers occur within three years of their installation, while stainless steel diffusers have a minimum 10 year life. Advanced diffuser erosion causes poor performance, surging, and internal damage. Some OEMs already adopted stainless steel diffusers as their standard.
5. **Inlet Filter Housing Dirt Leakage Causes Impeller Fouling & Vibration:** Ensure that only self-cleaning pulse type filters are used. Check that all pulsing air solenoids are operating and replace air filters every two years or more frequently. The pulse type filter housings have gaskets on their walls and silicon rubber seals, which can fail and allow outside, dirty air to enter compressors and cause frequent fouling (Figure 15.5). This issue was found to be occurring by all users. Disassemble each filter housing and renew all seals.
6. **Aftercooler Shells Internal Corrosion:** The aftercooler shells are typically carbon steel, as this reduces cost and poor company specifications allow such materials. The shell wall corrodes heavily and causes leakage at the tube-sheet to shell interface. This may lead to filling of the third stage volute with water during shutdowns, which can result in shaft shearing and heavy vibration failures. The solution to this

Figure 15.5 Signs of impeller damage from liquids and solids.

issue is to purchase new aftercoolers of full stainless steel or fabricate an identical size aftercooler shell from 304 or 316 stainless steel, using the original aftercooler detail drawings. Request a 1.25 design factor increase of aftercooler shell wall thickness to reduce stress on the stainless steel wall as compared to the existing carbon steel shell wall.

7. **Intercooler-Aftercooler tubing is typically fouled** with corrosion and deposits leading to poor heat transfer. This leads to higher inlet air temperature and surging. The intercooler aluminum fins can be damaged. Use medium pressure water jetting or chemical treatment to clean.
8. **Renew the intercoolers' bundle rubber air seals** in addition to the flange gaskets. Many users have leaking internal seals, which then allow hot air to bypass Intercooler and enter the next stage as hot air that is uncooled. Overhaul the intercooler bundles based on field inspections.
9. **High Wear of HS rotor thrust bearing** leads to seizures due to rotor axial movement into the volute face and impeller rub. Install 10-micron oil filters, replace filter housing rubber seals, and do not allow more than 1000 ppm water contamination of lube oil (see Figure 15.6). Replace thrust bearings at 5-year maximum intervals. The superior thrust design pinons are those which thrust directly against the bull gear face (see lower image in Figure 15.7). Select these designs for best reliability.
10. **The Aftercooler and Liquid Separator are defective,** leading to high water contamination in dryers. Also, drain lines

Figure 15.6 Rub and seizure: HS shaft impeller with part of broken shaft inside impeller.

can be plugged with pipe corrosion scale and water, which will lead to surging. It is recommended that all complete drain lines be replaced with new ones made of 1.50-inch stainless steel. Also, remove the air to the water separator and clean it.

11. **Intercooler Shell Liquid Drainage:** For some manufacturer designs, the drainage holes and piping is poorly designed and does not always drain properly. This causes liquids to enter the next stage impellers and mechanical damage occurs. It is a silent process that can lead to seizures and the erosion causes unbalanced vibration of pinions, which eventually results in shutdowns. The piping must have a 5% downward slope toward the automatic drain traps. Remove all existing piping and replace it with stainless steel piping of minimum 1.0-inch size.
12. **Water Slugs Entering & Damaging Impellers: Fabricate New Slug Vaporization Screens** Frequent liquid slugs escape from the coolers and enter the compressor stages causing erosion, unbalance, and major seizures with shaft shearing from large slug impact. Install vaporization screens that diffuse the slugs of water entering the stages. The screens should be made from 0.25-inch-thick 316 stainless steel alloy. The center portion should be perforated with 5/16 inch drilled holes. Insure a 7/16 inch between drilled hole centers for strength and integrity. Note: At the flange area, reduce thickness to 1/8 inch to allow easy installation.
13. **Inlet Air Filter Elements:** These should be of the pulse type, self-cleaning filter housing, and filters design only. Calibrate

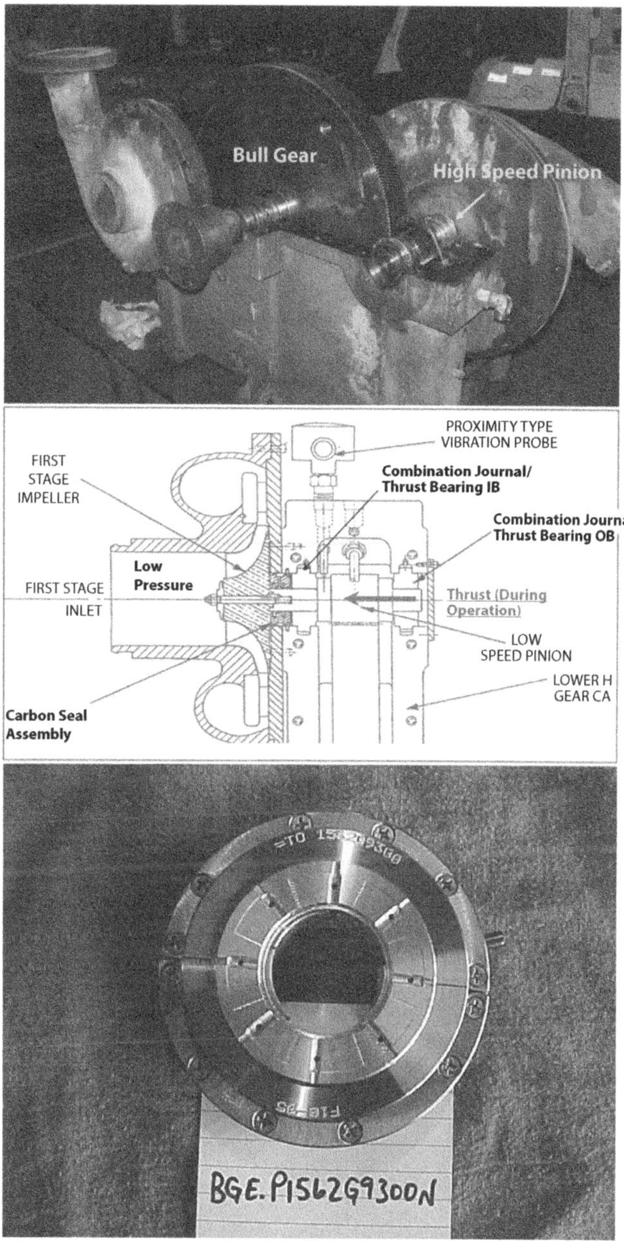

Figure 15.7 The top image shows the configuration of the high speed pinion and bull gear. The middle image shows the locations of the combination radial and thrust bearings for the high speed pinion and the direction of the thrust load under normal operating conditions. The lower image is a closeup of one of the combination radial/thrust bearings with an improved thrust face design.

the inlet air filter differential pressure transmitter. Replace all self-cleaning air filters on the fixed PM program, with a maximum period of 1.50 years.

14. **Normal Condensate Drainers and V-Notch Drain Valves Fail:** Water condensate that is not removed in the intercoolers causes erosion damage to the impellers as carryover and corrodes the entire system. This erosion causes high vibration due to unbalance. For intercoolers-aftercooler piping, replace drain valves and install automatic ball valve type condensate traps. For redundancy, you may also add SS solenoid valves with a timer. Drainers must be of a heavy duty design and construction, such as the Enesystem 2500 B Model (seen in Figure 15.8).
15. **Lubricating Oil Viscosity-Type:** The pinon bearings frequently accumulate varnish with overheating, so sites should utilize low-varnish hydrotreated turbine oils only. The only suitable viscosity lubricating oil giving long operating life with minimum vibration is ISO-VG-46 turbine oil. It is not recommended that the standard OEM selection of ISO-VG-32 oil be used. Higher high speed and low speed rotor vibration will result due the lower bearing damping properties of the ISO-VG-32 oil. **(See "Why The Oil Viscosity Type is Critical to High Speed Air Compressors" insert below for more details.)**

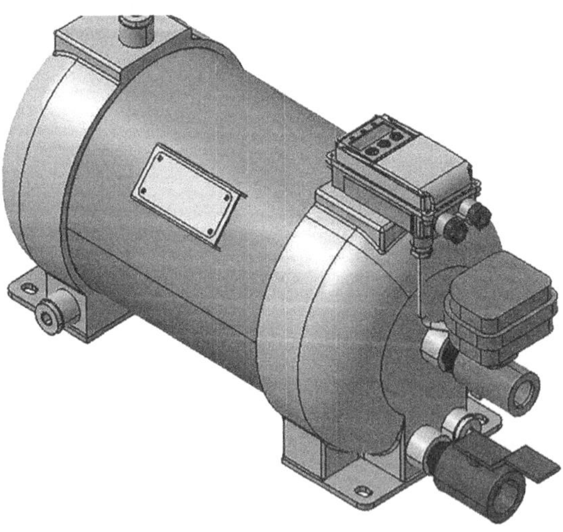

Figure 15.8 Auto drain trap DM-2500B electric actuator SS ball valve type.

Why The Oil Viscosity Type is Critical to High Speed Air Compressors:

1. Thinner oil causes reduced oil film damping for the pinion rotor journal bearings, which then increases rotordynamic vibration. We have direct experience of this occurring on high speed integrally geared air compressors at numerous facilities. At a Lubrication Plant facility using 750 HP x 52,000 RPM HS pinions in the year 2002, changing from new, non-contaminated ISO-VG-32 oil at 130 °F, to a new ISO-VG-46 oil, with no other system changes, bearing changes, or rotor removal, led to an immediate drop in HS rotor radial vibration from 0.95 mills, down to 0.60 mills pk-pk. No overheating occurred.
2. Thinner oil causes increased abrasion of shaft and bearing from fine particles. When operating with normal oil film thicknesses, these abrasives (which pass through the filtration system unless the oil is filtered to 5 μms and less) do not contact the rotating parts. However, with thin oil films, light rubbing contact occurs between shaft and bearing, resulting in wear.

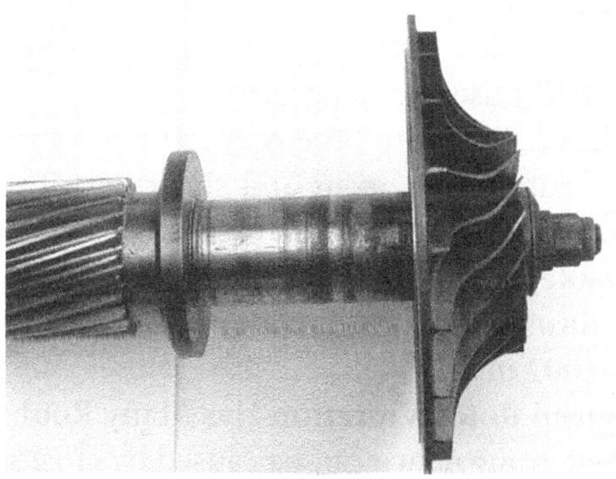

See **Figure 15.9** Pinion damage due to poor lubrication bearing.

3. Reduced reliability of machinery as any unusual pinion shaft loading will overcome the extremely thin oil film thickness and metal to metal rubbing contact results at the bearings (see Figure 15.9).
4. Note: Set the lube oil temperature control valve permanently at 120 oF when using ISO-VG-46.

16. **The third stage discharge air check** valves are usually made of carbon steel design which fails frequently from corrosion, causing sticking and high backpressure on the machine. This causes a high-pressure drop in the discharge line, low performance, surging, and vibration. Install new check valve design: The upgraded check valve should be either a piston type or dual disk type. Specify that the check valve be a non-slam, all stainless steel design with a 150-pound rating.
17. **Gearboxes Incorrect Design Footings:** Some manufacturers install the gearboxes above the skid or intercooler top baseplate using individual steel footing blocks at each gearbox foot. This design can cause flexing movement, low damping, and vibration. For those machines with frequent high speed and low speed rotor vibration, they can be modified with single integral one piece or fabricated welded block supports under the gearbox.
18. **Compressor Vibration or Seizure; Misaligned inlet-outlet piping**. Diffuser casings become distorted from inlet or outlet air piping flange strain. The compressor alignment to diffuser is checked at the shop, however when inlet piping is connected with pipe strain in the field, this force twists the diffuser housing bringing the impeller closer to the diffuser. At startup, the reduced axial clearance allows impeller rubbing to occur and shearing with severe damage. Before connecting the inlet and discharge piping to each volute, apply two dial indicators to the front and side of the diffuser casing. Maximum allowable piping deflection is 0.002 inch.
19. **High Speed Rotor Vibration Has Many Root Causes:**
    a) Impeller unbalance can be caused by erosion, fouling, or FOB damage.
    b) Other sources of imbalance include incorrect balancing of the rotor, high speed shaft journal wear, or bearing wear leading to excess clearance.
    c) Also, radial bearings loose in the gearbox casing housing can cause vibration. This bearing must fit with a 0.001 or 0.002-inch interference.
    d) High speed bearing misalignment to shaft axis will result in low rotor damping. Perform a bluing contact to confirm you have good contact.

e) Check for soft foot under the gearbox feet.
f) Finally, poor gear mesh oil spray coverage causes uneven rotor-gear overheating and high vibration.

16

Failure Analysis & Design Evaluation of a 500 KW Regeneration Gas Blower

By Abdulrahman Alkhowaiter

Introduction

The centrifugal single stage gas blower analyzed in this chapter is a highly critical unspared centrifugal blower for the CCR Reformer producing gasoline at a new refinery commissioned in 1999 (Figure 16.1). The blower is located high up on a seventh floor platform close to the reactor with difficult access and experiences high operating temperatures of 900 °F. The blower, which is directly driven by a 500 KW motor, operates at 1800 to 3600 RPM. During review of past failures and historical problems with this machine, the data revealed that this machine has operated unreliably for many years, resulting in site maintenance expenses exceeding USD 1.70 MM. The average blower Mean Time Between Failures (MTBF) from 1999 to 2013 was only 6 months, which was deemed to be completely unacceptable. The reliable lifetime between failures for such a machine in this service should be 4.0 years.

Historical blower failures include:

1. **Early bearing failures** of 4500 hrs. average life compared to required 40,000 hrs.
2. **Frequent overheating** of bearings.
3. **Premature shaft seal failures** with gas leakage and shaft sleeve damage at seal zone.
4. **High blower vibration** and sometimes noise due to rubbing of parts.

Present Situation:
The blower failed in March 2014 and the spare machine was sent to a local approved shop for overhaul repair. Refinery maintenance clarified that the blower has had a long failure history and had undergone bearing

Robert X. Perez (ed.) Condition Monitoring, Troubleshooting and Reliability in Rotating Machinery, (341–352) © 2023 Scrivener Publishing LLC

Figure 16.1 Centrifugal gas blower in a CCR reformer.

modification in 2007 (see Figure 16.2). During the recent 2014 failure investigations, several observation were made by the refinery:

1. The wrong cylindrical bearing number was being used. The wrong bearing has riveted brass and steel cages instead of a one-piece solid brass cage. The Vendor was claiming this was cause of recent failure.

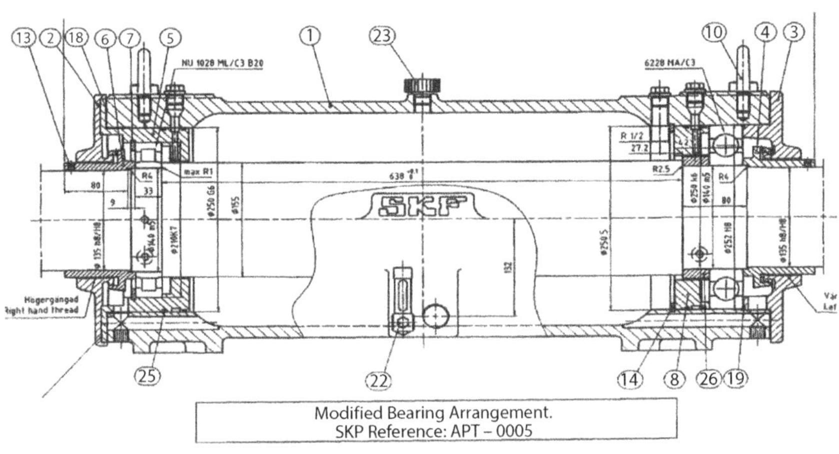

Figure 16.2 Modified bearings by OEM Year 2007 using new cylindrical roller bearing with oil injection.

2. Blockages on the lubrication nozzles, which indicates a lubrication quality issue was present. The practice was to just clean the filters if oil filter elements were not available. This will be corrected by enforcing the replacement of filter elements and to replace the oil annually.
3. Alignment issues such as concentricity of the shaft within the seal housing were found, resulting in internal rubbing.
4. Assembly issue: It was observed that different types of O-rings were utilized during the last repair. One of the O-rings was noticed without crush. This impacts bearing lubrication by internal leakages.
5. The oil level adjustment was found to be inaccurate but was subsequently adjusted to the correct level.

Detail Design Analysis

The following design analysis is based upon studying the equipment manual, drawings, and failure history, interviewing users, refinery engineers, and central repair shop craftsmen, and finally, evaluating every design detail on the blower. The result of applying a *Universal Failure Mode* list approach to the blower design with machine design fundamentals provided the following:

> The original manufacturer design rating Single Row deep groove bearings SKF 6228 MA/C3 are heavy duty and can operate for more than 40,000 hrs. under the operating conditions. This applies to radial and thrust loading. The bearing rating was properly chosen and at 165 KN the Dynamic load rating capacity is high.

In other words, there was no need to change bearings to cylindrical roller type as the failure modes experienced were not caused by a lack of load rating of original bearings. The root causes are numerous and will be defined. Proof of this is that multiple frequent failures have occurred in 2011 to 2014, even with the new cylindrical bearing. The failure analysis found a large number of errors leading to excess bearing loading above design values and when these are rectified, the bearings will operate smoothly as they will be lightly loaded by the combined gravity weight of the rotor acting radially and axial thrust force arising from gas differential pressure at thee impeller.

1. **Loss of Bearing Clearances due to Shaft Diameter Growth:** The blower operates at 900 °F and with such an elevated

temperature, the designer has properly chosen an optimum shaft material, i.e., High Nickel alloy - **Inconel-800**, to resist corrosion and maintain its strength at high temperatures. However, the designer has neglected to consider the shaft expansion due to heat conduction. Under steady state operations, there is about a 250 °F minimum average cross sectional temperature for the two bearing locations. Since the shaft OD is 140 mm or 5.50 inches in diameter at the bearings, the resulting thermal diametrical shaft growth from shop ambient of 80 °F to the operating temperature is: 5.5 inch x [250-80] x 0.0000073 inch/°F= 0.007 inch. This thermal growth leads to overstress at the bearing inner race since it has already been shrink-fitted to shaft with about a 0.002 inch interference fit. Total stretch of 0.009 inch is reduced by an inner race thermal growth of: 5.5 x [210-80] °F x 0.000006 = 0.004 inch. A differential is 0.005 is not acceptable as this distorts the raceway. The next problem is that this inner race expansion without equivalent outer race growth leads to loss of the critical ball to raceway clearance. This is one reason why the bearings were failing prematurely. **Solution:** a) Reduce total bearing shrink fit to 0.0015 inch or b) Allow bearing to operate with higher bearing temperatures using Synthetic oil. The oil reservoir temperature must be maintained at a 160 to 170 °F temperature range. This means that inner race temperature will be above 200 °F, approximately between 210 to 230 °F. There is no harm for such bearings to operate at metal temperatures in the range of 200 °F to 250 °F using synthetic oils. Gas turbine rolling element bearings operate at 350 to 400 °F conditions. The only point is that oil viscosity and type has to be chosen properly for correct lubrication at operating temperature.

2. **Loss of Bearing Clearances from Outer Race Confinement:** The SKF bearing housing has a housing fit to OD of bearings of only 0.002 inch, as checked at the central repair shop. This is a design error. The diametrical thermal growth of the ball bearing outer races (diameter 250 mm or 9.84 inch) at even a conservative metal temperature of 210 °F leads to a differential growth between the housing (in the 140 °F winter) and outer race of (210–140) x 9.84 x 0.000006 = 0.004 inch. This changes the bearing geometry by reducing internal clearances and overstressing the bearing race at contact with

balls. **Solution:** Machine the housing to allow a new dimension of 0.004 inch diametrical clearance to the bearing outer race for the radial bearing located at the non-drive end. The outer race still needs to be held axially by other mounting methods. The thrust bearing at the drive-end however must have greater clearance to allow free floating, so it requires 0.006 inch clearance to the housing.

3. **Overstressing of Thrust Bearing From Shaft Axial Thermal Growth:** The high temperature of shaft during steady state operation will equal approximately 250 °F due to the 900 °F temperature in the fan casing. The total shaft axial growth arising from this growth must be considered at the thrust bearing outer race, allowing it to float axially an amount equal to this plus 0.005 inch. The Inconel shaft growth from adjacent bearing to fan is not as important, as this is free uninhibited growth. The critical dimension is the 600 mm or 23.5 inches distance between bearings, as the bearings are held axially in the housing. Thermal Growth = [250 °F – 140 °F housing] x 23.5 x 0.0000073= 0.020 inch. Adding to this, a required 0.005 inch leaves a requirement for a 0.025 axial clearance at the thrust bearing to free the bearing from the thermal growth loads. **Solution:** The central repair shops to provide a minimum axial clearance at a drive end thrust bearing OD of 0.025 inch. This clearance is achieved by adding shims between bearing housing and the housing flange plate.
4. **Lube Oil Viscosity & Type is Not Correct:** The bearings operate at a DN (diameter x speed) value of 500,000. In addition, the bearing metal temperatures will operate continuously at 220 °F during summer months since we need to maintain correct metal temperatures. The manufacture selection of ISO-VG-46 oil is not suitable for single row Conrad bearings acting as radial and thrust bearings, i.e., the fixed bearing. This low oil viscosity leads to poor oil film, meaning rubbing and fatigue of the ball surface, which is why oil viscosity at these operating temperatures is critical. Please see below the NTN bearing recommendation chart (Figure 16.3). For these high DN value bearings operating at 100 °C (212 °F), the ideal oil viscosity is ISO-VG-68. **Solution:** Change lubricant to synthetic oil of viscosity ISO-VG-68.

Technical Data

Selection standards for lubricating oils

Operating temperature of bearings °C	*dn*-value	Viscosity grade of lubricating oil: Ordinary load	Viscosity grade of lubricating oil: Heavy or Impact load	Bearing type
−30 to 0	Up to the allowable revolution	22 32	46	All type
0 to 60	Up to 15,000	46 68	100	All type
	15,000 to 80,000	32 46	68	All type
	80,000 to 150,000	22 32	32	Except thrust ball bearings
	150,000 to 500,000	10	22 32	Single row radial ball bearings, cylindrical roller bearings
60 to 100	Up to 15,000	150	220	All type
	15,000 to 80,000	100	150	All type
	80,000 to 150,000	68	100 150	Except thrust ball bearings
	150,000 to 500,000	32	68	Single row radial ball bearings, cylindrical roller bearings
100 to 150	Up to the allowable revolution	320		All type
0 to 60		46 68		Spherical roller bearings
60 to 100		150		

Notes: 1. In case of oil drip or circulating lubrication
2. In case the usage conditions' range is not listed in this table, please refer to NTN.

Figure 16.3 Lubricant oil viscosity selection chart for NTN bearings.

5. **Thermal Expansion Causes Seal Rubbing to Sleeve:** The fan, even if insulated, has high thermal growth at the fan casing trapezoidal feet. Since the fan casing is at 900 °F, then with conduction plus convective heat transfer, the footing rough calculations show a minimum thermal growth difference of 0.025 inch between fan casing and bearing housing location. This means that the seal housing bolted to the fan cover will rise 0.025 inch, leading to rubbing with the shaft. **Solution:** Refinery Craftsman: During installation, adjust seal housing clearance to shaft with cold offset of 0.025 inch below shaft centerline. This means that the shaft is set higher than the seal bore centerline by 0.025 inch, which will compensate for seal thermal rise during operation. Note: Alignment of seal bore to shaft is extremely critical for reliability. Therefore, field personnel should contact a central engineering representative to attend and advise technicians during the final connection of fan shaft to fan casing at the site.
6. **Seal Sleeve is damaged by Carbon Ring Rubbing:** The carbon rings (Figure 16.4) are rubbing against the sleeve lightly by design, however, this sleeve should have a high hardness weld overlay to withstand this intense high temperature rubbing action. **Solution:** Therefore, either use an existing sleeve or fabricate a new Inconel sleeve. Then, apply very

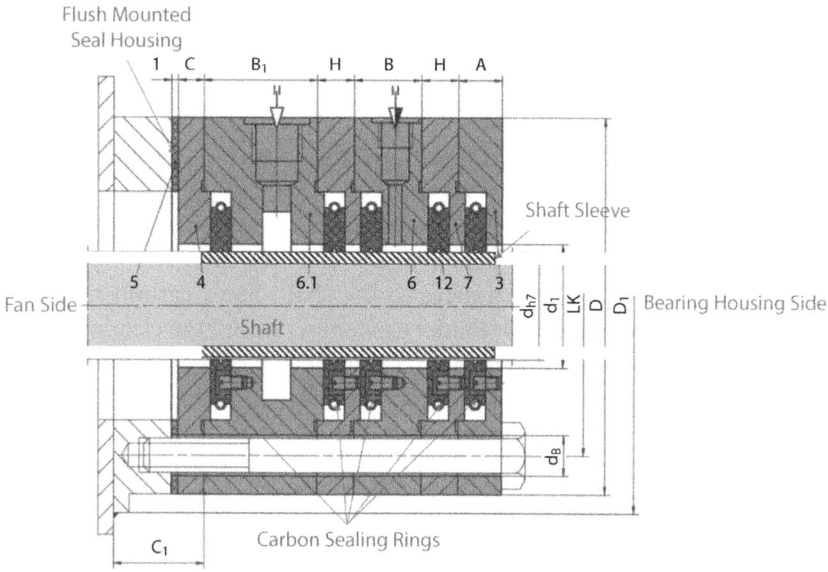

Figure 16.4 A typical cross sectional drawing of a flush-mounted, multiple carbon ring shaft seal with a sleeve. Note that this drawing shows 5 carbon rings, while the actual assembly contained 10 carbon sealing rings. Similar to the seal assembly mentioned in this chapter, this drawing depicts a seal assembly that is flush-mounted to the outside of a fan casing.

Hard Stellite-12 welding rod overlay. Minimum Stellite weld overlay final thickness after grinding shall be 0.090 inch. Stellite-12 achieves a hardness of Rockwell-C 50. This hard surfacing will protect the sleeve from any rubbing damage and hot corrosion from gas particles.

7. **Seal Housing Not Fitting Accurately to Fan Cover Plate:** The seal housing is poorly designed in that it has no register fit to housing (see Figure 16.4). This appears to be an attempt by OEM to give flexibility to the user to align the cover plate at the site, then move the seal housing as needed to the center shaft. This is a major design mistake, as when this is done the inner bore of the plate comes very close to the shaft. With thermal growth in service, the cover plate then grows upward, thus rubbing heavily with the shaft. **Solution:** The seal housing has to be fixed perfectly relative to the centerline of the fan cover's bore. The register fit should be machined into the Fan Cover plate with 0.100 inch depth. Machine the cover plate bore lightly by 0.010 inch to clean up bore. Allow

0.003 inch diametrical clearance at register. Refinery craftsmen note: Any centering of fan shaft to cover plate bore at field installation should only be accomplished by loosening the foot bolts of the skid and moving the skid in various horizontal moves. Applying shims to raise the skid upward will provide any vertical moves needed.

8. **Seal Housing High Gas Leakage:** This is both a safety hazard and a reliability problem. The manufacturer has incorrect carbon ring clearances to the shaft. For a carbon ring seal operating at 900 °F inlet gas temperatures, the clearance to shaft should be 0.005 inch diametrical ***at hot conditions*** to allow for normal shaft orbit and vibration. However, according to the central repair shop, the OEM is using 0.027 to 0.032 inch cold clearance. With such limits, it would be better to just apply labyrinth seals, as large clearances defeat the purpose of Carbon Rings. Calculation: The rate of growth of carbon is half that of the Inconel shaft which reaches 0.029 inches expansion at 800 °F differential temperatures. The Carbon Graphite expands 0.016 inch at the same temperature. By subtracting them, the differential growth is 0.013 inch. Adding 0.005 inch for clearance gives 0.018 inch. **Solution:** Machine the new seal sleeve with a step so that new Carbon Ring Clearance to shaft is 0.018 inch for the first 3.0 inches on the sleeve toward the hot end. The next sleeve clearance dimension to remaining colder rings stages is 0.014 inches. In addition, the present Nitrogen purge buffer regulator pressure of 3 psig is not suitable since the gas pressure is 36.0 psig. Therefore, the nitrogen pressure must be raised at 10 to 12 psig to give sufficient cooling and isolation.
9. **The Fan Rotating Deflector Plate is two pieces of split Aluminum:** This is fine for a 100 °F application, but not for a 900 °F fan. The OEM deflector is a two piece aluminum disk that can easily loosen through its bolting and will also loosen due to the very large thermal growth of the deflector compared to shaft growth. This causes unbalance. **Solution:** It is recommended that the site change this part to a fabricated stainless steel 316, solid one-piece deflector that is shrink-fitted to the shaft with only 0.002 inch shrink fit total. The central repair shop can fabricate the piece. Weld SS 316 ribs of 0.50 inch height and 0.25 inch thickness. The width of the deflector hub shall be 1.50 inches.

10. **Fan Bearing Design Reliability is Higher with Two Deep Groove Bearings:** Return the fan bearing housing back to the original design of two single row deep groove bearings, as these are superior to the modified design using a Cylindrical roller bearing. At the DN values of 500,000, a cylindrical roller bearing is not applicable due to excessive surface speeds and as long as all above design factors are applied, the original bearings will provide 40,000 hrs. of service. The cylindrical bearing also has a weakness in that it cannot be operated with splash lubrication at this operating speed, so it must have a spray jet of oil, but the spray nozzle tends to get blocked from debris, which is just another failure mode that does not apply to the original design.
11. **The Shaft Coupling is a Metastream Type Causing Unbalance Vibration:** The Metastream couplings are general purpose couplings which are not balanced at the factory and have excessive clearance fits at the hub to coupling spool register fit. This leads to non-concentric operation of the heavy spool and high vibration based upon many incidents. **Solution:** Send the motor and blower coupling hubs along with the spacer assembly to the shop. Request that the shop perform a repair to the hub to spool register fit to ensure there is a 0.003 inch loose fit. Also, balance full coupling assembly.

Conclusion

The reasons for the poor reliability of the CCR Reformer blower are two-fold: a) the difficult operating conditions at 900 °F and b) the fact that there are a large number of inherent design defects incorporated by the manufacturer. In some cases, workmanship errors did occur during assembly of the fan at the site, however, these errors were caused by poor design practices that did not provide automatic centering of components, such as seal housings, leading to internal misalignment issues, rubbing, and seal leaks. We found no other user or maintenance related deficiencies. This means that 95% of all problems are inherent in the design, as the manufacturer did not properly design this blower reliably at operating conditions, i.e., at 900 °F. No major reliability improvements will occur unless all the recommendations listed below are fully implemented.

Needed Action by Repair Shop

1. Return the bearing housing to the original design of two single row deep groove bearings.
2. Measure bearing fit areas on shaft and grind or polish shaft to achieve 0.0015 inch shrink fit dimension between bearing inner race and shaft OD.
3. Machine the SKF bearing housing bore to achieve 0.004 to 0.005 maximum inch diametrical clearance to bearing OD for radial bearing. The thrust bearing housing bore must be machined to achieve 0.006 to 0.007 inch maximum diametrical clearance.
4. Axial Float Clearance at Thrust Bearing: The thrust bearing must be given a 0.025 inch axial clearance between bearing outer race and axial cover. Add one single stainless steel fabricated shim to achieve this.
5. Seal Sleeve: Apply Stellite-12 weld overlay on the sleeve OD with minimum final weld thickness of 0.090 inch after grinding. Stepped Shaft Sleeve: Dimensional clearances to Carbon Rings: First 3.0 inches at Hot end: 0.018 inch diametrical clearance to carbon rings. Remaining diameter of sleeve at cooler rings is 0.014 inch clearance dimension.
6. Seal Housing: Machine a register fit at the center of the fan cover plate. Depth of register is 0.100 inch. Clean up plate bore by removing 0.010 inch. Allow 0.003 inch for register fit total clearance.
7. Deflector Disk: Change this part to a fabricated stainless steel 316 solid one-piece deflector that is shrink-fitted to the shaft with only 0.002 inch shrink fit total. Weld six 316 SS Ribs of 0.50 inch height and 0.25 inch thickness. Width of deflector hub shall be 1.50 inches.
8. Metastream Coupling Assembly: The central repair shop to perform a repair to the hub to spool register fit reducing this to 0.003 inch. Also, balance full coupling assembly at the central repair shop.

Action Required by Refinery

1. Lube oil: Change from existing mineral oil ISO-VG-46 to Synthetic Oil ISO-VG-68 grade. Maintain oil reservoir

temperature at 160 to 170 °F temperatures only. Alarm: Oil Bath Temp 200 F, Shutdown 210 F.

2. Refinery craftsmen: During installation, adjust Seal Housing clearance to shaft with cold offset of 0.025 inch below shaft centerline. This means that shaft is set higher than seal bore centerline by 0.025 inch, which will compensate for seal thermal rise during operation. Any centering of the fan shaft to the cover plate bore at field installation should only be accomplished by loosening the foot bolts of the skid and moving the skid in various horizontal moves. Applying shims to raise the skid upward will provide any vertical moves needed.
3. Nitrogen Purge: The present nitrogen purge buffer regulator pressure of 3 psig is not suitable since the gas pressure is 36.0 psig. Therefore, the nitrogen pressure must be raised at 10 to 12 psig to give sufficient cooling and isolation.

Shaft Coupling: Send both coupling hubs and spacer element to the central repair shop for balancing as set.

17

Operating Centrifugal Pumps with Variable Frequency Drives in Static Head Applications

By Robert X. Perez

Most centrifugal pumps in the industry are driven by reliable, squirrel cage induction, or electric motors, which run at essentially a constant speed (Figure 17.1). Their speed depends on the number of rotor poles and the electric supply frequency, which is typically 60 Hz in the US and 50 Hz in Europe. The most commonly used fixed speed electric motor options are 3600 rpm for 60 Hz power (3000 rpm for 50 Hz power), 1800 rpm for 60 Hz power (1500 rpm for 50 Hz power, and 1200 rpm for 60 Hz power (1000 rpm for 50 Hz).

In reality, induction motors operating under a load will never run at their synchronous speed, which is the rotating speed of the magnetic field inside the stator windings. They will always operate at a speed slightly less than the synchronous speed due to slip. Slip is defined as the difference between the rotating magnetic flux speed and rotor speed expressed in terms of per unit synchronous speed. For example, a 200 hp, 3600 rpm electric motor may operate at 3550 rpm at its rated load. This would mean there is (3600-3550)/3600 or 1.1% slip at full load. In practice, most designers and users assume that the magnitude of slip is minor and assume the motor speed is relatively constant from no load to full load condition.

While fixed electric motors are the norm in processing applications, it is increasingly more common for pumps to be operated by motors that are controlled by variable frequency drives (VFDs). VFDs control motor speed by varying the frequency of the power being sent to the motor. So, a 60 Hz motor rated for 3600 RPM can have its speed reduced to 3000 RPM if the supply frequency is reduced to 50 Hz. In theory, users can have any pump speed they want simply by turning a knob or manipulating a graphic user

Robert X. Perez (ed.) Condition Monitoring, Troubleshooting and Reliability in Rotating Machinery, (353–362) © 2023 Scrivener Publishing LLC

Figure 17.1 Due to their simplicity and reliability, motor-driven centrifugal pumps are frequently used to transfer liquids throughout industrial processes.

interface on a VFD. Not only can a user choose any desired speed, but they can also change it at will while the unit is running.

VFD Advantages

A major advantage of VFDs is their ability to start motors at zero voltage and frequency, which greatly reduces winding stresses and heat generation during a cold start. This "soft start" capability has a profound effect on motor reliability. Soft starting also 1) reduces power line disturbances, such as power line voltage sags, which can adversely affect voltage sensitive devices and 2) lowers the power demand at startup. Normally, if an industrial customer starts a motor at full voltage during peak hours, they are likely to be charged with surge energy prices. However, a VFD demanding less starting power can drastically reduce startup energy costs associated with soft starts.

Energy efficiency is another huge benefit of VFD driven centrifugal pumps. Instead of having to operate a pump at a constant speed and then throttling its outlet pressure down to meet the system's flow demands, the user can simply run the pump at the slowest speed required to meet the process flow requirements. These benefits are easily quantified in friction only pumping systems (see Figure 17.2) where the affinity flow law can be used to predict the relationship of flow versus speed, i.e., $Q_1/Q_2 = N_1/N_2$, where Q is flow and N is pump speed. A friction only system is a piping system where a pump only has piping losses to overcome and there are no differences in elevation or pressure differentials to contend with. Frictional

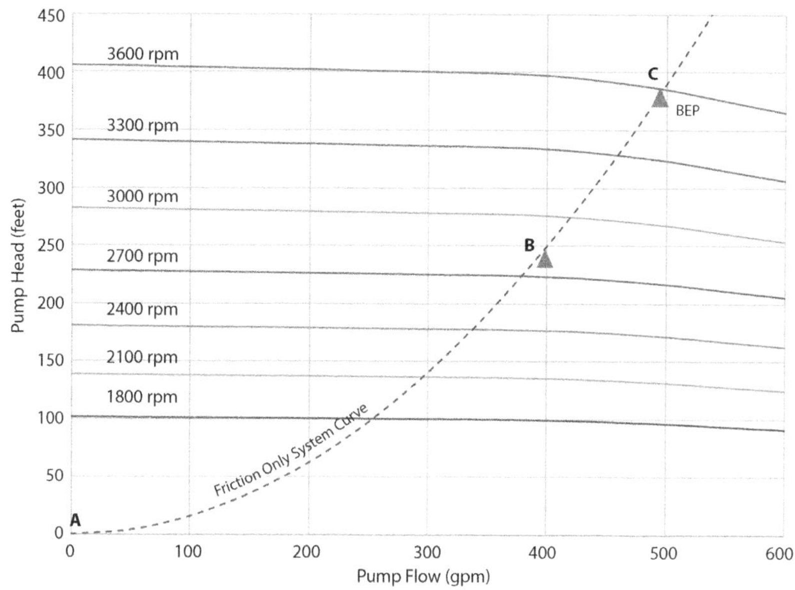

Figure 17.2 Friction only system curve superimposed on series of centrifugal pump speed curves.

losses (H_f) tend to follow the relationship $H_f = KQ^2$, where K is a flow constant and Q is flow. (Systems with static head and friction will be discussed later.) Remember that under steady state conditions, a centrifugal pump's operating point is defined by the intersection of the pump's performance curve and the system curve. For example, in the friction only piping configuration depicted in Figure 17.2, I have listed and labeled three possible operating conditions as points A, B, and C at the intersections of the system curve and the various performance curves.

At full speed (3600 rpm), the pump will deliver 500 gpm at point C. If only 400 gpm is required by the process, the pump can be operated at about 2850 rpm (Point B in Figure 17.2). At 2850 rpm, only 50% of the power needed to operate at full speed would be required (see calculation below), which represents huge power savings to the user:

$$\frac{Power_1}{Power_2} = \left(\frac{N_1}{N_2}\right)^3 = \left(\frac{2850}{3600}\right)^3 = (0.7916)^3 = 49.6\%$$

While this idealized scenario demonstrates the magnitude of power savings a VFD powered pump can offer in a friction only system, there are other factors to consider when evaluating the economics of VFD driven pumps. Factors such as system hydraulics, how well the centrifugal pump matches the pumping system, electrical power costs, the expected flow demand schedule, etc. are some of the design factors to consider when deciding whether a VFD makes economic sense for your specific application. The actual detailed economic calculations of variable speed pumping systems are beyond the scope of this chapter, so they will be left for another time.

Another advantage of VFDs is that they can readily be integrated into pump flow control systems, where they automatically control pump speed to control the flow or pressure to satisfy process requirements and eliminate the need for discharge throttling. Additionally, today most VFDs have the capability of reporting real-time centrifugal pump power measurements remotely and storing historical power information for trending and analysis. Power measurements can be used to continuously monitor the performance and condition of a motor driven centrifugal pump. All these capabilities make VFDs popular in centrifugal pumping systems. However, safeguards are required for using VFD driven pumps in static head systems.

Static Head Systems [1]

There are many types of static head pumping systems. Here are three commonly encountered categories of static head systems I have seen, along with some real world examples that you might encounter:

1. Pumping liquid from a lower elevation to a higher elevation. An example of this type of pumping system would be one where liquid is pumped from a tank at grade operating at atmospheric pressure into an overhead tank operating at atmospheric pressure. The difference in the liquid levels in the two tanks would represent the system's static head.
2. Pumping liquid from a lower pressure system to a higher pressure system.
    a. One example of this type of pumping system would be pumping into a pressurized pipeline. Let's say you are pumping from an atmospheric tank into a liquid pipeline running at 500 psi. The pump needs to provide about 500 psi of differential pressure minus the suction head in the tank before its discharge check valve could open and provide flow into the pipeline.
    b. Another example would be an arrangement of parallel pumps where several pumps on running and pressurizing

a common discharge header. If an idle pump needs to be started up, it will have to be able to create enough discharge pressure to open its check valve and establish a positive flow into the discharge header.

3. Pumping liquid in a system with a combination of elevation differences and a discharge that's at a higher pressure than the suction pressure. An example of this type of pumping system could be seen in a typical process setting where a pump is required to move liquid from a lower pressure process vessel grade into a higher pressure vessel at a higher elevation.

Figure 17.3 depicts a simple pumping system where a centrifugal pump is taking liquid from the tank on the left and pumping it into a discharge tank on the right. Notice that the liquid must pass through a check valve and then into an elevated discharge tank. If the pump is not running, i.e., rpm = 0, the check valve will remain closed and prevent liquid from the discharge tank from flowing back into the suction tank. If the pump in Figure 17.3 is driven by a VFD, for any flow to occur, it would first have to run at a speed required to generate sufficient head to open the check valve. Since check valves require some differential pressure to open, usually somewhere between a fraction of pound per square inch to a few pounds per square inch, the pump will have to develop enough pressure to overcome the static pressure plus a few more feet of additional pressure to start moving liquid into the upper tank. Then, for additional flow to occur, the pump needs to run at a speed to deliver sufficient head to overcome the

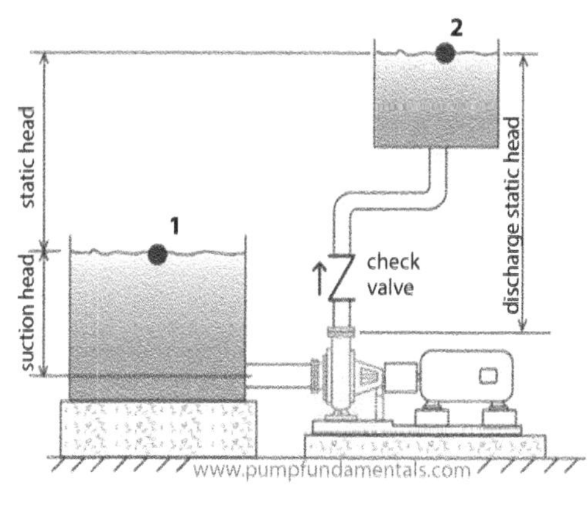

Figure 17.3 Centrifugal pump moving liquid in system with static elevation head.

friction losses in the piping plus the elevation head, i.e., the discharge tank level minus the elevation of the suction tank level.

Now, let's look at a graphical depiction of a realistic pump and system interaction scenario [2]. Figure 17.4 shows a hypothetical system curve made up of friction losses ($H_f = K \times Q^2$) plus static head (H_s) that is superimposed onto the family of speed versus flow pump curves. (Note: A simplifying assumption is to assume the piping system losses formula fits the form $H_{losses} = H_s + K \times Q^2$.) As mentioned earlier, the steady state operating point of a centrifugal pump is defined by the intersection of the pump's performance curve and the system curve. In Figure 17.4, I have labeled three possible operating conditions, Points A, B, and C, at the intersections of the friction and static head system curve and the family performance curves.

Notice in Figure 17.4 that for any flow to occur, the pump must generate 250 feet of head at about 2850 rpm (Point A) instead of zero head requirement for the friction only pumping system. In reality, the pump would need to generate 250 feet of head and some additional head just to open the discharge check valve to achieve a non-zero flow. Then, after the check valve has been opened and flow is established, pump flow versus speed would follow the relationship labeled "static head and friction system curve" in Figure 17.4.

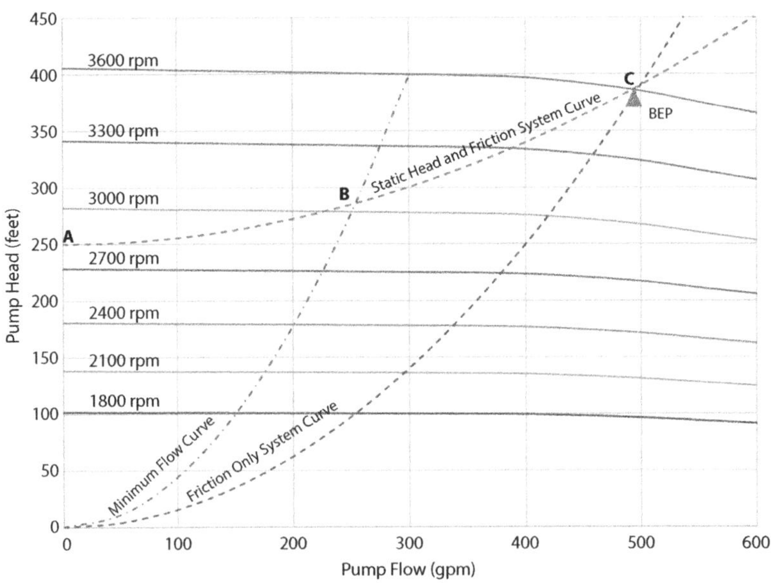

Figure 17.4 Superposition of systems curve and centrifugal pump speed curves.

This basic example illustrates how a static head pumping system requires a minimum speed to establish a positive flow. Why is this important? Because, it shows that if operators are not careful, they could inadvertently have a pump running dead headed, i.e., zero pump flow, for an extended period of time. Depending on the size of the pump and length of time the pump experiences zero flow, serious pump damage could be experienced and lead to major economic and environmental consequences.

Recommended Startup Sequence

We now arrive at the main purpose of this chapter, which is to describe a safe way of starting VFD driven pumps in static head applications. We have seen how a VFD driven centrifugal pump performs when used in different types of hydraulic systems. We learned that pumps in friction only systems create flow that is approximately proportional to speed. In contrast, we have seen that the flow realized in a friction and static head systems is highly dependent on the system's static head. In extreme cases, a VFD driven pump may have to nearly achieve full speed to simply overcome the static head and produce a safe, stable flow.

To safely operate a VFD centrifugal pump-motor set, users need to:

a. Operate them above the motor manufacturer's minimum speed. VFD manufacturers often recommend a minimum motor speed of 30% of their rated speed (18 Hz) to prevent motor overheating due to inadequate airflow. An inverter duty motor is allowed to have a lower minimum setting as low as 20% (12 Hz) of its rated speed.
   Advice: The minimum motor speeds listed here should be considered a rule of thumb. I recommend that you obtain the safe minimum operating speed from the motor manufacturer for your specific motor design and operating conditions.
b. Operate their centrifugal pumps above their minimum continuous stable flow (MCSF) and below their maximum recommended stable flow at all times. Refer to the pump manufacturer's recommendations for these flow limits. In the absence of manufacturer's recommendations, I recommend you always try to maintain a centrifugal pump's flow above 50% to 60% of the best efficiency point (BEP) versus speed line. For larger, high energy pumps, you should a maintain flow above 75% to 80% of the best efficiency point

(BEP) versus speed line. Additionally, users should try to keep maximum pump flows below 110% the best efficiency point (BEP) versus speed line.

c. Ensure the centrifugal pump has sufficient NPSHa to avoid cavitation throughout the expected operating range.

d. Always operate the pump so that the motor is not overloaded. Overloading may occur at the end of the pump performance curve or when operating above the rated speed. To prevent running out to the end of the curve and overloading the motor, provide sufficient back pressure to keep the pump from operating beyond 110% of the BEP flow.

e. Never allow the pump to operate beyond the pump manufacturer's maximum allowable speed. VFDs are capable of supplying electrical power above 60 Hz, which provides the capability of operating electric motors above their rated speed. Rotor components, seals, and bearings all have upper speed limits that must be considered when attempting to operate rotating machinery above design speeds.

Keeping all these precautions in mind, we are ready to discuss how to safely startup VFD driven pumps in static pressure systems. Let's first quickly review the centrifugal pump and system curve depicted in Figure 17.4. This figure contains:

- The family of pump head versus flow curves for a hypothetical centrifugal pump from 1800 rpm to 3600 rpm. Pump performance below 1800 rpm is not depicted in Figure 17.4, but we can assume it drops off continuously all way down to zero rpm.
- A hypothetical friction and static head system curve that intersects the y axis at 250 feet (point A) intersects the minimum flow line at point B and passes through the best efficiency point (BEP) at the rated speed curve (point C).
- A minimum continuous flow line for the pump. In this example, we are assuming the manufacturer has stated that the pump can operate safely at 50% of the best efficiency point for an extended period of time.

Assuming we are operating the pump and system seen in Figure 17.4, here is one possible starting sequence that could be used to safely put the pump into service:

1. Soft start the motor and accelerate the motor up to the manufacturer's minimum safe speed. Let's assume 30% of rated speed is the recommended minimum motor speed. For a 3600 rpm rated motor, this means we need to accelerate up to 1080 rpm. Since this is below our 1800 rpm pump performance line, we know that our pump will generate less than 100 feet of head and because our static head is 250 feet of head, we also know the check valve will be closed and that there will be no pump flow under these conditions.
2. You can hold this minimum motor speed condition for a few minutes without any issues.
3. Then, you should accelerate the motor to a speed that will ensure a flow above the minimum flow line. In this case, a speed of about 3050 rpm will generate a pump flow above the minimum flow curve. During this time, the pump's operating point will move from Point A to Point B on Figure 17.4. Once you have achieved the minimum recommended pump flow (Point B), you can remain at that flow indefinitely before needing to accelerate to full speed. This lull allows operators to check out the pump and motor for any potential issues before attempting to reach rated speed.

   Note: Running with a closed check valve puts the pump in a vulnerable situation because zero flow implies the pump's efficiency is zero. At zero efficiency, all the energy provided by the motor goes into generating heat, vibration, and noise. If gone unchecked, the pump will either lock up due to thermal expansion caused by internal heating, fail due to high vibration, or melt all the elastomers in the pump sealing element and result in a massive product leak.
4. Then, the pump can be accelerated from 3050 rpm (Point B) to rated speed to achieve the rated flow (Point C).

   Readers should keep in mind that all centrifugal pumps degrade over time due to internal wear. As pump wear ring clearances open up, a greater rotational speed will be required to generate the same head compared to when the pump was new. Horsepower requirements also increase due to the increase in internal recirculation. If gone unchecked, eventually the degraded pump may not be able to generate sufficient head to open its check valve due to a loss of performance, which is why

> pump performance should be regularly monitored. Don't assume just because you have a variable speed drive you will always be able to get a pump out of trouble by increasing speed.

An additional safeguard that I recommend for critical pumps in static head applications is the installation of some type of reliable flow indication to ensure there is sufficient speed to open the discharge check valve and establish a flow above the recommended minimum flow. Preferably, the flow indication should be tied into the startup logic and set up to shut down the pump if a prolonged low flow condition is detected. Blind flow switches are not recommended because they do not provide field personnel with any information about the actual flow and how close the pump is to a low flow trip.

Final Thoughts

Before designing your pump protection system and defining the startup sequence, make sure you understand the entire hydraulic system, pump characteristics, minimum flow requirements, and how the pump needs to be operated to satisfy the process. In most cases, your startup sequence should be able to be controlled automatically with a programable logic controller (PLC). I recommend that pump engineers work with their instrument control professionals to develop the required automatic sequences to operate and protect their pumps. It's also a good idea to get the centrifugal pump manufacturer's stamp of approval on your startup control sequence. A team effort is essential to implementing safe and reliable VFD control schemes for your critical centrifugal pumps.

References

1. Pump Fundamentals website, https://www.pumpfundamentals.com/
2. Perez, R. X., "How to Select the Right Centrifugal Pump," pp 20-22, AuthorHouse Publishing, 2015.

18

Estimating Reciprocating Compressor Gas Flows

Simple hand calculations can be used to evaluate reciprocating compressors in the field.

By Robert X. Perez

Reciprocating compressors, like the one depicted in Figure 18.1, are commonly used to move gases throughout the petrochemical industry. These types of compressors commonly employ one or more reciprocating pistons, which are driven off a common crankshaft, to compress gas and move it from a lower to higher pressure. There are multiple design and operating factors, such as piston and rod geometry, speed, and operating pressures, that affect the actual flow realized under operating conditions.

The purpose of this brief discussion is to explain to readers how they can estimate the flow capabilities based on readily available information. To calculate compressor flows, the users must understand a cylinder's swept volume, clearance volume, volumetric efficiency, and compression ratio.

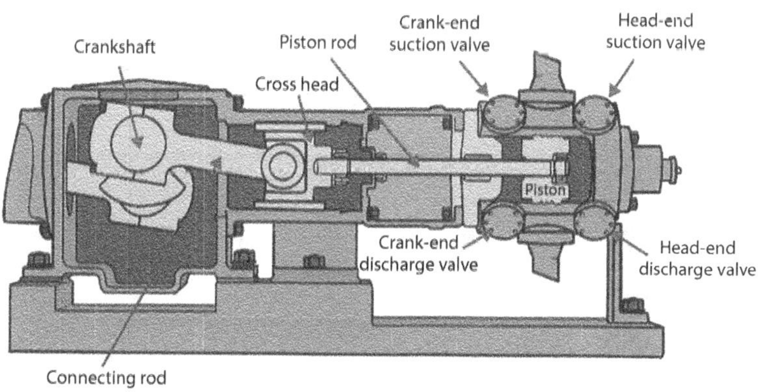

Figure 18.1 Single stage reciprocating compressor.

Robert X. Perez (ed.) Condition Monitoring, Troubleshooting and Reliability in Rotating Machinery, (363–372) © 2023 Scrivener Publishing LLC

Once these basic factors that affect flow are understood, then field troubleshooting and assessments are possible.

Swept Volume

The swept volume of a compressor cylinder is equal to the area of the compressor piston times the piston's stroke length, i.e., Swept volume = piston area x stroke length (see Figure 18.2). If a cylinder had a 100% volumetric efficiency, then the flow created by a cylinder would be its swept volume times the strokes per minute. The equation to determine the swept volume per minute is given below.

$$\dot{V}_{displacement} = \frac{\pi \times D^2}{4} \times S \times spm \tag{18.1}$$

Swept volume per minute = $\dot{V}_{displacement}$
D=piston diameter (feet)
S=stroke (feet)
spm=stokes per minute

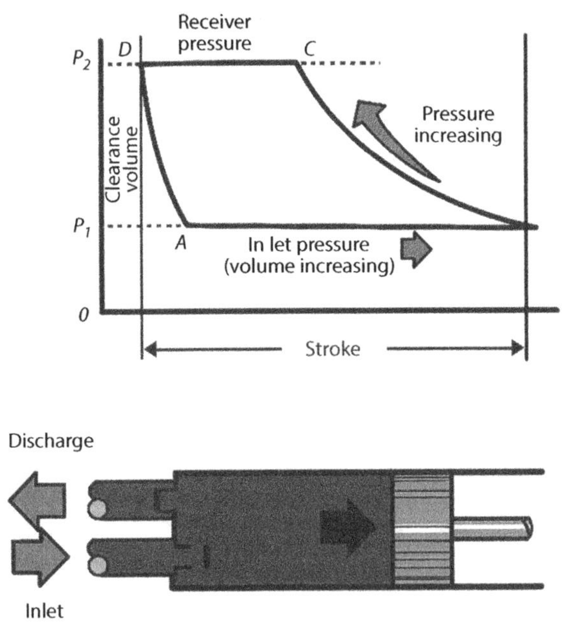

Figure 18.2 Typical pressure-volume diagram for reciprocating compressor.

The swept volume of the piston within the compression cylinder is a key factor in determining the gas flow a compressor cylinder can deliver. This is because the swept volume determines the amount of gas that enters the cylinder during the intake compressor stroke (A-B in Figure 18.2). By knowing the swept volume and suction conditions, the quantity of actual cubic feet and therefore the quantity of standard cubic feet of gas that enters the cylinder at the end of the suction stroke is known.

Clearance Volume

Clearance volume is the space remaining in the compressor cylinder at the end of the stroke. Clearance is made up of spaces in valve recesses and the space between the piston and cylinder end. At the completion of each compression stroke, the compressed gas trapped in the clearance space expands against the piston and adds to the force of the return stroke. Figure 18.2 is a pressure vs. volume (P-V) diagram that illustrates the effect of cylinder clearance.

From point A to B, the piston moves away from the cylinder head, which opens the suction valve, allowing suction gas to enter the cylinder. Compression takes place from point B to C as the piston moves toward the cylinder head and both the suction and discharge valves are closed. Then, the discharge valve opens allowing gas to exit the cylinder from point C to D. Finally, with both valves closed, the pressure drops down to suction pressure as the cylinder moves away from the cylinder head.

Expansion of the gas trapped in the clearance space occurs before the suction valve opens to admit new gas into the cylinder. As a result, part of the piston displacement occurs before the suction valve opens. The compression process in reciprocating compressors is nearly isentropic, so the energy required to compress the gas in the clearance space is recovered when the gas expands at the end of the compression stroke. For this reason, changes in the clearance space do not affect the compressor power.

Volumetric Efficiency

Volumetric efficiency (VE) is the ratio of the actual volume ($\dot{V}_{displacement}$) of gas (actual cubic feet/min) drawn into the cylinder to the piston displacement (cubic feet/min), i.e., swept volume (ACFM). This ratio is less than unity because of three fundamental effects. First, the gas is heated during admission to the cylinder. Second, there is leakage past valves and piston rings. And third, there is re-expansion of the gas trapped in the clearance volume from the previous stroke. Of these three, re-expansion has, by far, the greatest effect on volumetric efficiency.

$$VE = \frac{ACFM}{\dot{V}_{displacement}} \tag{18.2}$$

Rearranging terms, we get:

$$ACFM = VE \times \dot{V}_{displacement} \tag{18.3}$$

Compressor manufacturers have not reached a consensus on an appropriate calculation method because measurement of these effects is extremely difficult. However, users can estimate the volumetric efficiency with the following formula:

$$VE = 100 - r - c\left(\left(r^{1/k}\right)\frac{Z_s}{Z_d} - 1\right) \tag{18.4}$$

VE = volumetric efficiency (%)

r = compression ratio = P_d/P_s = Absolute discharge pressure/absolute suction pressure

c = cylinder clearance, % of piston-swept volume

Z_s = inlet compressibility factor

Z_d = discharge compressibility factor

k = ratio of specific heats, C_p/C_v

Note the term "-r" that has been added[1]. Instead of subtracting from 100%, this equation subtracts from 100 - r. This term is intended to account for the fact a real running compressor does not conform to pure thermodynamic theory. For example, the seals surrounding the compression chamber, specifically the compressor valves, piston rings, and packing, are not perfect. There is always some internal gas leakage. This means real VE will never agree with VE from theory. So, a "fudge factor" must be used, and "-r" is such a factor. Every compressor manufacturer has a unique method for adjusting the VE equation and "-r" is just one simple method. Compression ratio (r) typically varies from 1.3 to 3.5 so the -r term reduces VE (capacity) by 1.3% to 3.5%.

There are several ways to determine VE:

1. Calculate it with Equation 18.4
2. Obtain VE from recent field PV analysis
3. Get the VE value from the manufacturer

Actual cubic feet per minutes of flow can be converted to standard cubic feet per minute with the following formula:

$$SCFM = ACFM \times \left(\frac{P_s}{14.7}\right) \times \left(\frac{520}{T_s}\right) \tag{18.5}$$

Units:
T_s = Suction gas temperature, °R = 460 + °F
Ps = Suction gas pressure, psia
ACFM = Gas flow, actual cubic feet/minute
SCFM = Gas flow, standard cubic feet/minute
Note: The compressibility factor is assumed to be 1.00

Combining the relationship for the volumetric displacement of the cylinder with Equations 18.4 and 18.5, we can write the formula for determining the SCFM flow for a single acting compressor cylinder:

$$\dot{V}_{displacement} = \frac{\pi \times D^2}{4} \times S \times spm \tag{18.6}$$

$$\dot{V}_{displacement} = piston\ displacment / minute$$

$$ACFM = VE \times \dot{V}_{displacement} \tag{18.3}$$

$$SCFM = \frac{\pi \times D^2}{4} \times Stroke \times spm \times \left(\frac{P_s}{14.7}\right) \times \left(\frac{520}{T_s}\right) \times VE \tag{18.7}$$

Substituting equation 18.4 into this equation, we get:

$$SCFM = \frac{\pi \times D^2}{4} \times Stroke \times spm \times \left(\frac{P_s}{14.7}\right) \times \left(\frac{520}{T_s}\right) \times \frac{1}{100} \times \left[100 - r - c\left(\left(r^{1/k}\right)\frac{Z_s}{Z_d} - 1\right)\right] \tag{18.8}$$

If we neglect compressibility factor terms, we get:

$$SCFM = \frac{\pi \times D^2}{4} \times Stroke \times spm \times \left(\frac{P_s}{14.7}\right) \times \left(\frac{520}{T_s}\right) \times \frac{1}{100} \times \left[100 - r - c\left(\left(r^{1/k}\right) - 1\right)\right] \tag{18.9}$$

This annotated version of Equation 18.9 identifies the terms required to a) determine the swept volume of the piston, b) to convert from actual conditions to standard conditions, and c) to determine the cylinder's volumetric efficiency:

a. Swept volume

c. Volumetric efficiency

$$SCFM = \frac{\pi \times D^2}{4} \times Stroke \times spm \times \left(\frac{P_S}{14.7}\right) \times \left(\frac{520}{T_S}\right) \times \frac{1}{100} \times \left[100 - r - c\left(\left(r^{1/k}\right) - 1\right)\right]$$

b. Conversion to standard conditions

For a double-acting cylinder, the flow equation is:

$$SCFM = \frac{\pi \times (2D_c^2 - D_r^2)}{4} \times Stroke \times spm \times \left(\frac{P_s}{14.7}\right) \times \left(\frac{520}{T_s}\right) \times VE \tag{18.10}$$

$$SCFM = \frac{\pi \times (2D_c^2 - D_r^2)}{4} \times Stroke \times spm \times \left(\frac{P_s}{14.7}\right) \times \left(\frac{520}{T_s}\right) \times \frac{1}{100} \times \left[100 - r - c\left(\left(r^{1/k}\right) - 1\right)\right] \tag{18.11}$$

where

D_c=Piston diameter

D_r=Rod diameter

$$SCFM = \frac{\pi \times (2D_c^2 - D_r^2)}{4} \times Stroke \times spm \times \left(\frac{P_s}{14.7}\right) \times \left(\frac{520}{T_s}\right) \times \frac{1}{100} \times \left[100 - r - c\left(\left(r^{1/k}\right) - 1\right)\right]$$

In the colorized version of Equation 18.11, the red terms are related to the swept volume of a double acting piston, the blue terms are required to convert from actual conditions to standard conditions, and the purple terms are related to the cylinder's volumetric efficiency.

Note that these calculations assume there is adequate horsepower available for the required demand and that all the gas piping is sized to minimize pressure losses.

Calculating Flows in Multistage Cylinder Compressors

In multistage, reciprocating compressor applications, the net gas flow is largely determined by the flow capability of the first stage cylinder. This is because multistage compressor cylinders are typically configured in series, not in parallel. Regardless of the number of stages of compression a compressor has, the net flow through a compressor is limited by the quantity of gas the first stage cylinder can take in and then expel per stroke. The subsequent stages only add pressure to the flow stream. However, be aware that flow losses due to liquids dropping due to cooling or open bypass lines open between stages can result in a discrepancy between the calculated suction flow and the measured net flow.

Here are some additional thoughts to keep in mind when using Equations 18.9 and 18.11:

1. These flow equations assume that the compressor cylinder you are analyzing is fully loaded, i.e., no cylinder valves are unloaded and no volume unloaders are activated. This is important to keep in mind because unloading an end of a cylinder reduces the volumetric efficiency of that end of the cylinder to zero. Furthermore, activating a volume unloaded will dramatically reduce the head end of a cylinder's volumetric efficiency, which will in turn reduce its flow capacity.
2. These flow equations assume that all recycle lines are closed and that all the flow out of the 1st stage cylinder is sent to the next stage of compression or into the process. Leaking relief and control valves can also result in discrepancies between calculated and field measured flow values.

3. If two or more 1st stage cylinders are operating in parallel, then simply add the calculated flows for each cylinder to determine the total flow.
4. These equations are valid for any compression stage only if the actual interstage operating values of P, T, and r values are used. However, when calculating the second stage cylinder flow, instead of P_s use $P_{2nd\ stage\ inlet}$ and instead of T_s use $T_{2nd\ stage\ inlet}$, etc. In theory, the SCFM flow from the 1st stage to the last stage compression should be approximately equal if we assume that gas flow lost due to condensed liquids after interstage cooling is insignificant.

Flow Calculation Example

Let's run through a quick compressor flow example. Assume we are evaluating a single stage, single acting reciprocating compressor with a 13.5" cylinder bore and 7" piston stroke length operating at 700 rpm. The suction pressure is 100 psig and the discharge pressure is 250 psig. The clearance volume is 7% of the swept volume. Assume T_s=100 and k=1.31.

Given:

D_{piston}=13.5"

Strokes per minute= 700

k=1.31

P_s=100 psig

P_d=300 psig

T_s=100= (560 R)

C=10%

The compression ratio for this example is (300+14.7)/(100+14.7), which equates to 314.7/114.7 = 2.744. Plugging in the rest of the supplied values into Equation 18.9, we get:

$$SCFM = \frac{\pi \times (13.5/12)^2}{4} \times (7/12) \times 700 \times \left(\frac{114.7}{14.7}\right) \times \left(\frac{520}{560}\right) \times \frac{1}{100} \times \left[100 - 2.744 - 10\left(\left(2.744^{1/1.31}\right) - 1\right)\right]$$

After crunching the numbers, we find that the expected flow is 1799.16. In this example, the volumetric efficiency was found to be 85.65% of the swept volume.

Factors Affecting Compressor Flow

For simplicity, let's only evaluate the flow formula for a single-acting compressor cylinder. We see that the piston diameter and stroke, strokes per minute, suction and discharge pressure, cylinder clearance, and the suction and discharge compressibility factors affect the net flow though the cylinder.

$$SCFM = \frac{\pi \times D^2}{4} \times Stroke \times spm \times \left(\frac{P_s}{14.7}\right) \times \left(\frac{520}{T_s}\right) \times \frac{1}{100} \times \left[100 - r - c\left(\left(r^{1/k}\right) - 1\right)\right] \quad (18.12)$$

By inspection we can conclude that the:

1. Flow decreases as the piston clearance (c) increases
2. Flow deceases as the compression ratio (r) increases
3. Flow decreases as "k" decreases
4. Flow decreases as T_s increases
5. Flow decreases as P_s decreases
6. Flow decreases if Z_s/Z_d is less than 1

Final Words

Today, most compressor specialists perform these types of flow calculations using software programs. However, it doesn't hurt to understand the basic equation governing compressor flow. By understanding the key variables governing actual gas flows, users will be better equipped to assess field compressor performance. If a discrepancy between the actual flow from the expected flow is found, I recommend you first check the compressor speed, suction pressure and temperature, and compression ratio to see if they were correctly selected and entered into your calculations. If these field values are found to be correct, then you might expect an internal leakage or an open bypass to be present. Internal leakage at valves piston rings, etc. will lead to increased temperatures, which will preheat gas entering the cylinder and incrementally increase in the suction gas temperature. This preheating effect will affect capacity by increasing T_s in Equations 18.9 and 18.11 above.

If discrepancies between your calculations and field measurements cannot be explained either: a) double check your calculations, b) discuss your findings with the manufacturer to see if they can provide any insight, and

c) compare your results with a recent PV analysis. Most PV analyses automatically calculate the cylinder flows. PV analyses can also determine if internal cylinder leakage is present.

Phillippi, G, Basic Thermodynamics of Reciprocating Compression, Texas A&M Turbomachinery Symposium, September 2016.

19

Use Your Historical Records to Better Manage Time Dependent Machinery Failure Modes

By Robert X. Perez

Early astronomers used their astronomical observations and analytical powers to earn the favor of those in power. By being able to use their astronomical data to make predictions about the world around them, they were believed to see the future. Over the centuries, astronomers have been able to predict solar eclipses, the start of seasons, and when Haley's comet would reappear. Years ago, these predictors seemed to be magicians, but today we know they were using the scientific method to understand how the solar system works. We now know that these celestial phenomena were time-based events based on orbital mechanics. To understand their surroundings, they: 1) observed, 2) formed hypotheses, 3) tested their theories, 4) analyzed their results, and 5) made true conclusions based on their best evidence and intuition.

In the field of rotating machinery, instead of devising a new scientific theory, we want to make informed decisions concerning the condition of our machines. For example, we want to know: 1) When should we shut down and repair a machine that's in distress? and 2) When should we schedule a machine for inspection and repair based on the time between the last repairs? An efficient machinery reliability program employs a modified version of the scientific method. To make better machinery decisions, we need to:

1. Make detailed repair observations and take accurate field measurements (see Figure 19.1).
2. Store failure and repair data so it can easily be retrieved and reviewed. Repair data should include photos, measurements,

Robert X. Perez (ed.) Condition Monitoring, Troubleshooting and Reliability in Rotating Machinery, (373–378) © 2023 Scrivener Publishing LLC

Figure 19.1 A reliability technician uses an infrared camera to inspect an electric motor.

samples, and detailed observations to assist future investigators.

3. Capture and display field data so that it can be easily retrieved and analyzed.
4. Turn field data into useful information. Trend plots, scatter plots, and spreadsheets can all help make sense of your field data. Train your reliability technicians so that they are current on the latest condition monitoring techniques.
5. Decide on a path forward based on the data analysis and review of the historical records. Data should drive your decisions.
6. Periodically, review our decisions to see if we were right. Data is useless unless it is used and evaluated to determine its value.

There are several equipment failure distributions that reliability professionals often deal with. However, in this chapter, I will only discuss two common types of failure distributions related to rotating machinery: time-based failures, also called age related failures, and random or unpredictable failures. Time-based failures are those related to corrosion, erosion, wear, and creep mechanisms, which can be avoided or mitigated by preemptive maintenance activities. Here is an example of time-based machinery failures:

1. A gas turbine loses a hot section blade after a 45,000 hrs. run, resulting in a catastrophic engine failure. After inspection,

the blade failure was determined to be due to creep, which is a time dependent metallurgical failure mechanism. Historical records indicated that previously, two vibration trips caused by severe blade rubbing trips related to blade creep were noted in the history after engine runs of 43,000 hrs. and 44,000 hrs. This seems to indicate that creep related blade failures are strongly related to the turbine's service time. Based on the failure history, we should either refurbish the turbine well before 40,000 hrs. or redesign the turbine hot section to extend the service life of the blades.

2. A rod drop monitor on a reciprocating compressor shows a gradually downward trend. Historical records show that the rider bands are replaced every 2 years due to wear. Careful measurement of the rider band and cylinder liner wear during planned inspections could be helpful in determining their optimum replacement time.
3. A steam turbine shows an increasing temperature trend on the active thrust bearing pads. Past records indicate severe fouling has been seen due to poor steam quality after 3 years of operations. Your choices are to either improve the steam quality, try to clean the rotor online, or replace the rotor on a given interval.
4. A WPII (Weather Protected II) electric motor stator winding temperature has been trending up and reaching alarm levels during the heat of the day. Records indicate the motor must be disassembled and cleaned every 4 years due to the dirty environment. Perhaps better air filtration should be considered to extent the time between cleaning intervals.

Random failures are those controlled by human error, process upsets, or multiple unrelated external factors. Let's consider the expected life of a rolling element bearing. In a perfect environment, the life of a rolling element bearing is very predictable. However, in an industrial setting, its life is dependent on the quality of its installation, the cleanliness of the oil, and how the bearing is loaded during its lifetime. These factors can dramatically affect the useful life of the bearing in unpredictable ways. Random failure modes are typically managed using condition monitoring methods to detect early failures before secondary damage occurs. Examples of condition monitoring methods include: Oil analysis, vibration analysis, infrared thermography, temperature monitoring, ultrasonic monitoring, performance monitoring, etc.

However, for a condition monitoring method to be effective, it must be capable of reliably detecting the failure mode of interest in its early stages to allow time for maintenance planning. For example, vibration analysis techniques have been proven to consistently detect incipient mechanical issues such as imbalance, looseness, etc. In contrast, blade and shaft fatigue, cracks caused by thermal cycling, and mechanical seal wear are a few examples of a component failure modes that are not easily detected using traditional online condition monitoring methods. These types of undetectable failure modes require different management strategies to mitigate their associated risk.

Here are examples of failure events that are not time-based:

1. Tower bottom pump suction pressure and flow have both dropped after a processing tower upset. Records indicate that a suction strainer has plugged before and after tower upsets. Internal scale and coke are shaken loose during upsets and settle at the bottom of the tower, where they are pumped out by the bottom pump. Perhaps a differential pressure alarm on the strainer may make sense if this is a frequent problem.
2. Valves on a reciprocating compressor foul in an unpredictable manner due to changing upstream conditions. It makes sense to install valve cap temperature sensors to alert personnel when valve fouling has occurred.
3. Mechanical seals in process pumps fail in an unpredictable manner due to a number of factors: a) How the seal is repaired and installed, b) the reliability of its seal flush system, c) the level of the shaft vibration, d) and the pump's operating conditions. These multiple factors can erratically affect the operating life of the seal. Because incipient seal wear is difficult to detect on-line, we typically use seal leakage as an indicator of the seal's condition. However, there must always be a strategy in place to manage the possibility of a rapid and unexpected seal failure. Most process pumps are spared so that a pump with a bad seal can be taken out of service and replaced with an installed spare pump.

Machinery professionals are always trying to determine the optimum time to take their machines out of service for inspections or repairs. Machines with predictable failure modes should be well documented to aid in efficient planning and scheduling of outages. Remember that failure modes that are unpredictable and require the use of predictive maintenance methods to reduce the potential of secondary damage. We must trend carefully whenever we encounter machinery failure modes that are not detectable using proven condition monitoring methods. In the real world, machinery professionals use a combination of time-based inspections and condition monitoring to minimize the risk of unexpected machine failures.

Never underestimate the value of historical data and an experienced analyst who can interpret current and historical field data. The combination of careful observations, historical records, and technical analysis can lead to better machinery decisions for your organization. Each machine outage and failure presents an opportunity to learn about your machinery to reduce the risks associated with future failures.

Part 4
PROFESSIONAL DEVELOPMENT

20

Soft Skills and Habits that All Machinery Professionals Need to Develop

By Robert X. Perez

When you ask people that work around machinery, "How do you improve reliability?", they 'll often tell you need to buy better bearings, better seals, or better monitors or with some other hardware based improvement. Most reliability folks believe design improvements are the primary means of improving reliability. However, I believe that some non-technical skills and habits can also have a dramatic effect on the reliability bottom line. After over 30 years of machinery experience in operating plants, I have learned to appreciate the importance of employing effective communication and assessment techniques when interfacing with mechanics and operators in order to solve complex machinery problems.

Some of these techniques fall into the general category referred to as soft skills. Wikipedia defines soft skills as "a combination of interpersonal people skills, social skills, communication skills, character traits, attitudes, career attributes and emotional intelligence quotient (EQ) among others that enable people to effectively navigate their environment, work well with others, perform well, and achieve their goals with complementing hard skills." In this article, I will cover the following basic soft skills in addition to some good habits that all machinery personnel should strive to continuously develop:

Soft Skills:

1. Asking probing questions
2. Listening more carefully
3. Observing (see Figure 20.1 and Figure 20.2)

Robert X. Perez (ed.) Condition Monitoring, Troubleshooting and Reliability in Rotating Machinery, (381–390) © 2023 Scrivener Publishing LLC

Figure 20.1 We can learn a lot by simply walking around and observing the plant equipment and personnel behavior.

Figure 20.2 Keep an eye on your equipment.

Habits:

4. Continuously learning
5. Praising
6. Teaching

Asking Probing Questions

When attempting to retrieve historical information from operators or mechanics about a machine failure, it's important to ask probing questions instead of specific questions. For example, let's say you are investigating a seal failure that resulted in a major release. Don't ask, "What do you think caused the failure?" Asking operators or mechanics why they think a failure occurred before critical information is collected and analyzed will lead to the invention of possible scenarios that are not based on facts. When people are asked to speculate, they will create scenarios that may make sense to them but are not based on facts. Instead, ask those close to the failure, "Explain in your own words what led up to the seal failure. Start from days leading up to the failure." Let the interviewees talk freely and encourage an open and unforced narrative that will help you clearly understand what transpired during the lead-up to the failure. This approach will allow the interviewee to open up and improve the possibility of acquiring critical information that will help you determine the root cause of the failure. At this stage of the information, you should not try to focus on conclusions; you should focus on basic information.

Try to interview operators and mechanics individually, so that each interview is honest, thorough, and unbiased. Group interviews tend to stifle and possibly bias discussions. Try to allow the interviewees to talk freely with minimal interruptions or undue pressure. The goal is to encourage the interviewee to be comfortable and feel like their feedback is vital. You never know where key information is going to come from. Eyewitness testimonies should be considered "ore that could be turned into pure gold".

After the interview, look for commonality in the narratives. Each account may be slightly different, but the basic facts should be similar. Later, the testimonies can be transcribed so that they can be referred to at a later date. You may need to go back to some of the interviewees for clarification purposes.

You should only try drawing conclusions about what happened after all the data has been collected, compiled, and thoroughly analyzed. A root cause failure analysis may be required to eventually bring all the eyewitness accounts, machine data, and process data together in order to arrive at a final conclusion. Don't let management try to force you into a conclusion before you feel comfortable offering one you believe in.

Key Takeaway: Ask questions that stimulate conversation and keep the interviewee on point.

Listening More Carefully

When talking to operators and mechanics, we must learn to listen. Discussion should always be a two way affair. Technical people tend to lecture instead of engaging in meaningful dialogue. One trick I use to improve my listening skills is to force myself to stretch out the conversation with an operator or mechanic, which allows the person I'm talking to explain themselves fully and not feel rushed. Asking your questions multiple times in different ways provides more opportunities to listen and comprehend the content of the discussion.

I think that face time is the key. Take as much time as the interviewee will give you. Prolonging the interview can help calm the interviewee down and show them that what they have to say is important to you. If possible, move the interview to a quiet area where you both can sit and relax while you talk. Take careful notes so that you can refer to them later. You might have to return to the interviewee later.

Assume, the operator or mechanic is the expert in the topic they are explaining to you. It's likely they know the symptoms but may not understand the cause and effect relationships involved in the event you are investing. Operators know when events happened relative to process changes, but they may not always know the relevance of the events to the health of the machine. Similarly, mechanics know the history of a machine, but they may not understand how process changes can affect the machine.

When listening to interviewees, keep an open mind and listen for the unexpected. We tend to get jaded and sometimes hear what we want to hear. For example: Let's say you know that a pump has failed twice in the recent past due to a low liquid level issue. Don't assume that the pump failed again and you are trying to find out why. Listen carefully to the eye witnesses and try to understand what was going on around the time of the failure. There may be a subtle, unique issue that caused this failure that is completely different from the low level failures of the past.

I have worked at sites where everyone assumed machinery failures were either caused by operator errors or machine repair errors. Remember that each failure is different. Perhaps the latest failure is related to a different causal factor.

Key Takeaway: Force yourself to rise above the accusations and assumptions and keep an open mind at all times.

Observing

I had a boss who believed strongly in a management method called, "Management by walking around." The basic idea was that a manager can learn a lot by simply walking around and observing the plant equipment and personnel. I came to learn that it's amazing what you can discover by walking around a facility and just looking around.

In this same vein of thought, it is vital that when attempting to solve a problem, the responsible machinery reliability professional needs to go out to the machine site and walk about. It's amazing what you can learn by walking around and asking "dumb" questions. There is no substitute for boots on the ground. During site visits, I have found broken pipe supports, opened bypass valves that were supposed to be closed, and loosed hold down bolts by inspecting a machine (Figure 20.2). I recommend that during site visits, you:

- Inspect the piping into the machine and out of the process. Look for poor piping practices and unsupported pipe
- Look at the foundation and look to see if there is any broken grout
- Look at pressure gauges to see if there are any excessive pressure pulsations
- Look for signs of high vibration (alarms)
- Look for signs of non-functioning monitors or controls
- Look for product or lubrication leaks

I was once asked to review a seal oil pump that was causing a compressor to shut down due to a low discharge pressure. The pumps would start and provide adequate pressure, but after a while they would begin losing pressure and eventually force a compressor trip. The site personnel had been working with a pump supplier to solve the problem. They even tried new pumps but couldn't solve the problem.

I knew I needed to visit the site and look over the installation. When I inspected the pumps, I immediately noticed that the pump suction lines had major high points that potentially could accumulate gas vapor over time and affect pump performance. I theorized that when the pumps started up, the suction lines where liquid full, but as gas vapors in the seal oil gradually collected in the high points, they would eventually vapor lock the pump. When I talked to the pump supplier, they agreed with my theory. They were embarrassed that they didn't notice the potential of collecting process vapor in the suction high points. They assumed the suction piping was properly installed. If they would have carefully studied the piping, they would have spotted the problem.

One way to spot subtle problems is by asking others to perform what is called a "cold eye review." A cold eye review is when someone unfamiliar with a machine or process unit inspects the site and looks for visible issues such as high points in suction lines, unsupported piping, inaccessible valves, oil leaks, product leaks, unheeded vibration alarms, etc. Owners can become blind to problematic issues in front of their faces. Sometimes it takes someone with fresh eyes to spot potential problems.

When walking by a machine in the field, look up, down, and around and ask operators about anything that may be unusual or out of the ordinary.

Key Takeaway: Site visits can lead to surprise discoveries.

Continuously Learning

We live in a time when technology is rapidly changing and evolving. New analysis tools, such as laser alignment tools, infrared cameras, and advanced statistical data analysis tools, are now commonly used by maintenance work forces. These and other new high tech tools and methods are forcing machinery professionals to continuously learn new things to remain effective at their jobs and competitive in the job market. To keep abreast, we must force ourselves to follow the latest developments in the field and pursue all means of education at your disposal.

I think we would all agree that the shelf-life of technical knowledge is constantly shrinking. I would venture to guess that a significant portion of our machinery knowledge will become obsolete in 5 to 7 years. To keep up to date, we must develop a learning plan and then follow through with it. Here are a few ideas that you might want to consider:

- Continue to build your library with the latest books on machinery knowledge. Books on machinery construction and theory, lubrication and vibration analysis, reliability theory, and predictive maintenance methods are a few examples of books to consider acquiring. A large, diverse library is a sign that a machinery reliability professional cares about his or her job and is serious about learning. If you are not sure what books to acquire, ask your mentor or coworkers to recommend books that may be helpful.
- Explore all traditional means of learning such as reading technical books and periodicals and attending seminars. Nowadays, there are many online seminars and courses

available. There are also free webinars provided on a broad range of topics that can help those on a limited budget keep up with the latest developments in the field.
- Develop an annual learning budget that includes seminar fees, travel expenses, and books. First develop a "dream" budget and then negotiate with your boss in order to have him cover as much of the items on your list as possible.
- Whatever your boss won't support you should cover out of pocket. Remember that personal technical training is an investment in your future.

One of my favorite quotes is, "Everything is inferior to learning." This certainly was my mother's belief. She was always encouraging me to continue my education. She knew that education opened more doors of opportunity and was a means to career growth.

Key Takeaway: Technical learning should be a career long endeavor.

Praising

Something we never give enough of is praise. Praise is a positive form of feedback that can motivate those working around you to work harder and take pride in their work. Praise is a way to communicate to others that you appreciate their efforts and that they are doing the job correctly. Whenever we observe exemplary performance, we should offer sincere praise.

Simple phrases like "Great job" or "Way to go" can go a long way to encourage those around you to try harder.

Praise can also be put in writing. A memo or email to an operator's or mechanic's supervisor praising an employee or team is a great way of recognizing exemplary performance in your facility. A memo comprised of a short paragraph with details about accomplishments is a concrete way to pass on praise and provide particulars about why the recognition is deserved. The memo or email can be filed and referred to at a future performance review.

The flip side of praise is criticism. If an employee is not performing up to par, we must correct the bad habit or point out the mistake promptly and judiciously. The keys to criticism are that it must be:

- Constructive, meaning that it must be feedback that is clear and useful.
- Done in private. Never criticize anyone in public. Public criticism leads to embarrassment.

- Criticize succinctly. Provide the appropriate feedback and then drop the matter. Don't dwell on the issue.

Praise and criticism are valuable tools that guide staff to improve. Too much praise or criticism can be a problem. We must try to balance praise and criticism that we dole out.

Key Takeaway: Praise freely and criticize judicially.

Teaching

Machinery reliability professionals should be considered technical leaders. As leaders, one of our roles should be to teach those working around us about machinery and machinery reliability. Teaching can be a very satisfying part of our jobs, if done properly (Figure 20.3).

However, there is a difference between teaching and preaching. Teaching is the thoughtful and systematic transfer of knowledge from one person to another. Teaching can occur in a classroom, machine shop, or in field. Preaching, on the other hand, can be the result of someone trying to brag about what they know. Operators and mechanics know the difference between teaching and preaching and are usually turned off by someone talking over their heads and trying to show off.

Figure 20.3 Look for teaching opportunities.

Teaching benefits the teacher as well as the students he's teaching and can be considered a form of professional growth. The entire facility can benefit from formal and informal teaching. Here are just a few benefits for the organization:

- Teaching enlarges the teacher's knowledge base. Teaching forces you to learn the material in greater depth, which is good for you and the company.
- Teaching allows students to better perform their jobs by receiving instruction in the field where it can be immediately applied.
- Teaching can allow students to feel better about themselves. Operators and mechanics will think they must be important if you are taking time out from your routine to receive training.
- Teaching can be very rewarding if done properly. Everyone appreciates learning information that can make their job easier or possibly help those receiving new knowledge get a promotion in the future.

When applicable, use handouts that can be taken and saved by the attendees. Your training notes can be filed and then referred to again in the future.

To be a successful teacher you must always:

- Look for teaching moments. A teaching moment may arise in the field when a crew is working on a new machine, installing a new type of seal or bearing, or performing a new type of procedure.
- Present information at the appropriate technical level. For example, don't try to teach rotor-dynamics to a mechanic or mechanical seal design to an operator.
- Match the teaching content to the venue. In other words, if a topic requires graphics to get the point across, you probably need to do it in a classroom setting. If the topic can be conveyed in a few sentences and doesn't require graphics, then it can be taught in the field.
- Be careful to avoid teaching in a noisy environment. Noise can add to stress and frustration that can undermine your teaching efforts.

An excellent and inexpensive way to teach hourly workers is by employing "toolbox teaching sessions", which are 15 minutes sessions offered in a shop or field environment in an informal manner. Toolbox sessions can be provided on a weekly or monthly basis. Some examples of topics are alignment, bolt torquing methods, basic vibration analysis, etc. I am aware of a petrochemical facility that conducted these sessions for years and kept all the training notes in a binder in the shop for mechanics to refer to later.

Remember that most people learn better when they are actively involved in the training session. You can even get operators and mechanics involved in developing or presenting the training modules.

Key Takeaway: Always be on the lookout for opportunities to teach others. Hands on training, exercises, and short quizzes work best to help students retain key concepts and methods.

Closing Remarks

I hope I have convinced the reader that soft skills and good non-technical habits can be as important as technical skills when it comes to performing your job. Every failure should be considered an opportunity to improve machinery reliability at your site. To get the maximum information about a failure event, we need to learn how to question those closest to the failure and learn as much as you can about the failure. Failure investigations are where learning how to 1) ask probing questions, 2) listen more carefully, and 3) carefully observe the site can pay big dividends.

As a technical leader, you must be on the lookout for opportunities to teach and praise operators and mechanics as you work in the field. Remember that knowledge is power. The more we learn and retain about machinery technology the better we will perform and the more successful you will be during your career.

We must develop the habit of walking through the field and seeking out machinery issues and problems, as well as potential teaching topics. Machinery reliability professionals have a great responsibility to assist in improving machinery reliability so that the entire facility becomes safer and more profitable each year. However, this mission is everyone's responsibility not just yours. Everyone must work together to gradually chip away at reliability problems and optimize the site's overall reliability.

21

Developing Rotating Machinery Competency

By Robert X. Perez

Part I: Preparing Students to Work with Rotating Machinery

Rotating Machinery Related Job Functions

Rotating Machinery is a broad category of machinery (see Figure 21.1) found in the industry that includes:

- Fluid movers such as centrifugal and reciprocating pumps and centrifugal and reciprocating compressors
- Speed modifiers such as gear boxes, fluid couplings, and belts drives
- Drivers such as electric motors, steam turbines, and gas turbines

Why should you consider a career in rotating machinery? Most machinery professionals find working around rotating machinery to be very interesting and rewarding. As a machinery professional, you get to work on and sometimes solve very challenging technical problems. Here are a few examples of challenging problems that I (Perez) have faced:

1. A centrifugal pump is failing seven (7) times a year. I had to study the process conditions and pump's design to find the root cause of these failures.
2. Gas turbine blades are failing catastrophically. I had to examine the design of the blades and the operating conditions to determine the root cause of these failures.
3. New, multistage centrifugal pumps, operating in parallel were failing repeatedly. I had to carefully examine how

Robert X. Perez (ed.) Condition Monitoring, Troubleshooting and Reliability in Rotating Machinery, (391–402) © 2023 Scrivener Publishing LLC

Figure 21.1 Rotating machines, similar to those seen here, are highly engineered products found in most processing facilities. Starting with the upper left image and going clockwise, we see an image of a multistage centrifugal pump, reciprocating gas compressor, steam turbine, and a gear box. Their reliability performance requires that everyone involved in the design, selection, installation, and operation make sound technical decisions.

the pumps were operated to understand why these pumps where failing.

4. After installing a different coupling between an expander and electric motor, the motor is experiencing extremely high vibration levels. Consultants had to perform a rotor-dynamic analysis of the motor and coupling to understand the reason for the sudden increase in vibration. understand how operating

It's not enough to find the failed parts and replace them; machinery engineers must uncover the root cause of failures and attempt to eliminate the root causes of failure through redesigns or implementing new operating procedures.

The best feeling in the world is when you find a problem has been solved after your recommendations were implemented. As you grow in the field, the problems get increasingly challenging, which forces you to keep studying and learning. In many cases, rotating equipment professionals have to travel to field locations, shops, and manufacturing facilities. This mix of study, analysis, and travel leads to a varied and flexible work schedule.

As an added benefit, rotating machinery professionals to tend to be highly regarded in production settings and so they tend to be well paid.

End users and manufacturers require highly trained rotating machinery professionals to cover a wide range of job functions. There are those who design, market, select, operate, repair, and upgrade machinery. This diverse need for technical personnel requires a wide scope of formal and informal training programs to ensure they possess the right skills. Students in colleges, universities, and trade schools may not be aware of all the opportunities available in this challenging, yet rewarding, field. The purpose of this overview is to list the types of jobs that are available for graduates that involve rotating machinery and the types of courses they should consider if they want to gain an advantage over other candidates.

Here is a brief listing of some typical job functions found around the processing and machine manufacturing industry:

Operations: Operating personnel must be able to identify and operate all type of rotating machines in the field. They must also be able to perform basic machine inspections and troubleshoot field problems as they arise. When required, operators must be able to describe a field issue on a work order so that they can be addressed by the maintenance department. It is also important for Operators to understand how operating factors, such as flows, pressure, product composition, etc., affect machinery performance and reliability.

Maintenance Professionals: Maintenance personnel must be able to restore equipment that has failed back to "as new" conditions. During the repairs, they must also be able to perform basic failure analyses in order to determine the cause of failures. They must also be knowledgeable of design improvements that could improve machine reliability, such as seal or bearing upgrades.

Reliability Engineer: Reliability engineers look for ways to improve the reliability of machines. They look for design and repair procedural changes that could improve the reliability performance of their machinery. Reliability engineers need to be trained to look at historical failure data to determine if design upgrades make sense and are economically justified.

Project Engineer: Project engineers are involved in handling capital projects that will improve the profitability of a site.

They must specify, procure, and install all the mechanical, electrical, and control equipment required for the project. It is vital that project engineers have a firm grasp of how their equipment's reliability performance will affect the overall reliability of the plant.

Machinery Field Analyst: A machinery field analyst's job is to collect and analyze machinery field data, i.e., vibration, temperature, pressure, etc., in order to assess the condition of critical machines. They must also be able to sort through all the data and then summarize it in order to assess the conditions of machinery. A machinery analyst must be well versed in signal processing, patterns recognition, and troubleshooting complex systems.

Purchasing: Purchasing agents procure replacement parts and sometimes entire machines. They must understand what they are buying in order to make informed decisions. Purchasing the wrong part or machine can negatively impact operational reliability of a processing site.

Sales & Marketing: Original equipment manufacturers (OEMs) need to have an organizational arm that can market and sell their products. The more technical the product, the more knowledge the Sales & Marketing members must possess. They must understand why their equipment is superior to their competitors and be able to clearly communicate the benefits of their products to customers.

Design Engineer: Design engineers working for original equipment manufacturers customized existing products for customers. They must be very knowledgeable about their company's product line. When a new machine configuration is required, design engineers determine what specific metallurgy, geometry, bearings, etc. are required to fulfill the application requirements.

R&D Engineer: Research and development engineers are tasked with developing completely new machine designs for the market. If an OEM company sees a new opportunity in the market, R&D engineers will work to design, build, and even test prototypes to determine their viability. New designs offer unique challenges that must be overcome with innovative designs. Most large OEMs always have a full time R&D group working on new designs to keep their product offerings competitive.

The **Rotating Machinery Training Matrix** seen in Table 21.1 summarizes the required knowledge for the various job descriptions we have listed. The various job functions are listed at the top of the vertical columns and the required knowledge topics are listed on the left. "X" means only basic knowledge is required to properly perform the job function, "XX" means intermediate knowledge is required to perform the job function, and

Table 21.1 Rotating machinery training matrix.

		Job description								
		Operations	**Maintenance Professionals**	**Reliability Engineering**	**Project Engineering**	**Machinery Field Analyst**	**Purchasing**	**Sales & Marketing**	**Design**	**R&D**
Required knowledge	**Overview of Rot. Equipment**	X	XX	XX	XX	XX	X	X	XX	XX
	Operating Best Practices	XXX	XX	XX	XX	XX	X	X	XX	XX
	Fluid Mech. & Performance	X	XX	XX	X	X		X	XXX	XXX
	Lubrication	X	XX	XX		XX			XXX	XXX
	Signal Processing & Vibration Analysis	X	XX	XX		XXX			XX	XX
	Thermodynamics			XX		X			XXX	XXX
	Rotor Dynamics		X	XX		XX			XXX	XXX
	Maintenance Best Practices	X	XXX	XXX	X	XX		X	XX	XX
	Reliability Engineering	X	X	X	X	XXX	X	X	XXX	XXX
	Machine Design		X	X		X			XXX	XXX
	Materials Science		XX	XX					XXX	XXX

X - Basic knowledge
XX— Intermediate knowledge
XXX— Advanced knowledge

"XXX" means advanced knowledge is required to perform the job function. For example, the Project Engineer should have intermediate knowledge about general rotating equipment and operating best practices and basic knowledge about maintenance practices and reliability engineering.

To prepare for a career in the rotating machinery field, students should consider college courses related to the topics listed in Table 21.1. Here are some of the courses that we have found particularly useful:

1. Machine Design
2. Fluid Mechanics
3. Thermodynamics
4. System Dynamics
5. Strength of Materials
6. Materials Science
7. Economics

Once a student has these technical fundamentals mastered, they can move on to more advanced rotating machinery topics. Failure to learn these fundamentals will leave gaps in the essential body of knowledge required in the industry. Next, we will discuss how to keep improving your competency once you are working in the industry.

Part II: Steps to Improving Rotating Machinery Competency: Study-Practice-Share

It takes determination and focused effort to become competent in any discipline. This is especially true with rotating machinery. Because of the breadth of knowledge base required to be competent in rotating machinery, it can take years or even decades to be considered competent. So, where does a beginner start? I will spell out a methodology that I (Perez) used to grow technically in the field of rotating equipment.

Keep in mind that the road to rotating equipment mastery never ends. Those in the field learn every day and strive to continue learning. Each day will learn a new lesson or tool that we collect for the next field problem. The essence of continuous improvement is captured by the "study-practice-share" motto, which means that we first need to study the technical concepts, then we need to practice what we have learned, and finally we need to share what we have learned. Here, I will discuss this recipe for technical competency in more detail.

1. Study: Once you decide to work in the rotating machinery field, you need to ensure your academic coursework provides the proper foundation for more advanced training and activities. Refer to the recommended college preparation courses listed in Part I of this chapter for guidance. Next, you need to compile a solid library of books and then you must begin reading them. Here is a short list of books that you should consider purchasing. By no means is this a comprehensive list. These books should be considered a starting point for your post-graduate education.

a. Pump Books
   i. Pumps: Pump Handbook, Igor Karassik, *et al.*
   ii. Centrifugal & Rotary Pumps Fundamentals With Applications, Nelik, Lev
   iii. Operator's Guide to Centrifugal Pumps, Robert X. Perez
b. Compressor Books
   i. Compressed Air and Gas, Ingersoll Rand
   ii. A Practical Guide to Compressor Technology, 2nd Edition, Bloch, Heinz, P.
   iii. Gas Compression: A Primer On Compression Equipment & Technology, W. Norm Shade
c. Reliability Engineering Books
   i. Improving Machinery Reliability, Heinz P. Bloch (Consulting Engineer, Montgomery, TX, USA)
   ii. An Introduction to Machinery Reliability Assessment by Heinz P. Bloch and Fred K. Geitner
   iii. Fundamentals of Rotating Machinery Diagnostics (Design and Manufacturing) 1st Edition, by Donald E. Bently (Author), Charles T. Hatch (Author)
   iv. Oil Analysis Basics - Second Edition, Jim Fitch & Drew Troyer
d. In addition to technical reference books, you need to keep your rotating machinery knowledge current by subscribing to and reading trade publications. There are numerous publications available that deal with rotating equipment topics. Here are just a few to consider subscribing to:
   i. Pumps & Systems
   ii. Empowering Pumps & Equipment Magazine
   iii. Gas Compression Magazine
   iv. Compressor Tech 2 Magazine
   v. Machinery Lubrication Magazine (Noria)

2. Attend seminars: There are many seminars to choose from and limited training funds, so you need to choose wisely. Get the advice of other machinery professionals before deciding which seminars will provide the most bang for your training dollars. Seminars are useful because they provide a time-efficient means of getting exposure to the latest technical ideas and meeting experts in the field. Here are few of my seminar favorites:

a. Vibration Institute
b. Local Vibration institute Chapter meetings
c. Texas A&M Turbomachinery Symposiums
d. Noria Reliability Seminars

3. Talk regularly to in-house and outside experts. Take every opportunity to talk to experts that you have immediate access to. Most of these individuals are more than willing to share their knowledge. If an expert is busy, schedule a convenient time to get together and talk.
4. Strive to develop a well-rounded view of rotating machinery that includes knowledge about:

a. Machinery theory and design
b. Machinery operations
c. Reliability engineering
d. Troubleshooting and failure analysis

Try not to specialize too early in your career. Becoming too highly specialized too early in your career will limit your ability to analyze certain classes of problems and could inhibit your promotability. You can choose to specialize later in your career when you have determined the technical areas you enjoy.
5. Shop and field visits:
I cannot stress enough the importance of visiting shops and the field to gain practical technical insights. There's only so much you can learn from a book. You need to actually see how repairs are being accomplished, what internal parts and passageways look like, and what different failure modes look like. Most supervisors are willing to show you around their shops and explain the different repairs they are working on.

Field visits also provide opportunities for learning valuable lessons about machinery. In the field, you can learn how machines are installed and operated. You should study the layout of the process piping to see how machines fit into the process and analyze the auxiliary equipment and piping to see how they function. When you are in the field, you need to look up at the piping and supports and down to examine the foundation, baseplate, and hold-down bolts. Field audits are a valuable part of developing a holistic view of your machinery.

When you are in the field, talk to operators about the problems they've seen and issues that they are most concerned about. They are a valuable source of information about what's really going on in the field. Operators are the eyes and ears of a plant.

6. Listen to mechanics and operators and ask questions. Both mechanics and operators are great sources of historical information. They know when machines were installed and repaired and the types of problems the various types of machines have experienced. Good lines of communication between mechanics and operators are invaluable to solving subtle and complex field issues.

7. Practice: A key part of learning practical aspects of a topic is applying what you've learned. The application of your newly acquired knowledge enables you to see what is useful, what is superfluous, and what doesn't work in the real world. A typical example of applying what you have learned is calculating the theoretical discharge temperature of a reciprocating compressor. The calculation is straightforward if you have all the values required for the calculation, i.e., compression ratio, suction temperature, and the ratio of specific heats (k). Let's say you want to know if a reciprocating compressor is healthy so you decide to calculate the theoretical discharge temperature. You will quickly see if the theoretical discharge temperature agrees with what you see in the field. If there's a disagreement, you then need to investigate the possible cause of the difference between the two values. The problem may be inaccurate input values or they may actually be a problem. Either way, you'll learn something from the exercise.

There's an old joke where a man asks, "How do you get to Carnegie Hall?" The second man replies, "Practice, practice, practice." The same thought is applied to the field of machinery. If you want to solve big problems, you need to practice your craft in the real world with real machines.

Here are a few ways you can practice what you have learned around rotating machines:

a. Work on field problems
b. Perform calculations and compare results to field observations
c. Modify equipment and observe results
d. Select new machinery and see how they perform under actual field conditions

e. Install new machinery and follow them all the way though start-up to see if any problems develop
f. Observe machinery tear-downs and determine the reason for the failure.
g. Write failure analysis reports
h. Study failure data to identify trends and patterns

8. Write: Writing forces you to capture, in words, what you have learned and observed. Technical writing forces you to think logically and to be concise. Some people find writing to be painful, but I encourage you to look for writing opportunities to develop your writing skills. Here are some opportunities to write:

a. Maintenance work orders. This is normally how most people begin writing.
b. Internal machinery reports that include:
    i. Problem definition
    ii. Finding
    iii. Conclusion and recommendations
    iv. Economic justification of recommendations
c. Articles for trade journals. After a few years of practice, you can begin to submit articles for publication.

9. Share and Teach: Richard Feynman said, "If you want to master something, teach it." Anyone who has ever taught knows that to teach a topic you must know the material backwards and forwards. Furthermore, every time you teach a topic, you will glean an extra insight or two about the topic. Never believe it's beneath you to teach an operator or mechanic something basic. You should consider every teaching opportunity to be a chance to find a new and better way to explain the topic. Here are some ways to share your knowledge and experience:

a. Provide informal in-house sessions, such as toolbox huddles in the shop or in control rooms
b. Formal in-house training sessions in the shop and field
c. Present at technical conferences
d. Organize company technical exchange meetings

10. Become an advocate for machinery reliability in your facility, company, and in the industry. After you have achieved a renowned level of competence, it is your duty to lead the reliability efforts in your company and the technical community. These efforts could include organizing seminars, discussion groups, internal technical exchanges, etc. Involvement in these activities also helps you by allowing you to make new contacts and improve your organizational skills.

While I have known very knowledgeable machinery experts, I have never found anyone who knows everything about every rotating machinery topic, which is why it is important to make contacts at conferences and seminars. By joining forces with a group of experts, you will be able to tackle the toughest problems. Always be on the lookout for a new connection to your band of specialist.

About the Editor

Robert Perez is a mechanical engineer with more than 40 years of rotating equipment experience in the petrochemical industry. He has worked in petroleum refineries, chemical facilities, and gas processing plants. He earned a BSME degree from Texas A&M University at College Station, an MSME degree from the University of Texas at Austin, and holds a Texas PE license. Mr. Perez has written numerous technical articles for magazines and conference proceedings and has authored 5 books and coauthored 6 books covering machinery reliability. He collaborated with machinery reliability expert, Heinz Bloch, on his most recent book *Pump Wisdom: Essential Centrifugal Pump Knowledge for Operators and Specialists, 2nd Edition* (*Wiley 2022*). He is also the technical editor of *Kane's Rotating Machinery Dictionary.*

About the Contributors

William D. Marscher, P.E. is CEO and Senior Consultant for Mechanical Solutions, Inc. His career of 50 years has involved design, development, and troubleshooting of pumps, fans, compressors, turbines, and motors. He has BSME and MSME degrees from Cornell University, where he was a NASA fellow, and an MS in Applied Mechanics from RPI. He has held senior positions at Honeywell, Pratt & Whitney, Worthington, Dresser Industries, and CNREC. For ASME, he was Chair of the Predictive Maintenance committee and was Organizing Chair for two rotating machinery conferences. For STLE, he was the 1998/99 President, is a Fellow, and was Associate Editor of Tribology Transactions magazine. For ASTM, for 40 years he has been on the standards voting committees for wear, erosion, and metal fatigue. He is one of the U.S. voting members of the ISO/ANSI machinery vibration & noise standards committee TC108 and is an 18-year member of the Advisory Committee for the annual Texas A&M Pump Symposium. He is past Board Chairman for the Machinery Failure Prevention Technology Society of the Vibration Institute and is the recipient of the Vibration Institute's Jack L. Frarey Award for excellence in machinery diagnostics. He is also a Standards Partner for the Hydraulic Institute, where he was recently Vice Chair of the Vibration Committee, and in 2019 was named the HI/ANSI Standards Partner of the Year. Mr. Marscher has written 10 handbook chapters, has published over 80 technical papers and proceedings articles, and holds six patents.

Jim Fitch is the CEO and a co-founder of Noria Corporation. He has a wealth of "in the trenches" experience in lubrication, oil analysis, tribology, and machinery failure investigations. Over the past two decades, Jim has presented hundreds of courses on these subjects and has published more than 200 technical articles, papers, and publications. He serves as a U.S. delegate to the ISO tribology and oil analysis working group and has been awarded numerous patents. Since 2002, Jim has also been a director and board member of the International Council for Machinery Lubrication.

Bennett Fitch is the Chief Strategy Officer (CSO) for Noria Corporation. His years of experience at Noria include director of development for Lubrication Program Development (LPD) services and LubePM, the lubrication program management platform. Additionally, he serves as a senior technical consultant and is an instructor for Noria's Machinery Lubrication and Oil Analysis courses with an interactive, yet relaxed approach to engage his audience. His expertise is established from years within oil analysis and machinery lubrication as the core of his career. He received his bachelor's degree in mechanical engineering from the Georgia Institute of Technology with a concentration in applied tribology and a master's degree in business administration from the University of Tulsa. Bennett is also certified through the International Council for Machinery Lubrication (ICML) as a Machinery Lubrication Engineer (MLE), Machinery Lubrication Technician Level II (MLT II), and Machine Lubricant Analyst Level III (MLA III).

Abdulrahman Alkhowaiter (BSME) is a Rotating Equipment Consultant & Instructor with over thirty two years of experience covering a wide variety of machinery engineering and troubleshooting experience. His experience includes:

Maintenance Reliability positions covering Reliability Centered Maintenance of compressors, steam turbines, gas turbines, diesel engines, pumps, electric motors, gearboxes, blowers, and fans for seven years. In addition, he has worked 22 years in engineering consulting for a major petroleum company specializing as a compressor, gearboxes, couplings, and steam turbine specialist. Abdulrahman teaches international courses in Machinery Failure Analysis, Process Plant Compressors, Steam Turbines Field Troubleshooting, and Machinery Reliability Improvement.

Index

Also of Interest

Also in the series, "Rotating Machinery Fundamentals and Advances"

DESIGN, MODELING AND RELIABILITY IN ROTATING MACHINERY, Edited by Robert X. Perez, ISBN: 9781119631682. This broad collection of current rotating machinery topics, written by industry experts, is a must-have for rotating equipment engineers, maintenance personnel, students, and anyone else wanting to stay abreast with current rotating machinery concepts and technology.

MAINTENANCE, RELIABILITY AND TROUBLESHOOTING IN ROTATING AND PROCESS MACHINERY, Edited by Robert X. Perez, ISBN: 9781119631644. For Rotating equipment professionals, maintenance personnel, project engineers, process engineers, mechanical engineering students, and anyone working around critical rotating equipment.

Other books by the same editor

Troubleshooting Rotating Machinery: Including Centrifugal Pumps and Compressors, Reciprocating Pumps and Compressors, Fans, Steam Turbines, Electric Motors, and More, by Robert X. Perez and Andrew P. Conkey, ISBN 9781119294139. A must-read for anyone whose job depends on operating or maintaining safe, reliable, and efficient machinery in a process facility, this volume shows the newcomer and veteran alike how you can start troubleshooting rotating equipment problems immediately by implementing this book's common-sense methodology. *NOW AVAILABLE!*

Operator's Guide to Process Compressors, by Robert X. Perez, ISBN 9781119580614. The perfect primer for anyone responsible for operating or maintaining process gas compressors. *NOW AVAILABLE!*

Operator's Guide to General Purpose Steam Turbines: An Overview of Operating Principles, Construction, Best Practices, and Troubleshooting, by Robert X. Perez and David W. Lawhon, ISBN: 9781119294214. A must-read book for anyone operating or maintaining general purpose steam turbines in a process setting, this is a practical reference that includes a collection of field-tested procedures aimed at ensuring safe general purpose steam turbines start-ups, shutdowns, and over-speed tests. *NOW AVAILABLE!*

Check out these other titles from Scrivener Publishing

Dictionary of Industrial Terminology 2nd Edition, by Michael D. Holloway and Emma Holloway, ISBN 9781119363446. Greatly expanded and revised, this second edition is the most comprehensive and in-depth dictionary of industrial terms available, a must-have for any engineer, scientist, technician, or operator working in industry. *NOW AVAILABLE!*

Basic Electrical and Instrumentation Engineering, by P. Sivaraman, C. Sharmeela, A. Thaiyal Nayagi, and R. Mahendran, ISBN 9781119764465. Offering the most up-to-date coverage on the subject available, this new volume provides detailed concepts of AC circuits and power systems, transformers, DC machines, AC machines, measurement and instrumentation, and many other practical concepts for the engineer or engineering student. *NOW AVAILABLE!*

Electrical and Electronic Devices, Circuits, and Materials: Technological Challenges and Solutions, edited by Suman Lata Tripathi, Parvej Ahmad Alvi, and Umashankar Subramaniam, ISBN 9781119750369. Covering every aspect of the design and improvement needed for solid-state electronic devices and circuit and their reliability issues, this new volume also includes overall system design for all kinds of analog and digital applications and developments in power systems. *NOW AVAILABLE!*

Printed and bound by CPI Group (UK) Ltd, Croydon, CR0 4YY
17/09/2026
14964245-0004